Principles and practice of electron microscope operation

Practical Methods in ELECTRON MICROSCOPY

Edited by
AUDREY M. GLAUERT
Strangeways Research Laboratory
Cambridge

NORTH-HOLLAND PUBLISHING COMPANY
AMSTERDAM · NEW YORK · OXFORD

PRINCIPLES AND PRACTICE OF ELECTRON MICROSCOPE OPERATION

ALAN W. AGAR
Agar Aids
Stansted, Essex

RONALD H. ALDERSON
Tattersett
King's Lynn, Norfolk

and DAWN CHESCOE
Department of Materials Science
University of Surrey
Guildford, Surrey

NORTH-HOLLAND PUBLISHING COMPANY
AMSTERDAM · NEW YORK · OXFORD

Library of Congress Catalog Card Number 73-94298
ISBN series 0 7204 4250 8
volume 0 7204 4255 9
0720442540

1st edition 1974
2nd printing 1978
3rd printing 1980
4th printing 1982
5th printing 1985
6th printing 1987
7th printing 1990

Published by:
ELSEVIER SCIENCE PUBLISHERS B.V./BIOMEDICAL DIVISION,
P.O. BOX 211,
1000 AE AMSTERDAM, THE NETHERLANDS.

Sole distributors for the U.S.A. and Canada:
ELSEVIER NORTH-HOLLAND INC.
52 VANDERBILT AVENUE
NEW YORK, N.Y. 10017

This book is the laboratory edition of Volume 2 of the series 'Practical Methods in Electron Microscopy'.

Printed in The Netherlands

Titles of volumes published in this series:

Volume 1 Part I Specimen preparation in materials science
by P.J. Goodhew

Part II Electron diffraction and optical diffraction techniques
by B.E.P. Beeston, R.W. Horne, R. Markham

Volume 2 Principles and practice of electron microscope operation
by A.W. Agar, R.H. Alderson, D. Chescoe

Volume 3 Part I Fixation, dehydration and embedding of biological specimens
by Audrey M. Glauert

Part II Ultramicrotomy by Norma Reid

Volume 4 Design of the electron microscope laboratory
by R.H. Alderson

Volume 5 Part I Staining methods for sectioned material
by P.R. Lewis and D.P. Knight

Part II X-ray microanalysis in the electron microscope
by J.A. Chandler

Volume 6 Part I Autoradiography and immunocytochemistry
by M.A. Williams

Part II Quantitative methods in biology
by M.A. Williams

Volume 7 Image analysis, enhancement and interpretation
by D.L. Misell

Volume 8 Replica, shadowing and freeze-etching techniques
by J.H.M. Willison, A.J. Rowe

Volume 9 Dynamic experiments in the electron microscope
by E.P. Butler, K.F. Hale

Volume 10 Low temperature methods in biological electron microscopy
by A.W. Robards and U.B. Sleytr

Volume 11 Thin foil preparation for electron microscopy
by P.J. Goodhew

Volume 12 Electron diffraction: an introduction for biologists
by D.L. Misell and E.B. Brown

Contents

Acknowledgements

Much of the presentation of this book was developed during a long series of training courses in the use of the transmission electron microscope. We have been greatly helped over the years by the many 'students' who have asked awkward questions and stimulated new lines of thought. We must acknowledge the debt we owe to Dr. M. E. Haine, Prof. T. Mulvey and Mr. R. S. Page who unstintingly shared their experience and knowledge.

We particularly wish to thank the Editor for her many suggestions for the improvement of the presentation.

We are indebted to the following for permission to reproduce micrographs and figures: Dr. D. L. Allinson (9.8), Dr. V. E. Cosslett (9.3, 9.4), Prof. J. M. Cowley and Dr. S. Iijima (8.4), Dr. S. L. Cundy (9.13), Prof. G. Dupouy (2.27, 3.21. 9.11), Dr. A. M. Glauert (8.1), Dr. K. H. Hale and Miss M. Henderson-Brown (9.5), Dr. B. Hudson and Dr. M. J. Makin (6.3, 6.4), Prof. M. Iwanaga (7.8), Prof. J. B. LePoole (9.7), Dr. P. R. Swann (2.24, 9.9, 9.10), Prof. R. C. Williams (6.6), Dr. D. B. Wittry (9.13), AEI Scientific Apparatus Ltd. (2.4, 2.8, 2.10, 2.11, 2.18, 2.19, 2.20, 2.23, 2.32, 3.3, 3.5, 3.6, 3.7, 3.9, 3.19, 5.1, 5.6, 5.7, 5.11, 6.1, 6.2, 8.7, 9.2, 9.6, 9.12, 9.15), Ilford Ltd. (7.6), JEOL Ltd. (2.1, 2.31), Ebtec Inc. (2.6, 2.7), N.V. Philips (2.2, 2.9, 2.13, 2.17, 2.22, 2.25, 2.29, 5.3), Siemens Ltd. (2.3, 2.21, 2.26, 2.28), Smethurst High Light Ltd. (2.12), Zeiss Ltd. (9.1).

One of us (AWA) is also grateful for the use of electron microscope facilities at the Strangeways Research Laboratory, Cambridge, and at Pye-Unicam Ltd., for obtaining a number of micrographs used for illustration.

Chapter 1

The basic principles of the electron microscope

The aim of this book is to provide the background needed for an operator to obtain the best results from an electron microscope. To this end, the basic principles of the instrument are first described, followed by the physical construction of the microscope. The details of alignment and operation of the instrument follow, so that an operator can gain a clear idea of the functions of the controls available to him. The foregoing chapters should also have explained some of the potentialities as well as the difficulties of the design of an electron microscope, so that new possibilities and basic limitation become apparent.

Apart from understanding the operation of the instrument the operator must record the information which the instrument renders visible and important photographic techniques which are not always properly observed are detailed in one chapter. The basic mechanism of image formation (quite different from that in the light microscope) is described so that there is a background for intelligent interpretation of the image structure that is obtained. Chapters are also devoted to routine maintenance and fault-finding.

Optical microscopy has suffered greatly because it always appeared to be easy to achieve results without understanding how the instrument should be used. Electron microscopes are perhaps fortunate in appearing to be sufficiently complex so that they induce prospective operators to attempt to understand them before proceeding with their research programmes. It is the hope of the authors that this book will aid this process.

Since there are now many different models of microscope in service throughout the world, it is impracticable to describe constructions and procedures which are completely applicable to each one – and indeed the instruction manual will describe the best procedure for operating a particular instrument. This book aims to create an informed background for the operation of electron microscopes so that they may be used more intelligently and at a higher level of performance. Since the majority of those using the electron microscope are in the biological sciences, an attempt has been made to present the information in a descriptive rather than a mathematical form.

It will be apparent from the aims of this book that it has no pretensions to be a complete survey of the electron optics of the instrument. Those wishing to look deeper into this subject can consult a number of useful books (Zworykin et al. 1945; Cosslett 1951; Hall 1953; Haine 1954, 1961; Siegel 1964; Grivet 1972; Hawkes 1972).

1.1 The use of electrons for microscopy

A microscope is an instrument designed to render visible objects which are too small to be seen by the unaided eye. For particles greater than a few tenths of a micrometre in diameter, the light microscope is adequate, but for very small objects the light microscope fails because the wavelength of visible light is large compared with the objects to be examined.

When light rays emanating from a point, pass through a lens of semi-angular aperture α (Fig. 1.1), they form an image which is no longer a point but with the intensity spread out in what is known as an Airy disc (Fig. 1.2). The distance between the two minima on either side of the main intensity peak is given by $D = 1.22\lambda/n \sin\alpha$, where λ is the wavelength of the light, and n the refractive index of the material in which the object lies. When two

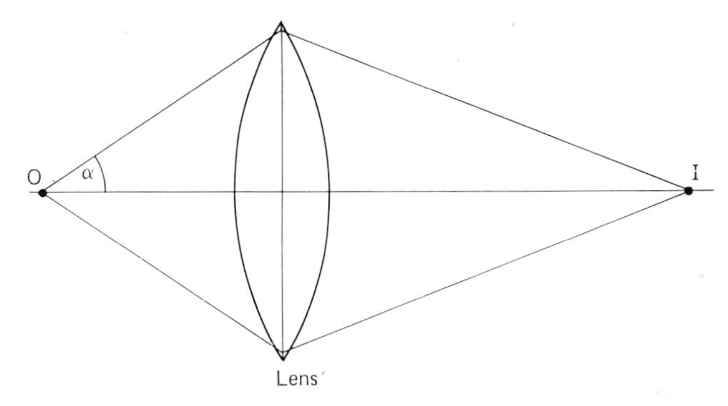

Fig. 1.1. Lens aperture defining Airy disc. The semi-angle α of the lens subtended from the object point O defines the image disc diameter at I (see Fig. 1.2 for intensity profile).

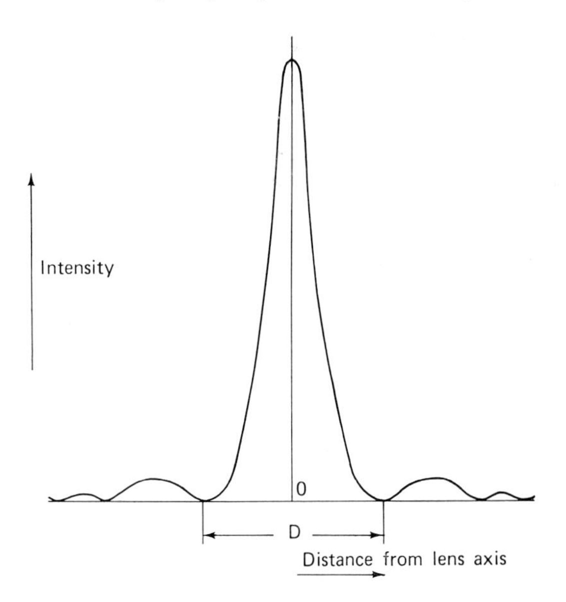

Fig. 1.2. Intensity profile of the Airy disc at the image point I. The ordinate represents image intensity. O represents the lens axis. The abscissae are distances from the lens axis. D is the diameter of the central intensity disc, given by $D = 1.22\lambda/n \sin \alpha$.

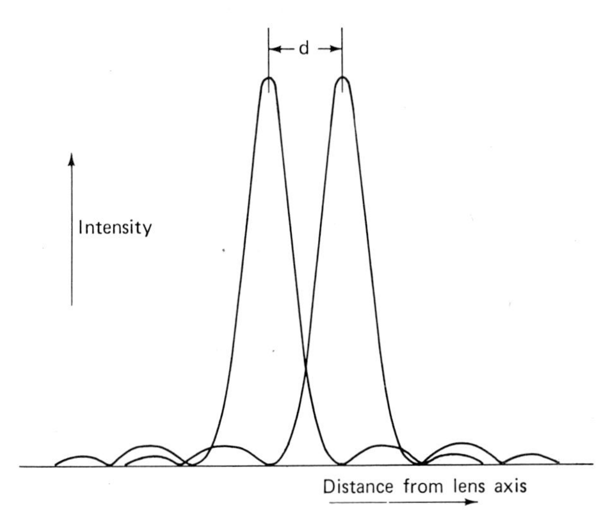

Fig. 1.3. The definition of resolution in terms of Airy disc separation. The central maximum of one image point coincides with the first minimum of intensity of the adjacent (just resolved) image point, distance $d = 0.61\lambda/n \sin \alpha$.

emitting points of the object lie very close together (Fig. 1.3) the intensity patterns in the image will overlap. The resolution of the system is defined as the distance between the maxima when the maximum intensity from one point is coincident with the first minimum from the other point. The separation of the two points is then

$$d = \frac{0.61\lambda}{n \sin \alpha} \tag{1}$$

Note that this separation does not depend on any property of the lens except its semi-angular aperture.

In a light microscope, using an oil-immersion objective lens, the value of $n \sin \alpha$ may be as large as 1.4, and taking a wavelength of 500 nm (1 nm $= 10^{-9}$ metres) the resolution limit is seen to be about 200 nm (2000 AU in the more familiar Ångstrom Units). Because the designers of glass lenses have over the years learnt how to correct most of the lens defects, it has proved possible to build instruments which approach close to this limiting performance. This limitation, called the diffraction limit, is due to the size of the wavelength of light and no further improvement can be expected without using a different illumination of shorter wavelength.

It was shown by De Broglie (1924) that an accelerated electron beam has an effective wavelength λ (in nm) given by

$$\lambda = 0.1 \sqrt{\frac{150}{V}} \tag{2}$$

where V is the accelerating voltage. For $V = 60{,}000$ V, $\lambda = 0.005$ nm, a wavelength shorter by a factor of 10^5 than visible light. In principle, therefore, an electron microscope should be capable of imaging atomic structures, if one considers only the wavelength (diffraction) limitation. As will be seen later other factors limit the achievable resolution to a considerably poorer figure.

1.2 Electron lenses

Although one can in principle use either electrostatic or magnetic lenses to focus a beam of electrons, practical instruments exclusively employ magnetic lenses as they can be made with smaller defects than electrostatic lenses, so only magnetic lenses will be considered here. The magnetic field required to form an electron lens is a locally strong axial field in the direction of the electron beam. The envelope of the electron beam is exactly analogous to that of a light beam passing through a converging lens, except that the electrons travel in a helical path through the lens; this additional rotary motion is a characteristic of magnetic lenses (Fig. 1.4). Since the rotary motion does not affect the focusing behaviour of the lens, the ray diagrams for a beam through an electron lens are identical with those for a light optical lens, and the lens formulae for light optics apply.

In a magnetic electron lens the focal length is determined by the field strength in the lens gap and by the speed of the electrons (determined by the accelerating voltage). With a given lens geometry, the strength of the field in the pole piece gap is proportional to the excitation of the lens, usually

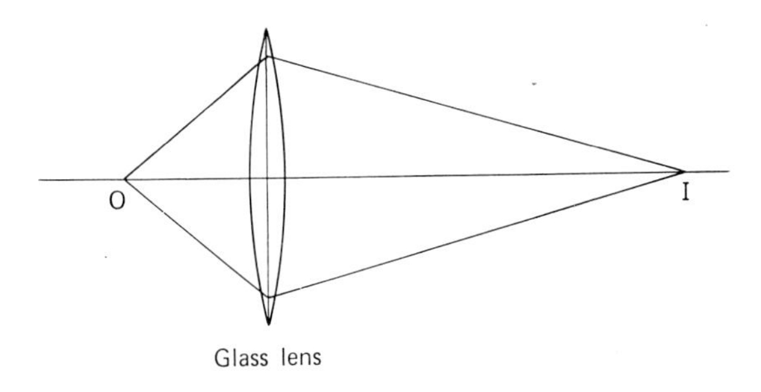

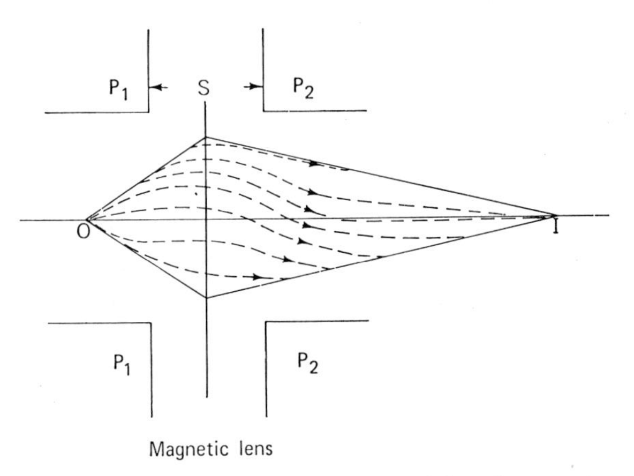

Fig. 1.4. Comparison of light and electron lenses. Each lens forms an image I from the object point O, so that the geometrical image formation is identical, but the electrons rotate in a spiral trajectory about the lens axis as they pass through the magnetic field formed between the pole pieces P_1 and P_2. The dotted lines indicate electron paths over the cone surface.

defined as the product of current in the windings and number of turns in the coil. Thus

$$f = K \frac{V_r}{(NI)^2} \tag{3}$$

where

f is the focal length of the lens;
K is a constant;
V_r is the accelerating voltage, relativistically corrected; and
NI the ampere turns in the excitation coils.

In practice, an electron microscope requires an accelerating voltage of at least 40 kV, since for lower accelerating voltages the electron penetration through the object is too small, and would result in a requirement for such thin specimens that there would be insuperable difficulties in specimen preparation. For accelerating voltages of 40 kV and above, the electrons are already travelling at an appreciable fraction of the velocity of light, and it is therefore necessary to use a relativistic correction to give the effective value of accelerating voltage V_r.

If the nominal voltage is V_0,

$$V_r = V_0\left(1 + \frac{eV_0}{2mc^2}\right) \tag{4}$$

where c is the velocity of light and e and m are the charge and mass of the electron, respectively, or

$$V_r = V_0(1 + 0.978 \times 10^{-6}\, V_0) \tag{5}$$

The great majority of electron microscopes operate in the voltage range 60–100 kV. A few are now operating at 1000 kV and above (§ 9.2).

It can be seen from Eq. (3) that the strength (focal length) (f) of a magnetic lens can be varied by altering the current (I) flowing through the windings, a great convenience in operation. It will also be seen that for a beam of more energetic electrons the lens current has to be increased in order to keep the focal length constant.

Following the analogy between an electron lens and a glass converging lens, a ray diagram can be constructed to show where an image will be formed (Fig. 1.5). (Throughout this book, the conventional converging lens of a light optical system will be used to denote a magnetic lens in the ray diagrams explaining the optical functioning of parts of the electron microscope.) The focal length is defined by the distance LF, F being the point to which all incident rays parallel to the lens axis (XX′) converge. A ray from a point on the object which passes through the centre of the lens will be undeflected. A ray parallel to the lens axis will pass through F on the lens axis. The intersection of these two rays will define the image point A′ corresponding to the object point A. Similarly the image point B′ corresponds

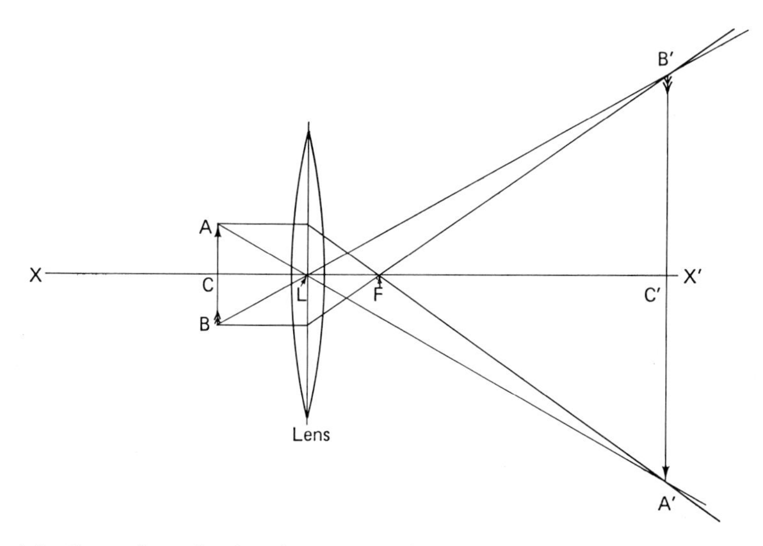

Fig. 1.5. Image formation in a lens. XX′ is the axis of the lens L. An object ACB is imaged at B′C′A′, with a magnification M where $M = v/u$; u is the object distance CL and v the image distance LC′. The focal length of the lens $f_0 = \mathrm{LF}$.

to the object point B. Thus the lens forms an image A′B′ of the object AB and, if image and object distances are u and v, respectively, the magnification of the lens in this mode of operation is M where $M = v/u$. It should be noted that the point C on the axis is imaged at the point C′ also on the axis of the lens. The plane through F normal to the axis is known as the *back focal plane*.

Following on from the analogy between light and electron lenses, there is a similar conformity in the geometrical optics of the light and electron microscopes, as shown in Fig. 1.6. The electron microscope is shown in its simplest form. The electron gun provides the illumination source. The illumination is then projected onto the specimen by a condenser lens. The specimen is imaged by the objective lens and the image is further magnified by a projector lens. (There is no direct comparison with an eyepiece of a simple light microscope, since the eye cannot perceive electrons directly. The analogy must therefore be with the projection microscope, where the image

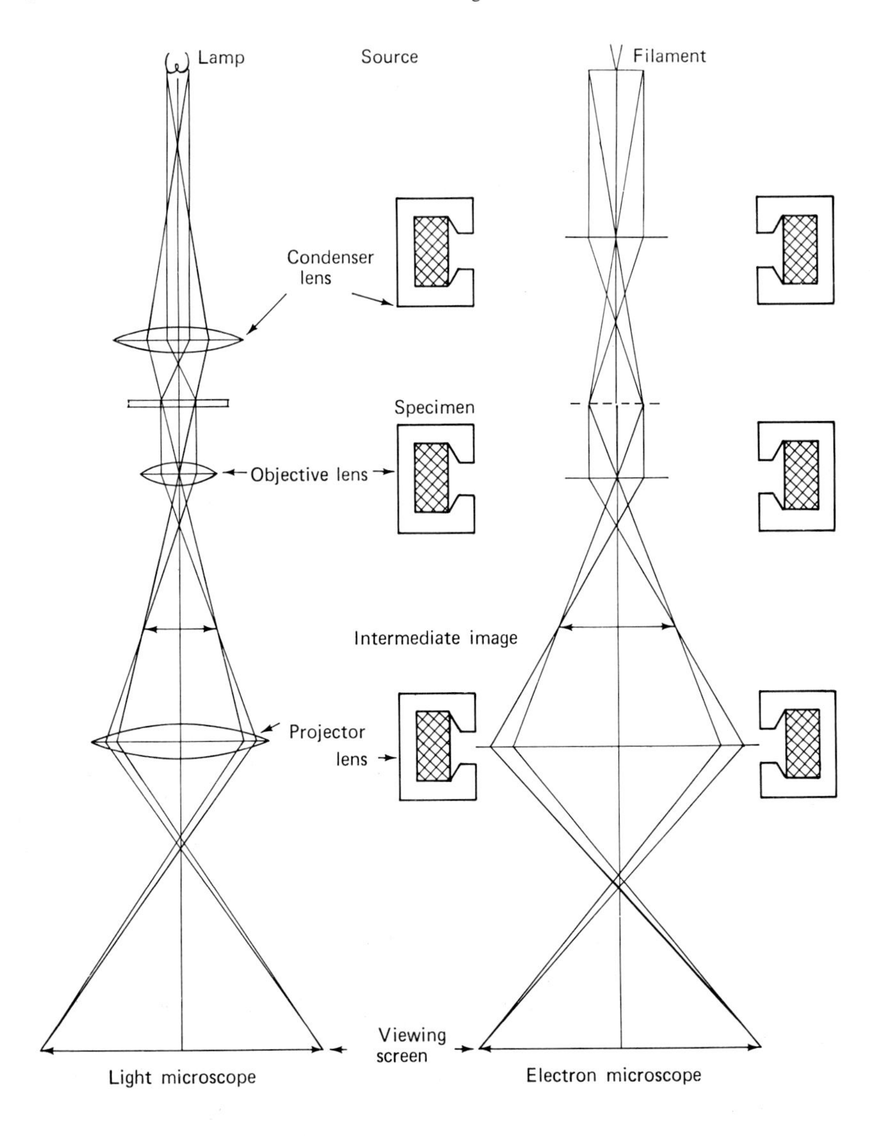

Fig. 1.6. Comparison of light and electron microscopes. In each instrument, illumination from the source (lamp, filament in the electron gun) is focused by the condenser lens onto the specimen. A first magnified image is formed by the objective lens. This image is further magnified by the projector lens onto a ground glass screen (light) or fluorescent screen (electrons).

appears on a ground glass screen.) The use of the same names for the lenses in the light and electron microscopes is a convenient reminder of their analogous functioning.

1.3 Defects of electron lenses

Electron lenses have similar properties and similar defects to those encountered in glass lenses. However, the solutions to the problems posed are often quite different.

1.3.1 SPHERICAL ABERRATION

Spherical aberration is the inability of a lens to focus all incident beams from a point source to a point. The outer zones of the lens have a greater strength, and light rays or electrons originating from a point are not imaged at a point (Fig. 1.7). F is the paraxial focus (the focus for rays very near the axis). It can be seen that at one point the envelope of the imaged rays has a minimum diameter known as the circle or disc of least confusion. This limiting disc has a diameter d_s given by

$$d_s = \tfrac{1}{2} C_s \alpha^3 \qquad (6)$$

where C_s is the spherical aberration coefficient and α the semi-angular aperture of the lens.

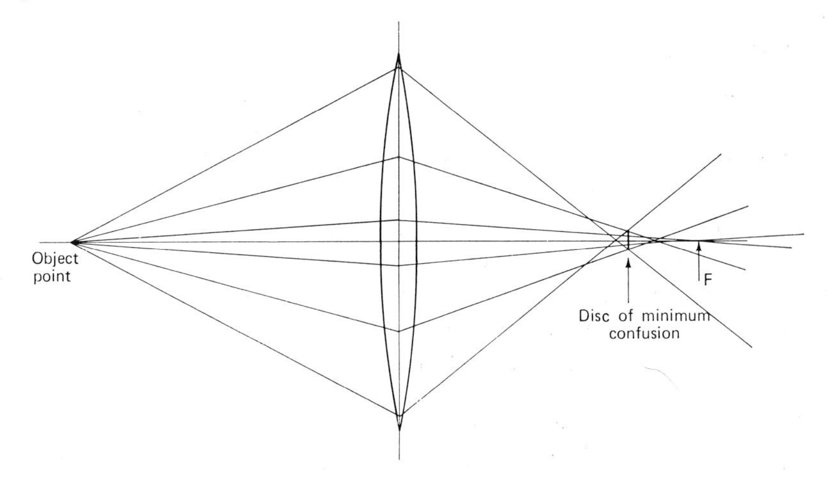

Fig. 1.7. Spherical aberration in a lens. The rays close to the lens axis (paraxial rays) are focused at the Gaussian focus F. Rays entering the lens at a larger angle are converged more strongly. The disc of minimum confusion is where the envelope of emergent rays has its smallest diameter.

In a light-optical system, spherical aberration is corrected by combining a diverging lens of different refractive index with the converging lens. Unfortunately, such a lens cannot readily be realised in an electron-optical system and the defect can only be minimised by reducing the angular aperture, α, of the lens. Camera enthusiasts will remember that a cheap lens with poor properties can yield good pictures if it is well stopped down. This is the procedure used with the objective lens in an electron microscope. There is a limit to how far this process may continue, since if α is made very small indeed the diffraction limit to resolution starts to become important. Eq. (1) for electrons can be written

$$d = \frac{0.61\lambda}{\alpha} \qquad \text{(for small } \alpha) \tag{7}$$

(the refractive index of a vacuum for electrons is unity).

There is consequently an *optimum aperture* at which the two limits to resolution, d_s and d (Eqs. 6 and 7) are equal and for which the effect of spherical aberration is minimised, consistent with the diffraction limitation.

One may then define a theoretical limit to the resolving power of the electron microscope (δ_0) where

$$\delta_0 = BC_s^{\frac{1}{4}}\lambda^{\frac{3}{4}} \tag{8}$$

where B is a constant, value $0.56 > B > 0.43$, λ is the electron wavelength and C_s is the spherical aberration coefficient.

For a 100 kV beam, δ_0 may be as small as 0.19 nm for the electron lenses at present available.

This value of δ_0 has been achieved in the symmetrical objective lens designed by Kunath et al. (1966) in which the specimen is placed at the centre of the lens field of a very strong lens, with the pole piece tips saturated magnetically. In principle, one could obtain much more powerful fields by the use of superconducting lenses, but they present very difficult technical problems both in the lens construction and in the space available for the specimen. It seems unlikely therefore that there will soon be a significant improvement in the resolving power of electron microscopes.

1.3.2 CHROMATIC ABERRATION

This is analogous to the chromatic defect in glass lenses. A cheap magnifying lens gives rise to images with coloured fringes due to the different focal lengths of the lens for red and blue light. The defect can be largely corrected by forming a compound lens employing glass of different refractive indices, but the highest resolution is obtained by using monochromatic illumination.

The chromaticity of the electron beam is determined by the energy spread of the electrons within the beam, since this gives a spread of electron wavelengths. Remembering from Eq. (3) that the focal length of an electron lens is

$$f = K \frac{V_r}{(NI)^2}$$

and differentiating,

$$\frac{\Delta f}{f} = \frac{\Delta V_r}{V_r} - \frac{2\Delta I}{I} \tag{9}$$

which shows that a variation in either accelerating voltage or lens current will cause a change in the effective focal length of the lens.

The way in which these variations affect the resolution is dependent on the sensitivity of the objective lens to such variations. The chromatic aberration coefficient C_c is the parameter which expresses this quality, and the resolution d_c as limited by chromatic effects is given by

$$d_c = C_c \alpha \left\{ \frac{\Delta V_r}{V_r} - \frac{2\Delta I}{I} \right\} \tag{10}$$

where α is the objective semi-angular aperture.

The chromatic aberration coefficient is similar in magnitude to the focal length, although it can be modified by perhaps a factor of two by choice of lens design. Eq. (10) leads to a requirement for a stability in the objective lens supply and the high voltage supply of a few parts per million for a resolution of 0.5 nm.

1.3.3 ASTIGMATISM

This is a defect of magnetic field asymmetry resulting in differing lens strengths in two directions at right angles. The result is that a point object is imaged at two focal lines (Fig. 1.8), and the smallest circular image of the point object is halfway between the focal lines. The amount of the astigmatism of the lens is defined as the distance between the focal lines. The astigmatism of an objective lens in a good microscope will be typically 1 μm or less.

The defect arises from lack of perfection in the machining of the lens pole pieces, in particular lack of circularity in the bores and flatness of the pole faces and also from asymmetry in the magnetic material of the lens itself. Fortunately, it is a defect which can be corrected without too much difficulty by using a compensating cylindrical lens field, at least to a level at which it does not significantly affect the final resolution of the microscope. The correction levels required and the mode of achieving them are described in § 4.9.

There are other imaging defects, but those enumerated above are the most important in determining ultimate resolution, and all that need be considered by the practising microscopist. For the reader who wishes to follow the development of the basic theory of electron lens design, the original papers of Glaser (1941) and Liebmann (1951a, 1951b, 1952, 1955) are recommended.

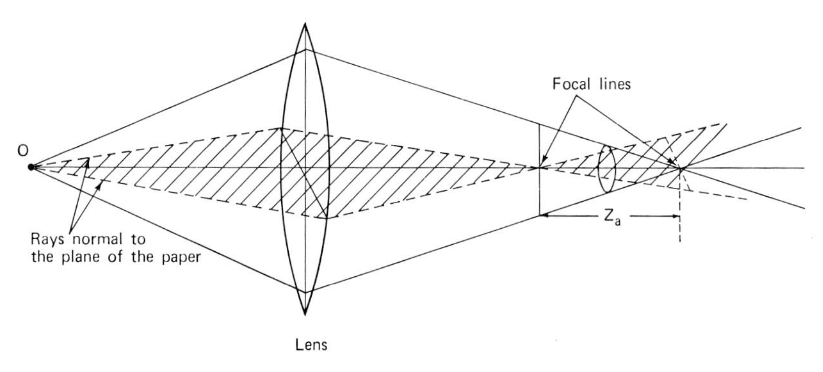

Fig. 1.8. Image formation with an astigmatic lens. The lens is stronger in a plane perpendicular to the paper than in the plane of the paper, so that a point object O is imaged into two focal lines. The astigmatism of the lens is measured by Z_a, the distance between the focal lines. A circular image is formed halfway between the lines.

The logic of how a practical design is evolved is discussed by Haine (1961) and Fert and Durandeau (1967). Many electron optics centres now have computer programmes for designing lenses which will produce accurate information provided the essential parameters are determined first.

1.3.4 CHROMATIC CHANGE OF MAGNIFICATION

This is a defect which is not always recognised, and, in fact can only be removed at the design stage, and is not under the control of the operator. The effect can be described as the variation of the lens strength with the distance from its axis, for electrons of different wavelength. This means that, near the edge of the field of view, the magnification of the final image will be different for the unscattered electrons and the scattered electrons which have lost energy in transit through the specimen. This results in an image which is blurred at the corners, although it appears sharp in the centre. In practice, the effect is only noticeable at low magnification (where the electrons travel far from the lens axis), and with relatively thick specimens (which cause large energy losses in the electron beam).

1.4 Depth of field

One other feature of cheap camera lenses, a great depth of field, is also characteristic of electron lenses with their small angular aperture. It is a property which turns out to be exceedingly useful, as will be seen later. It appears to be one of the few physical implications of the design which are useful as opposed to restrictive!

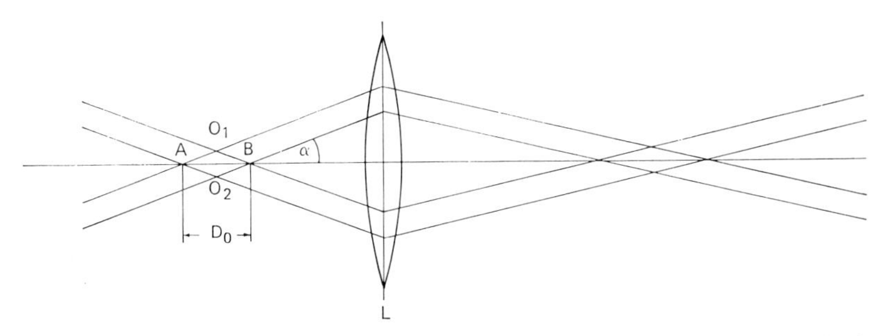

Fig. 1.9. Depth of field. Object points O_1 and O_2 are separated by the resolution limit d of the lens L. Rays from these points cross the axis at A and B. Hence points between A and B will look equally sharp and AB is the depth of field D_0 of the lens, for a semi-angular aperture α.

The depth of field D_0 of an electron lens is the axial distance over which the lens may be focused without a perceptible change in image sharpness. Let O_1 and O_2 (Fig. 1.9) be object points separated by the limiting resolution distance d for the lens. Parallel rays through O_1 and O_2 cut the axis of the lens at A and B, and any points within the distance AB will therefore appear to be equally sharp. The distance AB is then defined as the depth of field D_0 where

$$\frac{d}{2} = \frac{D_0}{2} \tan\alpha$$

Hence

$$D_0 = d/\alpha \tag{11}$$

where α is small.

Thus for $d = 1$ nm and $\alpha = 5 \times 10^{-3}$ radians,

$$D_0 = 200 \text{ nm}$$

This is considerably greater than the thickness of any specimens normally examined in a 100 kV electron microscope at this level of resolution, and thus the specimen appears equally sharp throughout its thickness. Note, however, that the depth of field is not an absolute figure; for a resolution of 0.5 nm, and an objective aperture of 10^{-2} radians, the depth of field is reduced to 50 nm and is becoming a limiting factor for some specimens. The effect is much more marked in the very high voltage microscopes where specimens with thicknesses up to 10 μm have been examined. This indicates the clear possibility of optical sectioning, by altering the fine focus control to bring different levels of the specimen into focus (Cosslett 1969). This is a technique in common use with the light microscope.

1.5 General features of electron microscope design

The foregoing paragraphs have shown that, at the present state of the art, the best attainable resolution of a 100 kV electron microscope will be about 0.2 nm; high performance commercial instruments now routinely operate at the 0.5 nm level, although the performance of a given instrument with a particular specimen may be worse, for a number of reasons which are discussed later.

This capability of very high resolution carries implications for the whole of the design of the instrument. Firstly, if one is to perceive detail so small, the magnification available must be very large. The top magnification of a high performance instrument must be at least 200,000 ×, and is often higher. This calls for at least three magnifying lenses, (including the objective lens). The image intensity is inversely proportional to the square of the magnification, since each doubling of magnification spreads the electron intensity over an image area four times as large. In order to maintain an adequate viewing intensity, therefore, as the magnification is increased, it is necessary to have an exceedingly bright source and an efficient illumination system for the electron microscope.

The high resolution calls for great mechanical rigidity of the microscope column, and isolation from outside vibrations, and the whole instrument must be maintained at a high vacuum in order to permit uninterrupted passage of the electron beam. Since in the transmission mode of operation (the normal mode for most specimens) the electron beam has to pass through the specimen, the specimen itself must be very thin (less than 0.2 μm for a 100 kV beam).

In the remainder of this chapter, the important features of the optical design implicit in the above outline are considered.

1.6 The electron gun

It is fortunate that an illumination source of very high intensity can be provided by employing a heated tungsten cathode as a source of thermally-emitted electrons, and by concentrating and controlling the electron beam by a biased cathode shield (or Wehnelt cylinder) between the filament and the anode.

The operation of electron guns was studied in a model system by Haine and Einstein (1952) and they were able to show that by an appropriate combination of operating parameters, a 'brightness' β (A/cm^2/steradian) closely approaching the maximum that is theoretically possible from a heated wire can always be obtained from the gun. This brightness is given by

$$\beta = \frac{\rho_c eV}{kT} \,. \qquad (12)$$

Where ρ_c is the cathode current density, e the electronic charge, V the accelerating voltage, k Boltzmann's constant (8.6×10^{-5} eV/°K) and T the absolute temperature.

It is seen from Eq. (12) that the brightness is proportional to the accelerating voltage – a fact apparent when the kilovoltage of operation is changed in the electron microscope. The effect of the temperature is more complex. ρ_c, the cathode current density, rises exponentially with T and therefore the brightness rapidly increases with the temperature of the filament, even though T also appears as a factor in the denominator.

Haine and Einstein (1952) showed that there are four important parameters in an electron gun (Fig. 1.10). If the distance of the filament tip behind the aperture in the Wehnelt cylinder is defined as h, the parameters are h, T the temperature of the filament, V_b the negative bias voltage of the Wehnelt

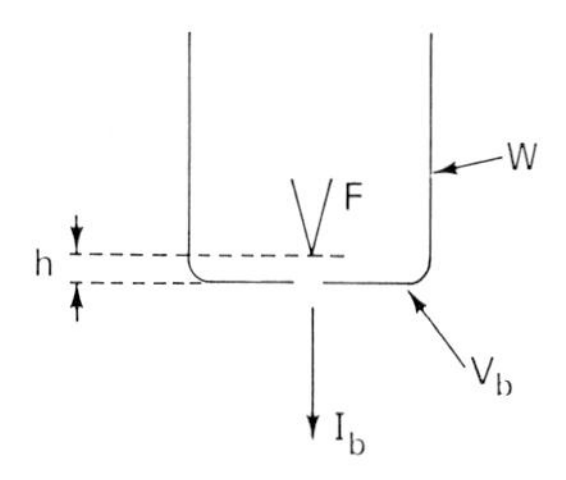

Fig. 1.10. Parameters in an electron gun. The filament F is contained in a Wehnelt cylinder W. The filament is at a height h with respect to the aperture in W (h is positive for a filament tip behind the aperture). The bias voltage V_b is the voltage between F and W. Electron beam current is I_b.

cylinder with respect to the cathode and β the brightness. Taking any shape of Wehnelt cylinder (flat or re-entrant) and an arbitrary value of h, there is in general some value of the bias voltage which will allow the theoretical value of β to be obtained for the value of T used. The effect of decreasing h, for instance, is to reduce the required bias and to increase the value of electron beam current at which the maximum brightness occurs. The angular width of the beam is also modified by both h and by the shape of the cathode.

These results might seem to indicate that there is no bar to obtaining a gun brightness as high as is desired. Unfortunately, the small increases in filament temperature, leading to large increases in brightness, also lead to a dramatic shortening of filament life, due to the higher rate of evaporation of the material of the wire at elevated temperatures. Most filaments operate on a constant-current stabilised supply, and once a slight thinning has taken place by evaporation, the temperature of that part of the wire rises further, thus increasing the evaporation rate. The failure of the filament will usually occur at about 6% thinning. Fig. 1.11 shows how the temperature of the filament is related to gun brightness and filament life. Therefore, in practice, there is

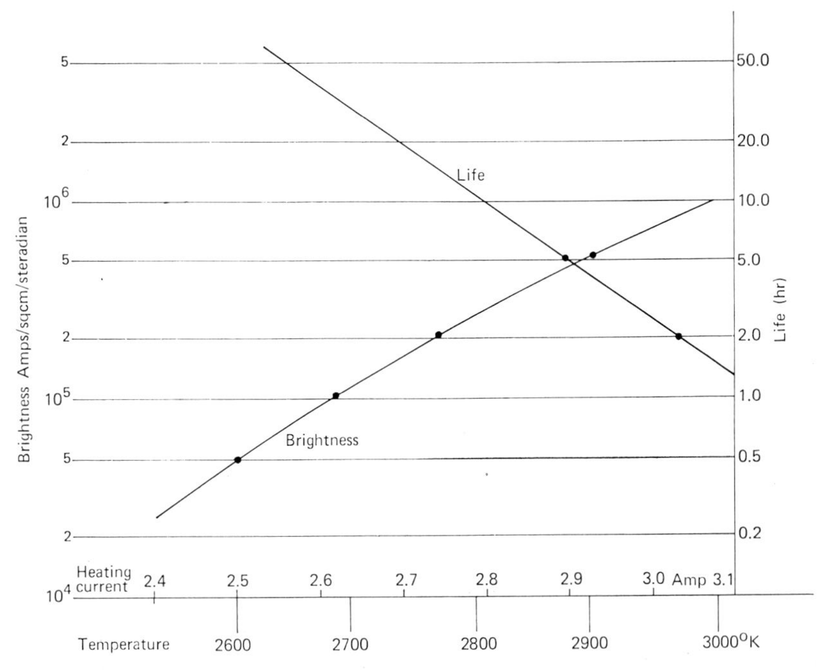

Fig. 1.11. Relationship between filament temperature, gun brightness and filament life. Note the logarithmic dependence of both brightness and life upon filament temperature.

a relatively narrow band of possible operating conditions. If the filament is not hotter than say 2700 °K, the brightness is insufficient for high magnification operation. If the temperature is greater than 2900 °K, the life of the filament is impractically short.

Because the cathode current density rises so rapidly with temperature, the gun emission has to be stabilised, and the common form of control is the self-biased gun. This is obtained by including a high resistance in one of the supply leads to the filament (Fig. 1.12) so that the electron beam current

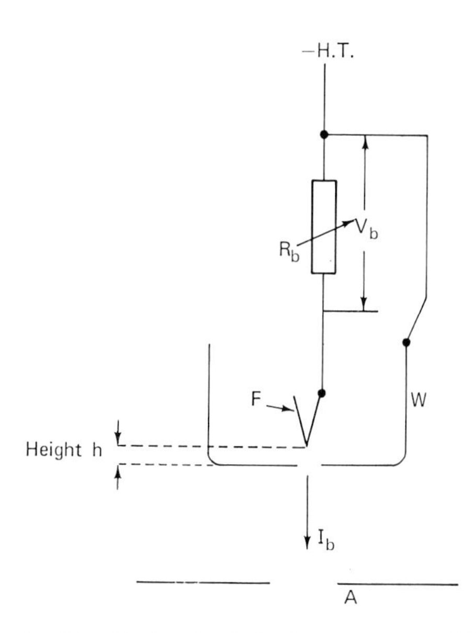

Fig. 1.12. The principle of the biasing circuit for the electron gun. The Wehnelt cylinder W is biased with respect to the filament F by the potential drop across the bias resistor R_b. The beam current is I_b and filament height is h. A is the anode.

generates a voltage across this resistance. This bias voltage is applied to the Wehnelt cylinder. Any increase in beam current causes an increase in the bias voltage which acts to reduce the beam current again. The use of this circuit modifies the electron gun characteristic so that it assumes the form shown in Fig. 1.13 (Haine et al. 1958). The reason for this modification of the characteristic is that the bias voltage is no longer an independent variable at the control of the operator, since it changes automatically with the beam current. Commercial microscopes are designed with other parameters pre-set or with recommended settings (e.g. filament height) so that the operator can normally achieve the maximum available brightness with the bias and filament controls provided.

The stages in the heating of the filament may be observed by using the condenser system to image the filament tip as the heater current is gradually increased from zero. The points marked (i) to (v) on the curve in Fig. 1.13 roughly correspond to different forms of the image of the filament. When the heating current is low (i), the beam current is low and there is hardly any bias on the Wehnelt cylinder. Electrons from a considerable area of the filament

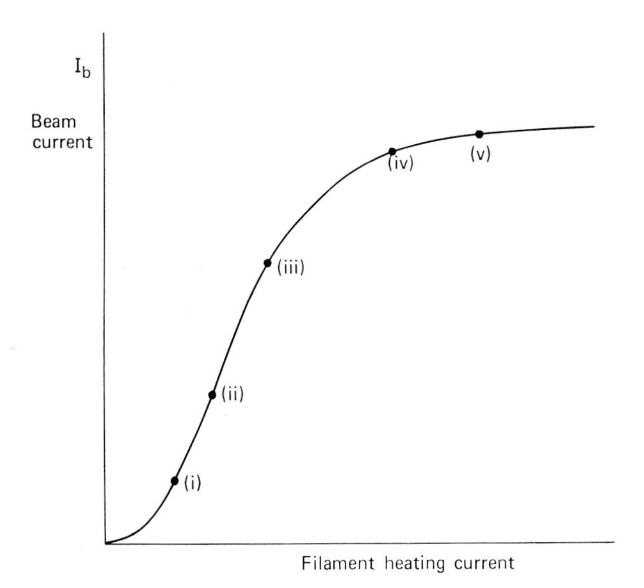

Fig. 1.13. Plot of electron beam current against heating current for an electron gun. The points (i) to (v) correspond to stages in 'saturation' of the filament illustrated in Fig. 1.14.

wire can therefore escape past the bias shield, even some from the rear of the filament wire, and multiple spots appear (Fig. 1.14a). Further heating of the filament (ii) and (iii) causes a roughly proportional increase in beam current and increased bias voltage, so that the area of the filament from which electrons can reach the anode is decreased (Figs. 1.14b and c). Fig. 1.14c shows that only the front of the filament wire is now able to emit electrons which will pass the bias screen. Further heating (iv) causes loss of most of the structure in the image of the filament wire (Fig. 1.14d). As 'saturation' is approached (v), the spot reduces in size and increases in intensity (Fig. 1.14e). The smallest beam size and maximum intensity are obtained by imaging an

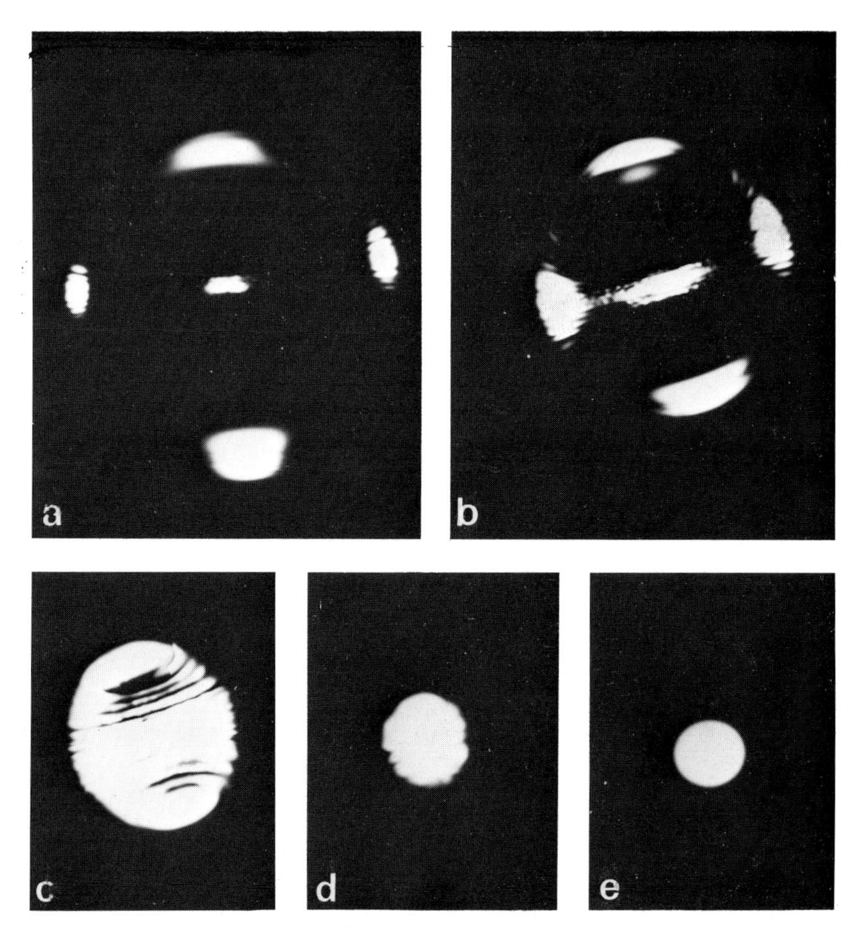

Fig. 1.14. Images of the electron source. (a) Very low filament current. (b) Increased current, spots collapsing. (c) Increased current; one large spot with structure. (d) Increased current; spot contracting, small amount of residual structure. (e) Operational condition, filament 'saturated'.

electron beam crossover just in front of the filament tip, and it is this beam crossover which is the effective source for the electron microscope. Fig. 1.15 shows the form of the electron beam trajectories under conditions of increasing beam current and bias voltage and illustrates how the crossover becomes the effective electron source. The size of this source is of the order of 30–50 μm in diameter. Further increase in heating of the filament does not result in any further increase in beam current or intensity. However, it does produce a higher filament temperature, with shorter filament life, so it is

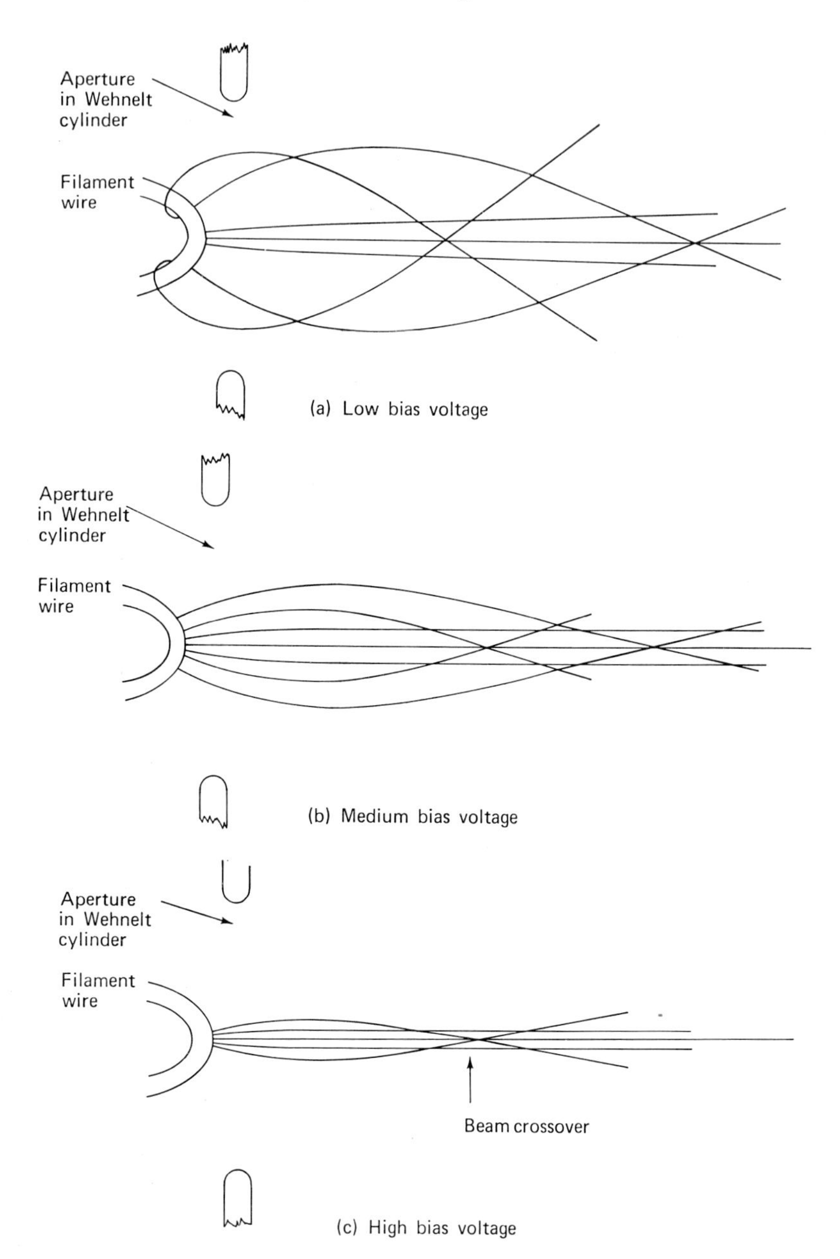

Fig. 1.15. Magnified diagrams of the filament and the aperture in the Wehnelt cylinder to show the general form of electron beam trajectories under conditions of increasing beam current and increasing bias voltage. (a) Low bias – multiple beam spots. (b) Medium bias voltage – spots converging. (c) High bias voltage – single high intensity crossover.

important in practice just to achieve this plateau of intensity and then to avoid any further increase in filament current.

Here it should be mentioned that a properly saturated gun may still result in some striation of the illumination spot, and it may not disappear with additional heating of the filament. The striations arise from poor quality filaments with strong drawing marks in the tungsten wire. This gives rise to zones of strong and weak emission within the small area from which electrons are drawn from the filament, and there is no way of avoiding the problem (except to change the filament). It can be minimised by working with a very high bias voltage, which will restrict the emitting area of the filament still further and may avoid the irregular region.

1.7 The condenser system

The simplest condenser system is a single, rather weak lens which projects the electron crossover in the gun to a concentrated image at the specimen plane. The geometry of the arrangement usually results in a focused spot of illumination of about 30 μm at the specimen (Fig. 1.16a). This is much larger than is needed for the illumination of the specimen for high magnification operation – at a magnification of 100,000 × on the plate, the diameter of the area of the specimen which is imaged is only about 1 μm on the largest plates used. The large area of illumination is therefore wasted for imaging purposes and only results in excessive heat dissipation in the specimen and unnecessary electron irradiation of regions not yet examined. For these reasons single condenser illumination systems are now rarely employed.

A double condenser system has a strong first lens to reduce the electron crossover to an image of about 1 μm diameter and a weaker second lens to project this crossover onto the specimen plane (Fig. 1.16b). The second lens produces a slight magnification, giving a final focused beam diameter at the specimen of 2–3 μm. It will be noted that for a given size of aperture in condenser 2, the cone of electrons accepted from the electron source is much reduced compared with single condenser operation, and is more likely to be uniform and of high intensity. When a greater area of the specimen needs illumination, as in low magnification operation, the second condenser lens is defocused, usually by overfocusing, that is by forming the crossover above the specimen (Fig. 1.17). In normal operation, the first condenser lens is not altered in strength, as it is advantageous to retain a small spot for working at optimum resolution and for minimum irradiation. If a more intense beam

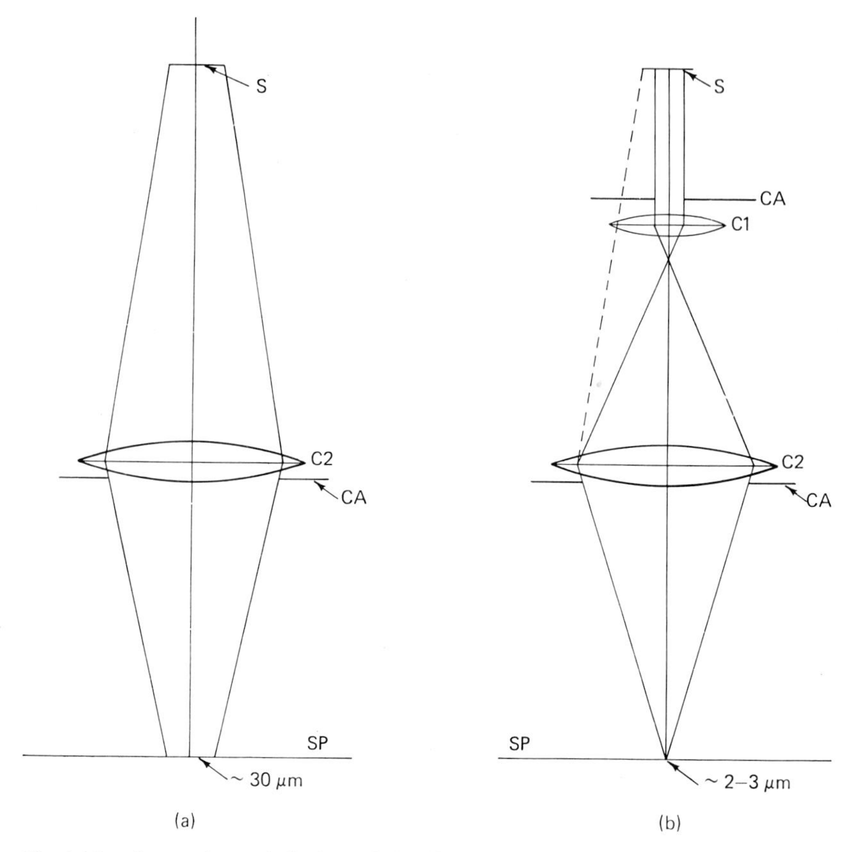

Fig. 1.16. Comparison of single and double condenser systems. (a) Single condenser system. (b) Double condenser system. The condenser 2 lens C2 projects an image of about 30 μm diameter on the specimen plane SP from the source S. When the condenser 1 lens C1 is also used, the projected image is 2–3 μm in diameter. Apertures CA limit the beam angle.

is required (for penetrating metal foils, for example) the focused spot size may be increased by weakening the first condenser lens.

1.7.1 CONDENSER APERTURES

The second condenser lens is equipped with a physical aperture which limits the beam striking the specimen. This is necessary both to protect the specimen from excessive heating and to limit the generation of X-rays from the microscope which might otherwise reach harmful levels.

The size of the aperture in the second condenser lens determines the maximum semi-angular aperture of the illumination, α_c, as viewed from the

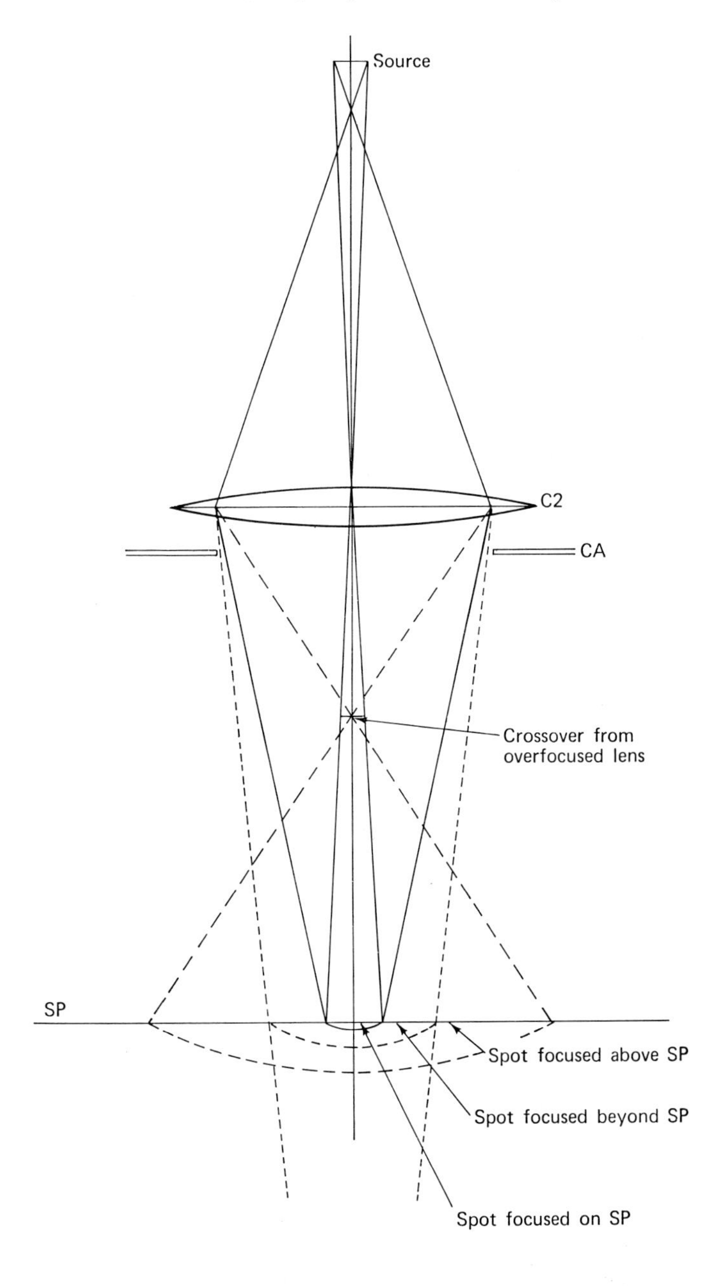

Fig. 1.17. Variation of beam intensity at the specimen by condenser lens variation. C2 is the condenser lens, CA the condenser aperture. The spot diameter at the specimen plane SP is determined by the focus of C2.

specimen (Fig. 1.18). This maximum angular aperture is achieved near the point of condenser focus, i.e. when the beam crossover is imaged on the specimen. The larger the aperture angle, the greater the maximum illumination intensity, but in general, the poorer the image quality. The loss in image quality may not be apparent at first, but it can usually be perceived easily if

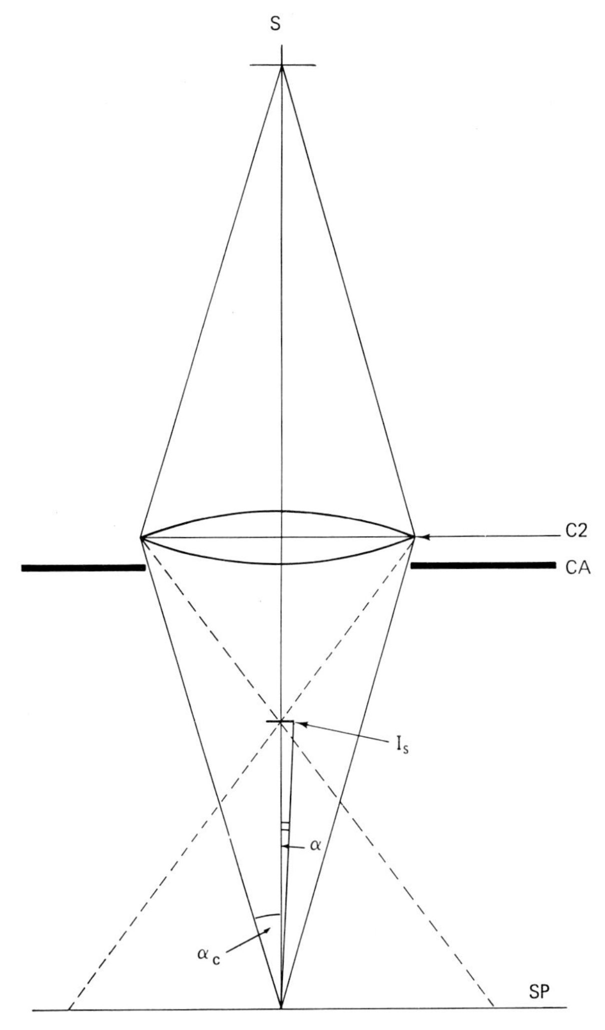

Fig. 1.18. Variation of illumination semi-angular aperture with condenser excitation – ray diagram. When C2 is focused on the specimen plane SP, the condenser aperture CA defines the limiting semi-angular aperture α_c. When the lens is overfocused, the source image I_s defines the semi-angle α.

the aperture size exceeds 250–400 μm in diameter (about 10^{-3} radian semi-angular aperture). If the second condenser lens is defocused, the illumination semi-angular α is defined not by the size of the condenser aperture but by the size of the crossover image and its distance from the specimen. A plot of semi-angular aperture against condenser lens excitation (Fig. 1.19) shows how the maximum aperture is limited by the physical size of the condenser aperture, but that the effective aperture falls very rapidly as soon as the condenser lens is defocused. Thus it is very easy to reduce the effective aperture even with a large physical aperture in position.

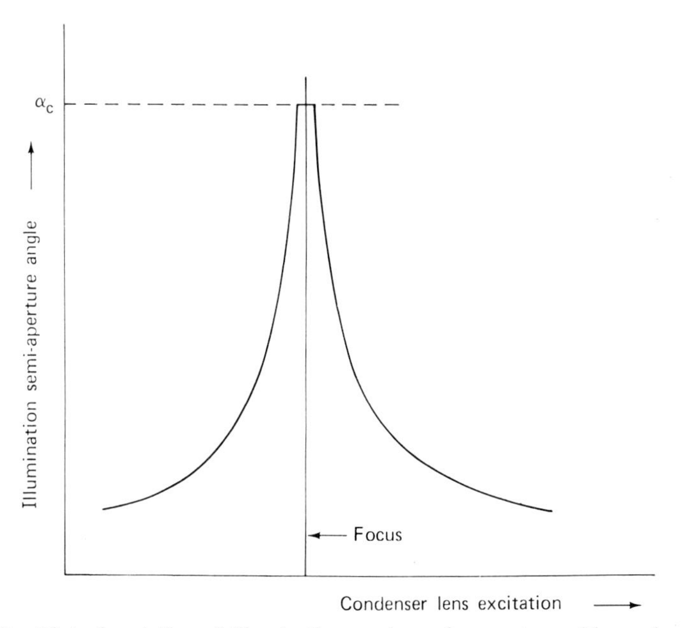

Fig. 1.19. Plot of variation of illumination semi-angular aperture with condenser lens excitation. The limiting illumination aperture angle α_c is determined by the diameter of the physical condenser aperture.

1.8 Beam alignment system

It is necessary to have a convenient system for aligning the electron beam as it leaves the condenser system and before it reaches the specimen. It proves desirable to be able to translate the beam with respect to the specimen without altering its inclination, and to tilt it about the point of observation without translation. Both these motions can be obtained by electrically excited deflector coils. The principle is illustrated in Fig. 1.20. If the lower

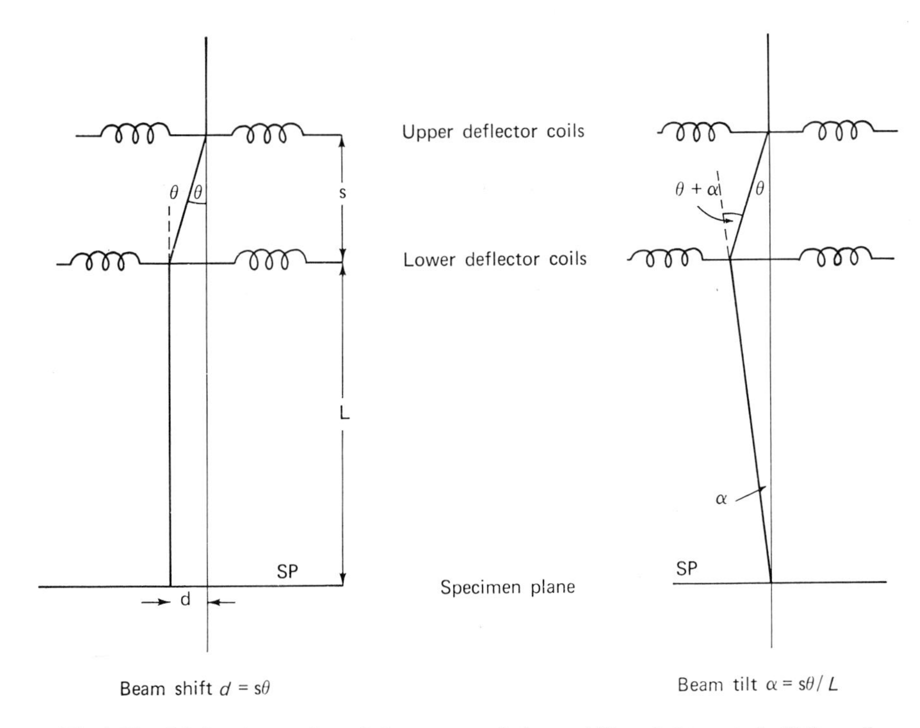

Fig. 1.20. Mode of operation of electromagnetic beam shift and tilt controls. If the coils introduce an equal and opposite beam inclination θ, there is a beam shift $d = S\theta$ at the specimen plane SP. If the lower coils produce a deflexion $\theta + \alpha$, the beam is tilted through an angle α at SP. $\alpha = S\theta/L$.

coils deflect the beam by the same angle θ as the upper coils, a pure translation of $S\theta$ is obtained, where S is the spacing of the coils. If the lower coils provide a larger deflection angle of $\theta + \alpha$, then the beam is tilted through an angle α with respect to its former direction, and provided $S\theta = L\alpha$, the illumination does not move.

1.9 The objective lens

The optical features of the objective lens which fundamentally affect the performance of the electron microscope have already been considered (§1.3). These demand that the lens be manufactured to the highest possible accuracy, but for the most part the operator cannot greatly affect the performance of the lens since its parameters are built into it. However, the lens is equipped with two important controls: the objective aperture and the astigmatism corrector. Fig. 1.21 shows how electrons are scattered through different angles by the specimen, and how the objective aperture, placed at the back focal plane of the lens, stops a number of these scattered electrons. By exchanging the aperture for one of a different size, the effective aperture of the objective lens can be varied, thus varying the proportion of electrons from any given object point stopped by the aperture; as will be seen in § 3.3, this varies the contrast of the image. The semi-angular aperture also affects the resolution (§ 1.3.1 and § 1.3.2). Typically, an objective aperture is 50 μm in diameter. For a focal length of 2.5 mm, the defined semi-angular aperture is 10^{-2} radian.

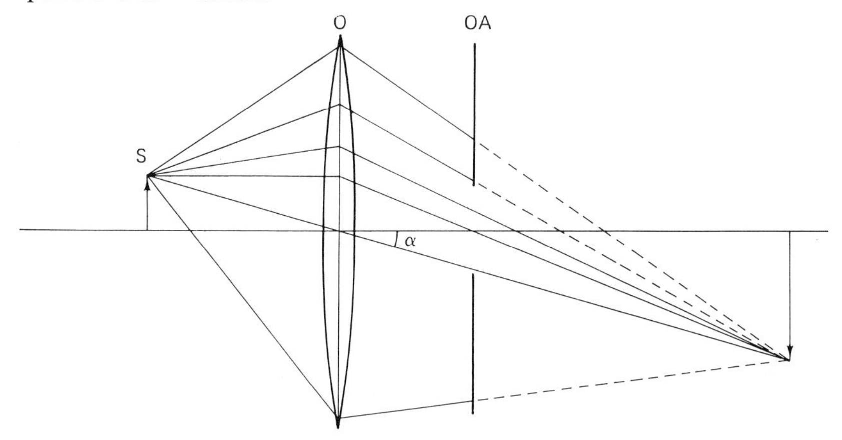

Fig. 1.21. Function of the objective aperture OA in stopping widely scattered electrons from the specimen S in front of the objective lens O; α is the semi-angular aperture of the lens.

The other important component in the objective lens is the astigmatism corrector. It was pointed out in § 1.3.3 that the residual astigmatism of the lens requires correction. The correction is applied by creating a cylindrical deflecting field in the objective lens (either by electromagnetic or electrostatic correctors). The corrector normally has eight poles, and by combining these in opposition, and varying the current applied to the coils, the correcting field may be adjusted to have any desired strength and orientation.

A typical arrangement of an electrostatic corrector is shown in Fig. 1.22, since it is slightly easier to visualise its action. The positive poles attract the electrons and so tend to distort the field as shown by the dotted ellipse. This is adjusted to counteract the original asymmetry of the lens field. An electromagnetic corrector has coil-excited poles in a similar configuration. The actual procedure for correcting astigmatism is described in § 4.9.

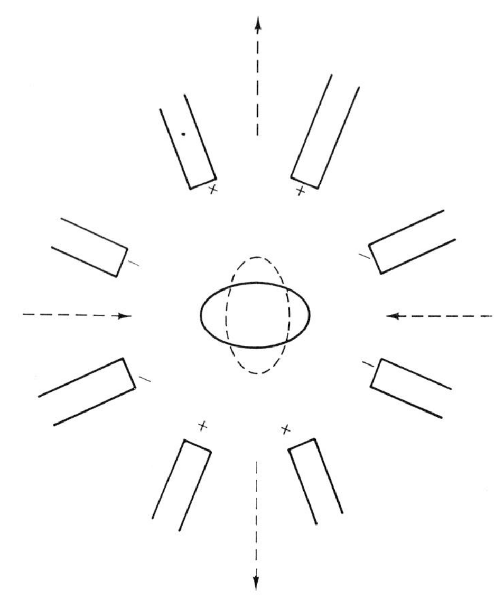

Fig. 1.22. Electrostatic octupole astigmatism corrector. The original lens field asymmetry is shown in full line, the asymmetry due to the corrector in dotted line. The dotted arrows show the direction of the forces on the electrons due to the corrector.

1.10 The projector system

1.10.1 MAGNIFICATION RANGE

The lowest magnification required of an electron microscope will be in the range achievable by a light microscope, so that structures may be directly compared before higher magnifications are used. There is much merit in having available a magnification of 100 × or 200 × for survey purposes as well as the normal imaging range of 1000 × upwards.

The top magnification required is determined by the best resolution attainable by the instrument, by the resolution of the photographic plate (or, for visual observation, the fluorescent screen) and the visual acuity of the human eye.

Considering first the visibility of detail on the fluorescent screen, the size of detail required is related to the image intensity, and the visual acuity of the eye. It has been shown (Agar 1957) that in viewing the screen there is a maximum acuity of about 350 μm, corresponding to a minimum detectable resolution on the screen of about 35 μm, when a 10 × viewing telescope is used. A rather high screen intensity is required to render such detail visible, and it is more efficient to work with a lower screen brightness (of the order of 30 milli foot lamberts, corresponding to a beam intensity of about 3×10^{-11} A/cm^2 at the screen at 100 kV. At this intensity level, the visual acuity is of the order of 700 μm (70 μm at the screen). The acuity falls quite sharply if the brightness is decreased further. On the other hand, the acuity does not improve rapidly enough with increased screen brightness to compensate for inadequate electron optical magnification. This in practice leads to an optimum magnification for examining 0.5 nm detail of about 150,000 ×–200,000 ×.

At a magnification of 200,000 × two points in the specimen separated by 0.5 nm will be imaged 0.1 mm apart on the photographic film. This is about the limiting resolution distance of the human eye in good lighting. If this is enlarged 10 × on to a print, however, the resolved particles are now 1 mm apart, enough to be easily perceived by the unaided eye. However, as will be shown in § 7.2.3 there are considerations of noise in the electron beam/ recording system which demand still higher electron optical magnifications for the highest resolution work. These lead to a desirable top magnification of 500,000 ×. These high magnifications are also useful in getting the best performance out of an image intensifier system.

1.10.2 HIGH MAGNIFICATION OPERATION

The magnification of the final image M_T is the product of the magnifications of the individual lenses, thus

$$M_T = M_0 \times M_1 \times M_2 \tag{13}$$

where

M_0 is the magnification of the objective lens;

M_1 is the magnification of the first projector lens (sometimes called the intermediate lens); and

M_2 is the magnification of the second projector lens.

Typical values of these magnifications would be

$M_0 = 50$

$M_1 = 16$

$M_2 = 250$

whence $M_T = 200{,}000\times$.

A ray path through the lenses for forming a high magnification image is shown in Fig. 1.23. It will be observed that each lens forms a real image, and there is an image inversion with each stage of magnification.

Variation of the magnification can be obtained by reducing the magnification of any one (or all) of these stages. In practice, most of the magnification

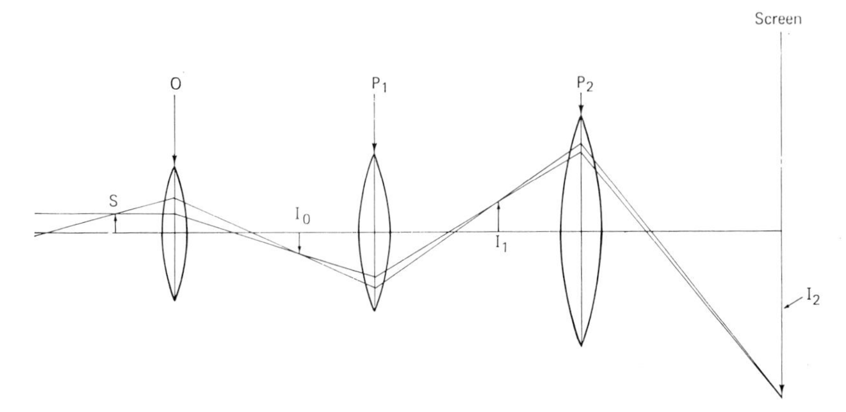

Fig. 1.23. Ray diagram for high magnification mode of operation. Note that each lens forms a real image, with image inversion. I_0 is the image formed by the objective lens O, I_1 is formed by the first projector P_1 and I_2 by the second projector P_2, on the screen.

change is obtained by altering the strength of the intermediate (projector 1) lens. As it becomes weaker, the object plane of this lens moves towards the objective lens (since, with a final projector at fixed strength, the image plane of the intermediate lens remains fixed). In order to retain the image in focus, the objective lens is refocused, and since its image distance is now reduced, both these lenses now have a lower magnification. The limit to this process is set by the onset of an unacceptable amount of image distortion.

1.10.3 IMAGE DISTORTION

The type of beam deflection caused by spherical aberration in an objective lens (§1.3.1) also gives rise to distortion of the image in the projector system. Consider a magnifying lens system forming a real image (Fig. 1.24). For an object ABC, an image A′B′C′ would be formed with a perfect lens system, but owing to spherical aberration the image points are displaced and the displacement of the image point C″ from C′ is greater than that of B″ from B′. This results in an image suffering from pincushion distortion: an extreme example is illustrated in Fig. 1.25.

When a lens is used in a demagnifying mode, the object is virtual, and the image is erect. In this case, spherical aberration causes contraction of the outer parts of the image, as shown in Fig. 1.26. The resulting barrel distortion is illustrated in Fig. 1.27.

The overall distortion of the image is the sum of the distortions in the projector lenses. At high magnifications the lenses are all strongly excited and

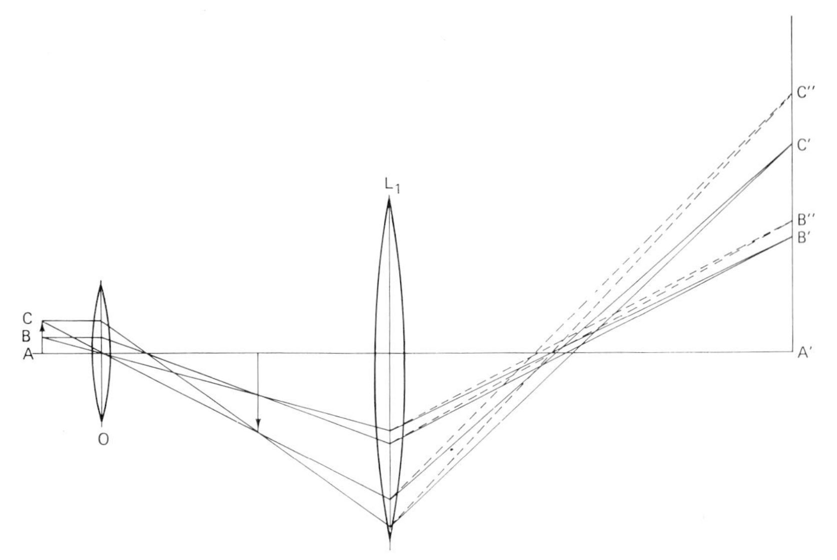

Fig. 1.24. The origin of pincushion distortion in a projector (intermediate) lens forming a real image. An object ABC is imaged at A′B′C′ in the intermediate image plane if the lens is perfect. The spherical aberration causes the actual image to be formed at A′B″C″.

Fig. 1.25. Gross pincushion distortion in an image of a specimen grid.

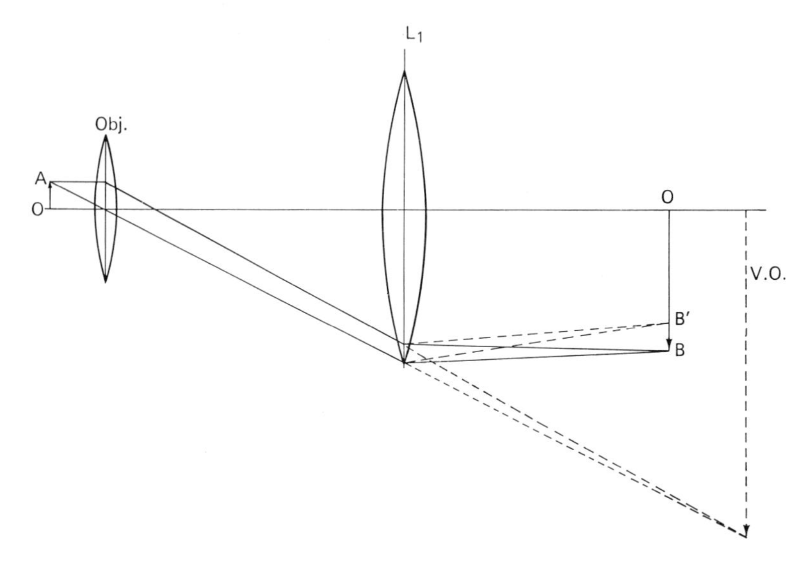

Fig. 1.26. The origin of barrel distortion in a projector lens with a virtual object V.O. The specimen at OA would be imaged at OB by a perfect lens. The spherical aberration of L_1 causes the actual image to be formed at OB′.

Fig. 1.27. Gross barrel distortion in an image of a specimen grid.

the overall pincushion distortion is low (only about 1.3%). As the intermediate lens strength is reduced to achieve lower magnifications, it becomes necessary to use a larger part of the aperture of the lens and the distortion starts to rise at an increasing rate.

1.10.4 LOW MAGNIFICATION

When the pincushion distortion rises in this way, the lenses are switched to a new, low magnification mode. The intermediate lens is now made weak, and the objective is also weakened so that the image from the objective falls behind the intermediate lens, which therefore has a virtual object (Fig. 1.28). With the lens in this condition, it introduces barrel distortion which can be made to balance out the pincushion distortion of the final lens. A significant additional range of magnification can be obtained in this way.

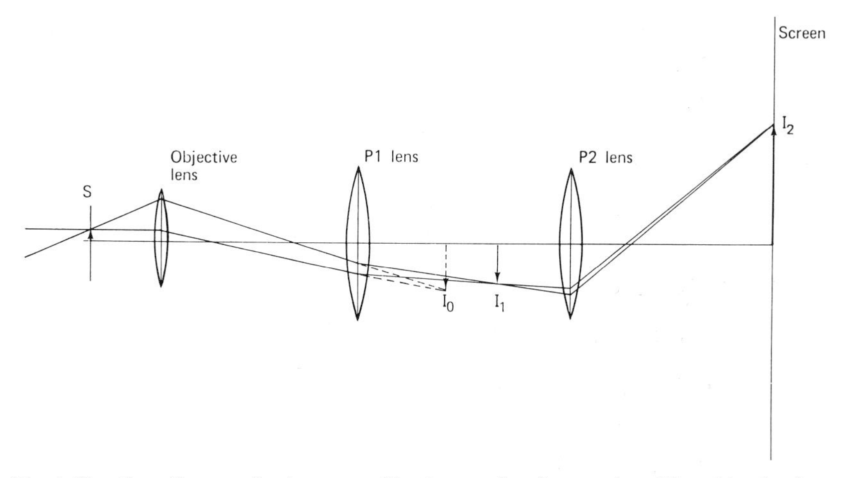

Fig. 1.28. Ray diagram for low magnification mode of operation. The objective lens forms a virtual image I_0 behind the intermediate lens P1, the image of which is at I_1. S is the specimen, I_2 the image on the final screen, and P2 the second projector lens.

It will be noted that, in this mode, the image suffers only two inversions through the system, and at the point where the changeover between high magnification and low magnification operation occurs, the image on the viewing screen will be seen to invert. Finally, as shown by Kynaston and Mulvey (1963) a further range of magnifications can be obtained by variation of the final projector lens strength. In spite of all these precautions it is quite difficult to achieve a large field of view at a magnification of say 1000 × in a three lens imaging system while retaining an adequate top magnification.

Fig. 1.29. Spiral distortion in the image: (a) with lenses set to have opposed rotations; (b) with all lenses having rotations in the same direction. Straight lines drawn on each micrograph show the distortion towards the edge of the picture in (b).

Apart from barrel and pincushion distortion, there is a further distortion caused by the rotation of the image as it passes through a magnetic lens – namely spiral distortion. This is minimised by arranging that the lens rotation directions are balanced against one another as far as possible. Fig. 1.29a shows an image of a cross grating replica formed in a microscope set up to minimise spiral distortion. In Fig. 1.29b, all the lenses are operating to give image rotation in the same sense, so that the spiral distortions add. The increase in distortion is perceptible. There is always some residual distortion at the edges of the image, but in a commercial electron microscope this should be difficult to perceive. The lens strengths are normally programmed to maintain conditions of low image distortion automatically. A distortion of 2% at the edge of the plate is quite difficult to detect visually.

1.10.5 SCAN MAGNIFICATION

It is very useful to be able to view a large area of the specimen in initial search operations. This can be achieved by switching off the objective lens (or leaving it working with a very low excitation) and using the intermediate lens to image the object directly. This mode of operation yields an image of about 200 × in many microscopes. There is of course only one value of magnification at which the image is focused. Some microscopes may be switched automatically to this mode by an appropriate switch. In an instrument with three projector lenses, however, it is possible to obtain a whole range of low magnification focused images by using the first projector as a long focal length objective, and by varying the magnification with the second projector lens. This very low magnification range, obtained in a three-

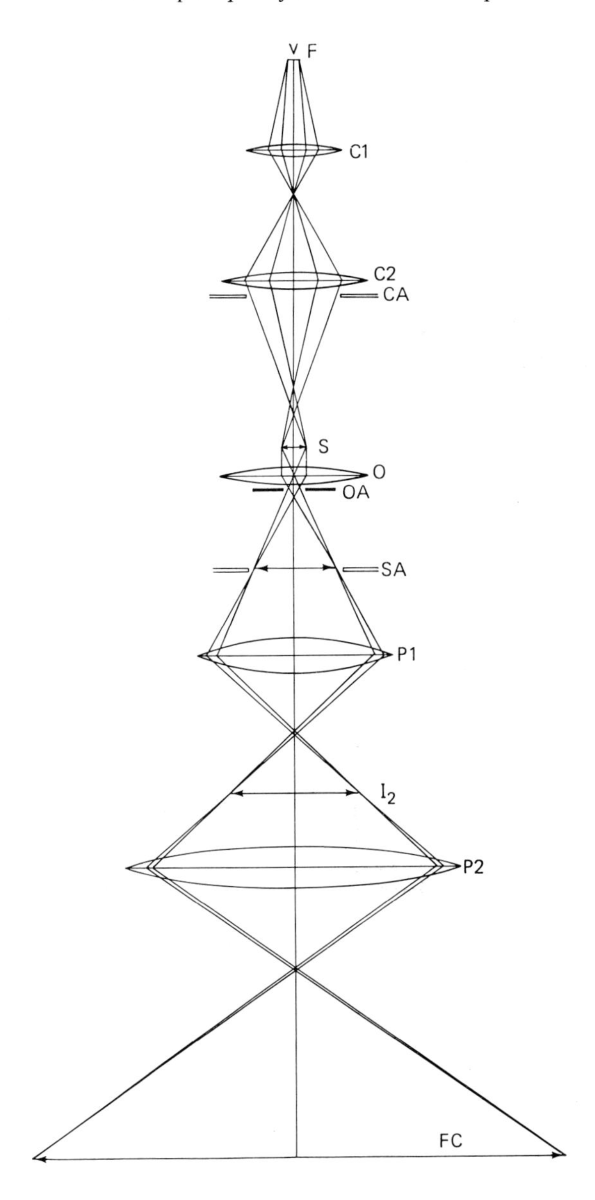

Fig. 1.30. Ray diagram for a complete electron microscope. Filament F, condenser 1 lens C1, condenser 2 lens C2, condenser aperture CA, specimen S, objective lens O, objective aperture OA (in the back focal plane). 1st intermediate image and selector aperture SA. Intermediate lens P1, second intermediate image I_2, projector lens P2 and final image on the fluorescent screen FC.

projector system enables the instrument to achieve a higher top magnification with the objective and two projectors since it is no longer necessary to achieve such a low bottom magnification in the normal imaging mode. (It is not usual to use all four lenses for magnification simultaneously.)

The further great utility of a third projector lens lies in the much greater flexibility of operation in the diffraction mode and this is discussed in more detail in § 3.10.1.

There are other modes of operation of the electron microscope (e.g. dark field § 3.10.3), but this introductory consideration is adequate as a background for the description of the physical appearance and construction of the instrument, described in the next chapter.

1.11 The complete optical system

Putting together all the elements of the design considered above, the optical design of the electron microscope yields a typical layout and ray path shown in Fig. 1.30. This is a microscope working with double condenser in the high magnification mode. The next chapter will describe how this design is realised in commercial electron microscopes.

REFERENCES

Agar, A. W. (1957), On the screen brightness required for high resolution operation of the electron microscope, Brit. J. appl. Phys. *8*, 410.

Cosslett, V. E. (1951), Practical electron microscopy (Butterworths, London).

Cosslett, V. E. (1969), High voltage electron microscopy, Q. Rev. Biophysics *2*, 95.

De Broglie, L. (1924), Dissertation (Paris, Masson) Phil. Mag. *47*, 446.

Fert, C. and P. Durandeau (1967), Magnetic electron lenses, in: Focusing of charged particles, A. Septier ed. (Academic Press, New York and London), Vol. 1, p. 309.

Glaser, W. (1941), Strenge Berechnung magnetischer Linsen der Feldform $H = H_0/(1 + (x/a)^2)$, Z. Phys. *117*, 285.

Grivet, P. (1972), Electron optics (2nd Ed.) (Pergamon Press, Oxford).

Haine, M. E. (1954), The electron microscope: a review, in: Advances in electronics and electron physics, Vol. V1, ed. L. Marton (Academic Press, New York) p. 295.

Haine, M. E. (1961), The electron microscope (Spon, London).

Haine, M. E. and P. A. Einstein (1952), Characteristics of the hot cathode electron microscope gun, Brit. J. appl. Phys. *3*, 40.

Haine, M. E., P. A. Einstein and P. H. Borcherds (1958), Resistance bias characteristics of the electron microscope gun, Brit. J. appl. Phys. *9*, 482.

Hall, C. E. (1953), Introduction to electron microscopy (McGraw Hill Publishing Co., London).

Hawkes, P. W. (1972), Electron optics and electron microscopy (Taylor and Francis, London).

Kunath, W., W. D. Riecke and E. Ruska (1966), Spherical aberration of saturated strong objective lenses, Proc. 6th Int. Congr. Electron Microscopy Kyoto *I*, 139.

Kynaston, D. and T. Mulvey (1963), The correction of distortion in the electron microscope, Brit. J. appl. Phys. *14*, 199.
Liebmann, G. (1951a), Imaging properties of a series of magnetic electron lenses, Proc. Phys. Soc. *B*, *64*, 956.
Liebmann, G. (1951b), The symmetrical magnetic electron microscope objective lens with lowest spherical aberration, Proc. Phys. Soc. *B*, *64*, 972.
Liebmann, G. (1952), Magnetic electron microscope projector lenses, Proc. Phys. Soc. *B*, *65*, 94.
Liebmann, G. (1955), A unified representation of magnetic electron lens properties, Proc. Phys. Soc. *B*, *68*, 737.
Siegel, B. M. (1964), Modern developments in electron microscopy (Academic Press, London).
Zworykin, V. K., G. A. Morton, E. G. Ramberg, J. Hillier and A. Vance (1945), Electron optics and the electron microscope (John Wiley, New York).

Chapter 2

The design of the electron microscope

In this chapter, the practical details of the design of an electron microscope are described in the light of the background considerations outlined in Chapter 1.

A typical high resolution electron microscope is shown in Fig. 2.1. The microscope column is mounted vertically, with the electron gun fed by a high voltage cable, at the top, and the various electron lenses below it. At the base of the column is the main viewing window and a viewing telescope. Specimen movement controls are brought down to knobs on either side of the viewing chamber. The heavy column is supported on a substantial desk which also contains the operating controls and their immediate circuitry. The knobs on the rear panels give control of focus, magnification, image brightness, etc., and there are meters to indicate beam current, state of the vacuum, and a display showing the operating magnification. Halfway up the column, on the left, is the liquid nitrogen reservoir for the anticontaminator. In the background is a large cabinet containing the high voltage and lens power supplies. This contains all the elements which give rise to magnetic fields which could disturb the stability of the electron beam. At the rear of the column is a vacuum pumping tube giving access to various parts of the column. Out of sight, on the floor behind the microscope, is the rotary vacuum pump which provides the first stage of vacuum pump-out.

2.1 General construction of the column

Since the outside of the instrument does not at first sight give much information about the make-up of the instrument, the various parts will be described from the cross section shown in Fig. 2.2.

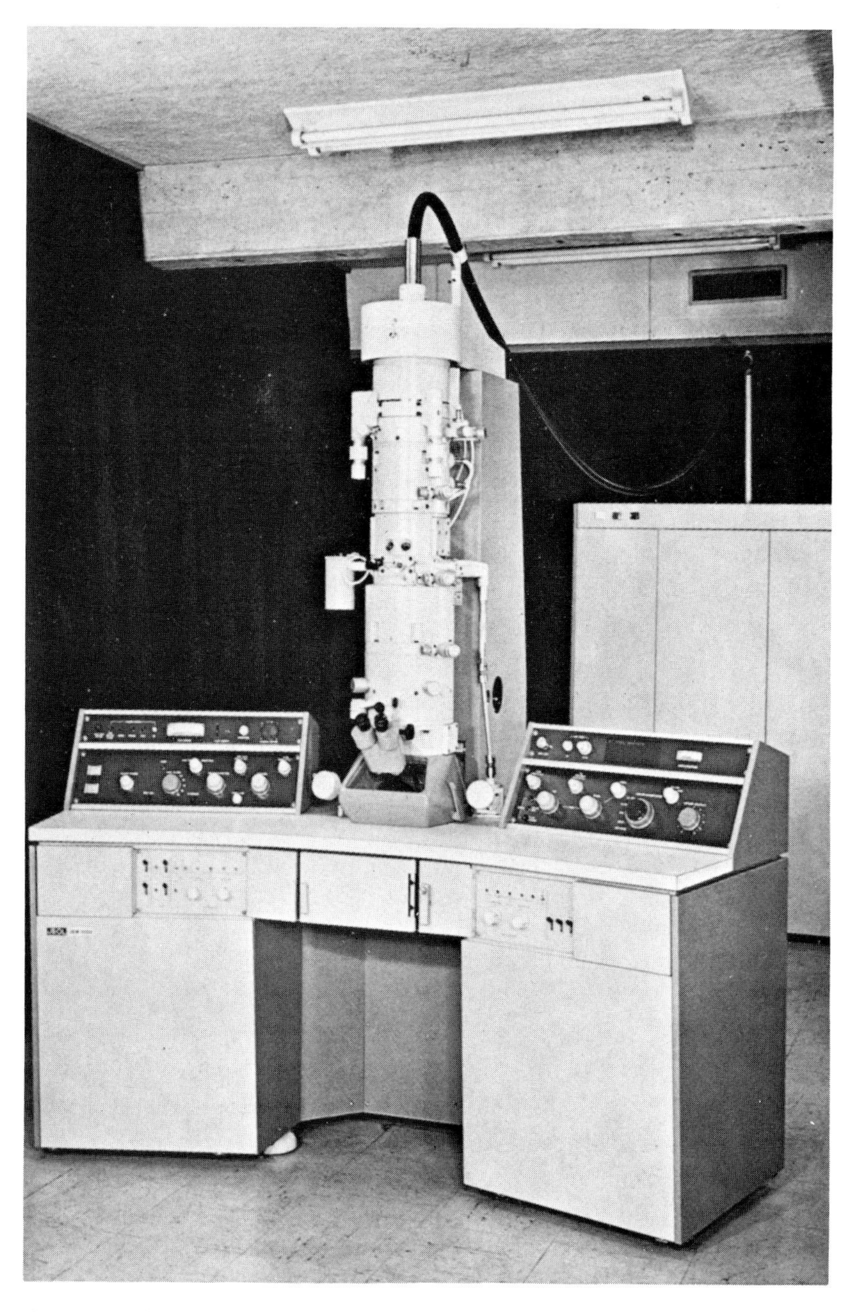

Fig. 2.1. A high resolution electron microscope. The JEM-100U. (Courtesy the JEOL company.)

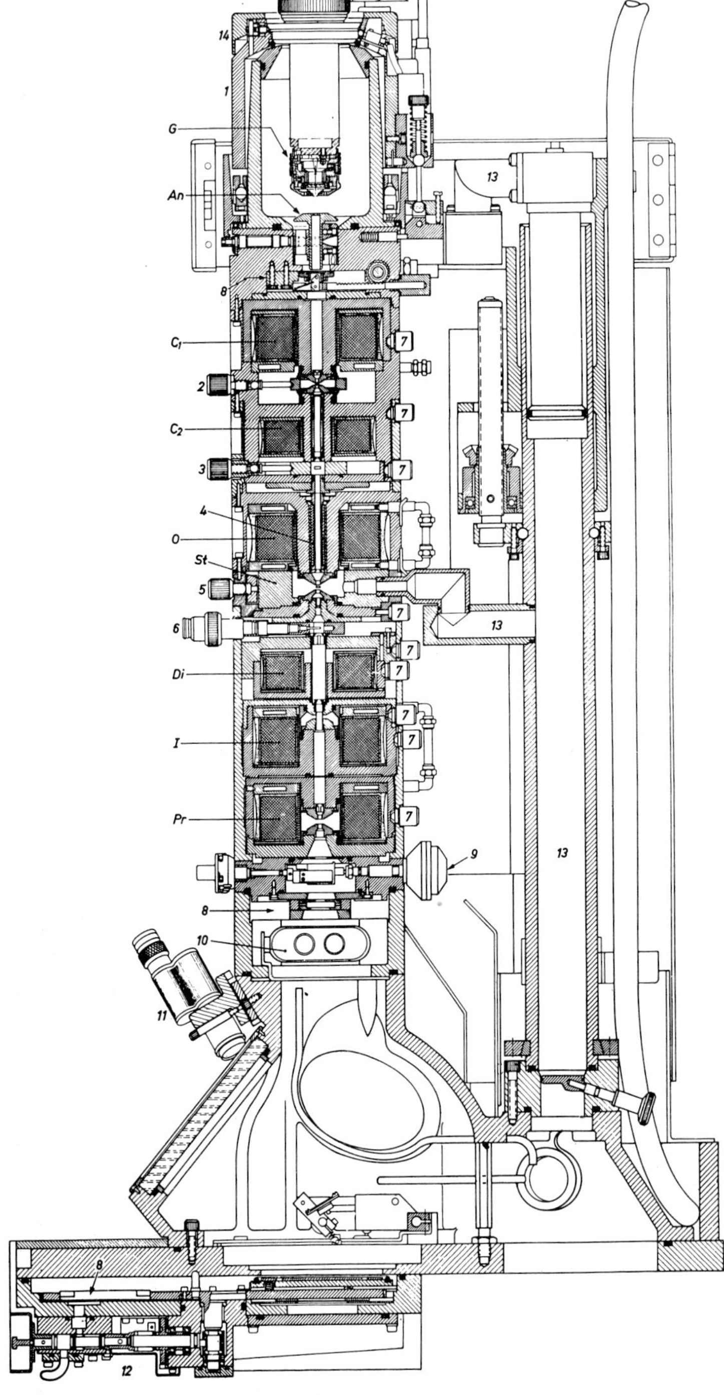

Fig. 2.2. Cross-sectional view of a high resolution electron microscope EM 300. (Courtesy Philips Ltd.)

A screened high voltage cable leads into the ceramic high voltage insulator which carries the cathode assembly (G), consisting of the tungsten filament and its mount, and the surrounding Wehnelt cylinder which is biased negatively with respect to the filament. The whole gun assembly is supported in the gimbal ring 1 and is held at a high negative potential with respect to the anode (An) which is at earth (ground) potential. Electrons emitted from the filament are accelerated towards the anode, and pass through a hole in its centre into the microscope column. The insulator assembly can be moved laterally with respect to the anode by micrometer drives on the top metal ring (14) of the column.

The electron beam now enters the condenser lens system – the strong first condenser lens (C_1), then the defining aperture in this lens (adjuster is marked 2) and into the second condenser lens (C_2). The adjustable condenser aperture is moved by knob 3. The condenser astigmatism corrector is on a plate immediately below C_2.

The objective lens O has the windings for exciting the lens above the lens gap. Along the upper part of the bore of the lens are the alignment coils 4 for deflecting the illuminating beam before it strikes the specimen. The focus wobbler coils are also at this point.

The specimen is inserted into the pole piece gap through an air-lock mechanism (not shown, for simplicity) and it is moved by a stage mechanism St (details omitted). The objective contrast aperture is also in the pole piece gap and is moved by the knob 5. The objective astigmatism corrector is an assembly pushed into the lower part of the lens bore.

Immediately below the objective lens is the selector (or intermediate) aperture centred by the knob 6. This is used for selecting a region of the specimen for electron diffraction analysis.

The beam then enters the diffraction lens (Di), also called the first intermediate lens and then the second intermediate lens I and the final projector Pr. All of the lenses are alignable within the microscope column by means of the adjusters 7.

Below the projector lens is the electrically controlled shutter (9) which permits accurate exposure timing. Below this is a vacuum valve 8 which allows the microscope column to be sealed off from the viewing chamber and camera. A 35 mm film camera may be inserted at 10, in the upper part of the viewing chamber, which is equipped with large lead glass viewing windows (to protect the operator from X-rays). The fluorescent screen on which the image is rendered visible lies flat on the base of the viewing chamber. It may be raised at an angle to permit normal viewing from the front of the instrument.

The binocular telescope 11 allows a small focusing screen (not shown) to be viewed at a magnification of 10 ×.

Below the viewing chamber is the camera for plates (or cut film) (12).

The whole instrument is pumped out to a high vacuum by rotary and diffusion pumps, via the pumping manifold 13 at the back of the microscope column. Apart from a large aperture pumping connection to the large volume of the viewing chamber there are separate pumping connections into the objective lens gap and to the electron gun.

The electromagnetic lenses are water cooled, and the pipework connections may be seen at the rear of the column.

The components of the instrument will now be described in more detail.

2.2 The electron gun

The elements of an electron gun were shown diagrammatically in Fig. 1.10, and a photograph of an electron gun terminal is shown in Fig. 2.3. The ceramic insulator I carries the high voltage cable termination, bringing the electrical supplies to the tungsten filament, contained within the Wehnelt cylinder W. The filament tip is near the small hole at the bottom of the re-entrant portion of this cylinder. The Wehnelt cylinder is held onto the insulator by the screwed stress-distributor ring SDR. An earthing bar EB prevents any residual voltage from harming the operator (electrical stresses in the insulation of the high voltage cable can give rise to quite a high voltage on the cathode assembly even when the high voltage supply is switched off).

Since the tungsten filament has only a limited life, it must be easily replaceable, and this is why the gun top is hinged, and why the Wehnelt cylinder can be easily removed. In the Siemens instrument illustrated, the filaments are supplied pre-centred and set to a correct height. In other instruments, the filaments are centred and adjusted in height relative to the Wehnelt cylinder by adjusting screws before being placed in the instrument. Fig. 2.4 illustrates a typical arrangement, with the filament shown removed from the Wehnelt cylinder.

A description of the correct procedure for replacing a filament will be found in the Instruction Manual of the instrument being used. The filament setting recommended by the manufacturer will normally yield the maximum brightness, when the bias conditions are correctly adjusted. For applications where low irradiation of the specimen is important, the filament may be set deeper within the Wehnelt cylinder (h increased, see Fig. 1.10). If this is

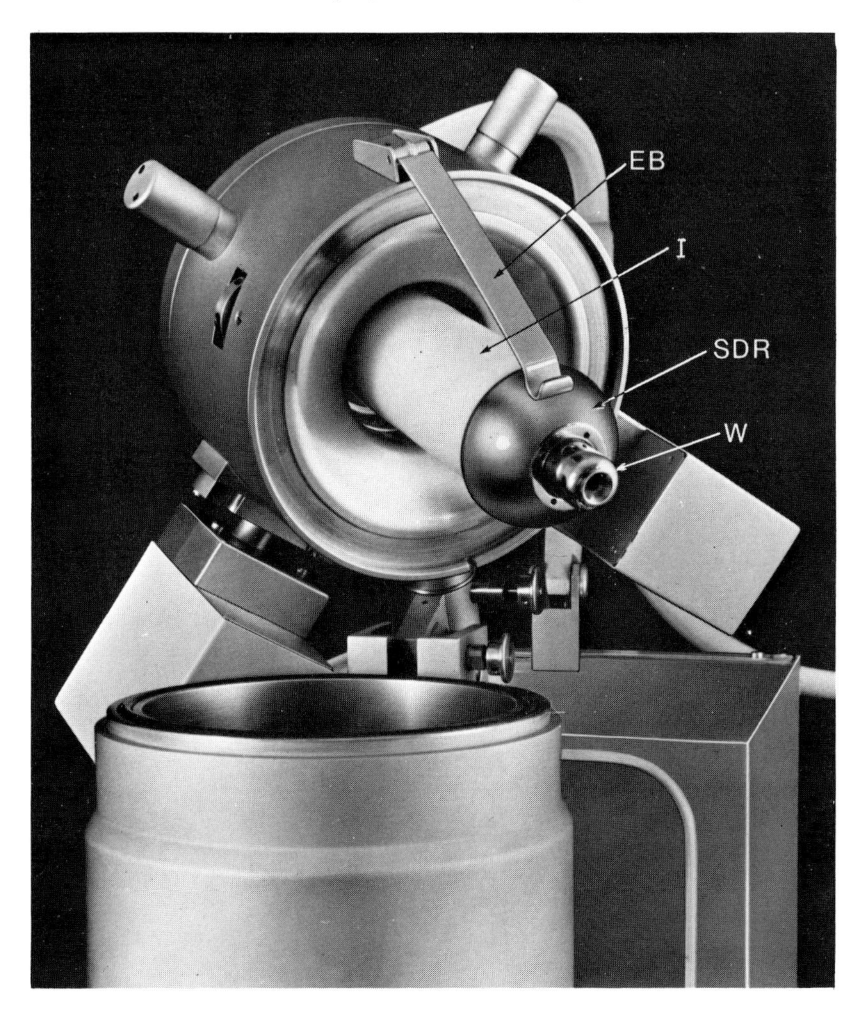

Fig. 2.3. An electron gun opened for filament changing, showing the high voltage insulator I, the Wehnelt cylinder W, stress distributor ring SDR and earthing bar EB. (Courtesy Siemens.)

combined with a fairly high bias voltage, the gun will yield a beam of low brightness and low beam current. It will incidentally yield a longer filament life because the operating point will occur at a lower filament temperature.

To achieve the maximum illumination intensity, the electron crossover in front of the cathode must be aligned with the hole in the anode. There is usually a mechanical lateral adjustment of the cathode or anode to enable

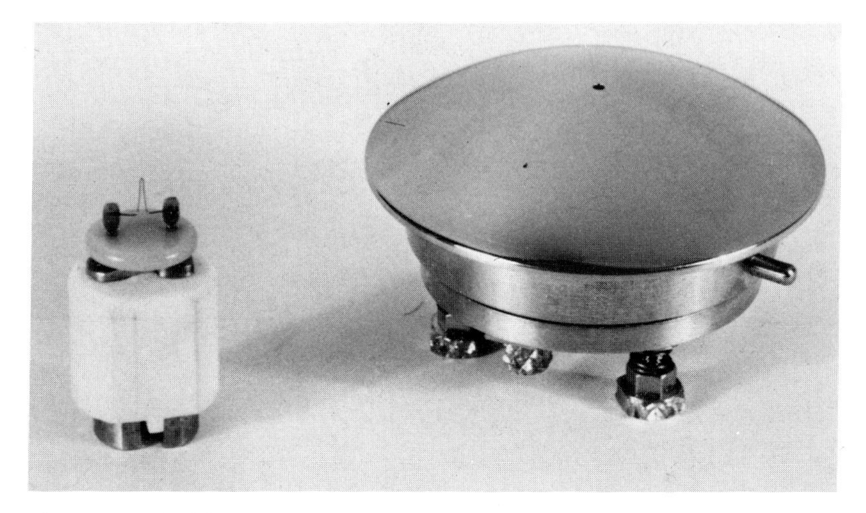

Fig. 2.4. At the left, an electron microscope filament in its ceramic mount; this normally fits inside the Wehnelt cylinder (at the right), shown with the filament tip near the aperture in the top surface. Note the levelling screws underneath the assembly for centring and setting the height of the filament. (Courtesy AEI Scientific Apparatus Ltd.)

this alignment to be made conveniently during operation of the microscope, as described in Chapter 4.

2.2.1 THE ELECTRON EMITTER

Although the great majority of electron microscopes employ a standard hairpin filament (Fig. 2.5) there are other filament designs in restricted use. If the filament tip is ground down to a sharp point (Fig. 2.6), it heats up locally and the emission is concentrated into a smaller area and gives rise to a reduced spot size (Bradley 1961). Such filaments tend to have a somewhat shorter life than the normal ones so they are not used routinely in many laboratories.

A more refined system is the pointed filament (Fig. 2.7) in which a small oriented single crystal of tungsten is welded to an ordinary filament tip and is then etched down to a very fine point (Hibi 1954; Fernández Morán 1966). Such filaments yield much finer beams than the normal ones and are more or less essential for critical work in imaging crystal lattice planes, where a highly coherent source is desirable (see § 3.8).

More recently, a cathode of lanthanum hexaboride has been used by Broers (1970). Such a cathode has a long life and yields high brightness. It requires a considerably better vacuum than is normally obtained in electron

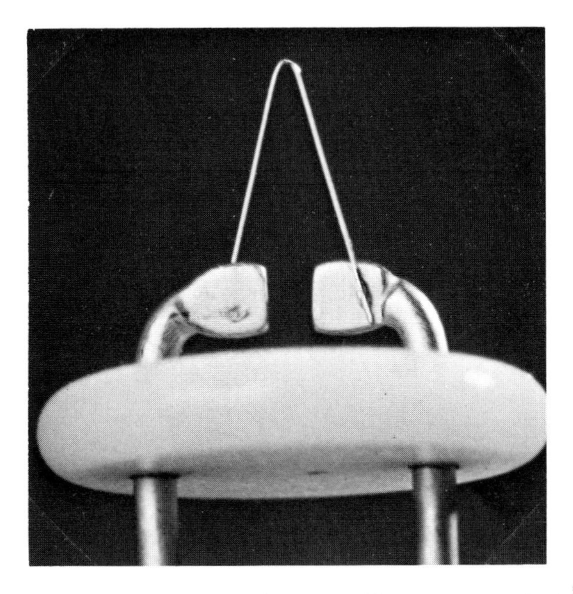

Fig. 2.5. Standard type of hairpin filament; uniform tungsten wire in V shape. (Courtesy Agar Aids.)

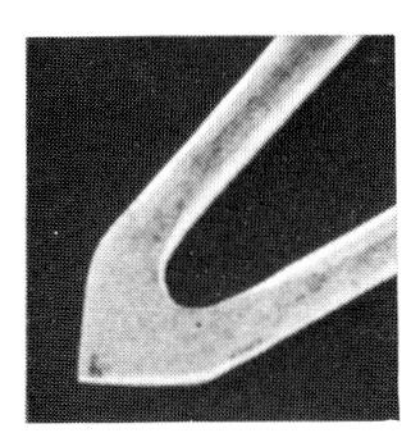

Fig. 2.6. Filament with ground point (after Bradley 1961). (Courtesy Ebtec.)

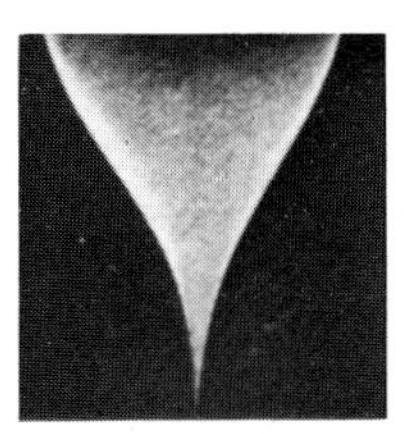

Fig. 2.7. Pointed filament. The tip is a single crystal of tungsten welded onto the heater wire. (Courtesy Ebtec.)

guns in microscopes currently available. The heat dissipation is also quite large and presents design difficulties. It is not yet used in commercially available transmission instruments, although it has been used in scanning microscopes.

Finally, there is the possibility of using a field emission cathode which requires no heating. However, it does require a vacuum of the order of 10^{-10} Torr and its adoption in conventional microscopes will have to await a radical redesign of the vacuum system. It is essential for ultimate performance scanning microscopes (Chapter 9) of the type developed by Crewe et al. (1968).

2.3 The condenser lenses

Fig. 2.8 shows a cross section through a condenser lens assembly (condensers 1 and 2). Since these are typical electron lenses, some general features of lens construction will be mentioned here.

The strong axial magnetic field is generated between the pole faces P_1, P_2, which are pierced by a cylindrical hole to permit passage of the electron beam. They are held at the required spacing by a non-magnetic spacer NS. A heavy iron circuit joins the poles and is designed to minimise the stray magnetic flux in other parts of the magnetic circuit. It encloses the windings which carry the current which provides the excitation for the iron. These windings are usually water cooled to maintain temperature stability in the lens. The lens may be fitted with a fixed defining aperture (as in condenser 1 shown AC1) or a movable aperture AC2 when it may be necessary to change the aperture (as in condenser 2). Such apertures are centred by micrometer controls outside the column.

The pole pieces of the lens are made of soft iron and can very easily be scratched or dented. They must be handled with extreme care whenever they are removed, because mechanical damage to the pole pieces can ruin the optical properties of the lens. The mechanical tolerances in a new lens are of the order of tenths of a thousandth of an inch (a few micrometres).

It is usual to keep the volume of the lens which is at high vacuum to a minimum, so the vacuum seals are arranged so that the windings and water circuits can be at atmospheric pressure. In some recent instruments, the whole lens is outside the vacuum, since the electron beam passes through a tube along the axis of the lens. This is not a feasible arrangement in the objective lens, of course, or anywhere where a movable aper ure has to enter the beam path.

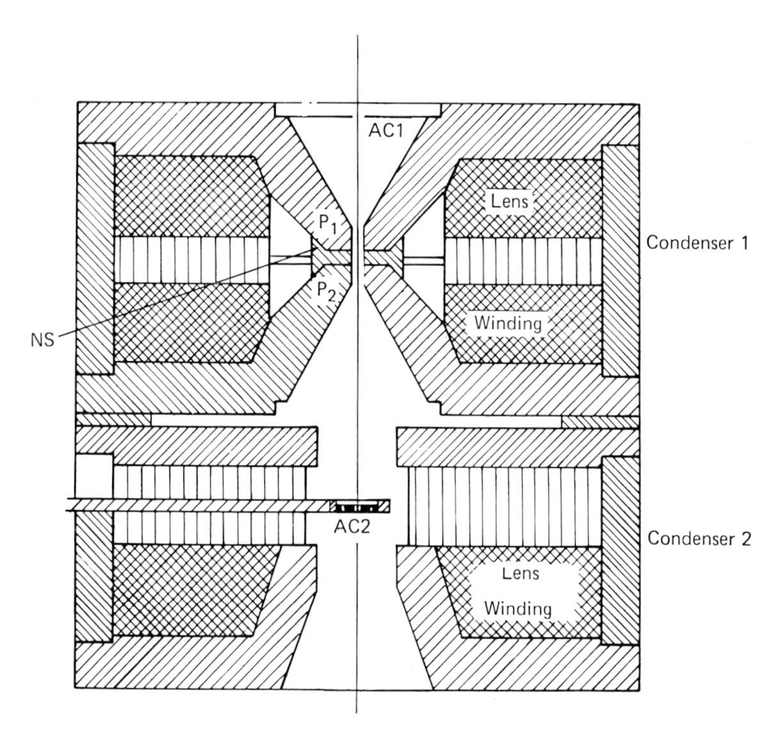

Fig. 2.8. Cross section through a double condenser lens assembly. The upper lens is condenser 1, with a fixed aperture. The lower lens is condenser 2 with an adjustable aperture. Pole pieces are indicated by P_1, P_2, and the nonmagnetic spacer by NS. (Courtesy AEI Scientific Apparatus Ltd.)

It will be noted that the condenser 1 lens has a relatively small lens gap and bore (i.e. it is a strong lens of short focal length) and that the condenser 2 lens is relatively weak, with large pole piece dimensions. It will be seen by reference to Fig. 1.16 that condenser 2 never operates in a short focal length condition.

2.3.1 CONDENSER ASTIGMATISM CORRECTOR

The condenser system usually introduces some astigmatism into the imaging beam, which is manifest as an elliptical spot at the specimen. This is disturbing, both because of uneven image quality and of loss of intensity. It is compensated by using an astigmatism corrector at the second condenser level. Such a corrector is constructed from four, six or eight electrodes, independently insulated, but capable of connection together in opposing pairs.

These can be excited so as to create a cylindrical lens (stronger in one axis than the other) and with the strengths of the two axes continuously variable in both magnitude and sense. A cylindrical field can therefore be established so as to compensate for any residual cylindrical field in the lens system. The corrector may equally well be made from a system of electromagnetic poles (Fig. 2.9) rather than electrostatic ones. In such a system there are fewer insulation problems, but the remanence of the magnetic material must be extremely low and it is usually necessary to use Mumetal. The electrodes or poles are generally interconnected electrically so that the control knobs give simple direction and strength variations. However, some instruments (e.g. Philips) favour X–Y controls of strength only. It is not found necessary to make very frequent corrections of illumination astigmatism during operation. The procedure is described in § 4.5.

Fig. 2.9. Condenser astigmatism corrector assembly. This is an octupole electromagnetic system. (Courtesy Philips.)

2.4 Illumination alignment coils

It is necessary that the illuminating beam should pass along the axis of the objective lens, in order to simplify operation and optimise performance. This is usually achieved by electromagnetic deflector coils which can impart both a lateral translation of the beam and a tilt centred on the centre of the objective lens as described in § 1.8. Two sets of coils are required in order to produce X and Y motions. Since the beam is deflected away from the axis by the first set of coils, the lower coils have to be larger. A typical coil arrangement for a metallurgical instrument (AEI EM 802), requiring large tilt angles, and with a top entry specimen stage, is shown in Fig. 2.10.

Fig. 2.10. Deflector coil assembly from EM 802. The larger lower coils are to accommodate the off-axis beam produced in the first deflection. (Courtesy AEI Scientific Apparatus Ltd.)

The arrangement is much more compact in a side entry specimen system (e.g. the Philips instrument shown in Fig. 2.2), although it is more difficult to obtain large tilt angles in the beam because of geometrical limitations. These deflector coils may also be used for other operational purposes (e.g. high resolution dark field operation as described in § 3.10.3).

2.5 The objective lens

This is the most important component of the microscope as it determines the ultimate performance of the instrument. The choice of the size of the pole pieces of this lens also has a profound effect on the specimen stage facilities which again affect the potentialities of the microscope. As a first approximation, the ultimate resolution of the microscope improves as the focal length of the objective lens decreases, since the spherical aberration coefficient and chromatic aberration coefficient both decrease approximately as the focal length decreases. Since the focal length depends, for a given kilovoltage of operation, on the magnetic field strength in the objective lens, a very strong lens is required. This means that the specimen has to be placed in the lens field to obtain the necessary short focal length, and this immediately poses the problem of introducing the specimen into this position. Originally, specimens were always held on the end of a cartridge introduced from above the lens, and indeed this arrangement is still used in several instruments. It is necessary to have a fairly large bore in the upper objective pole piece, so as to leave room for the specimen cartridge to be moved for survey of the whole area of the specimen support. An alternative system is to introduce the specimen on a rod from the side of the lens, into the pole piece gap. This avoids the problem of a large top bore but there is very little space in the lens gap to accommodate a specimen rod and a movable objective aperture.

The desirable characteristics of specimen stages are discussed in the next section (§ 2.6) in the context of applications of the microscope. For the moment, it will merely be noted that compromises have to be made between ultimate resolution and flexibility of operation, since the strongest lenses do not leave enough room for complex specimen movements.

The cross-section of a typical objective lens with a top entry stage is shown in Fig. 2.11. It will be seen that the specimen plane S lies within the pole gap, and that the objective aperture rod E and the anticontaminator blades (not drawn) have also to be accommodated. (The anticontaminator is cooled to liquid nitrogen temperature and its function is explained in § 6.14.) The

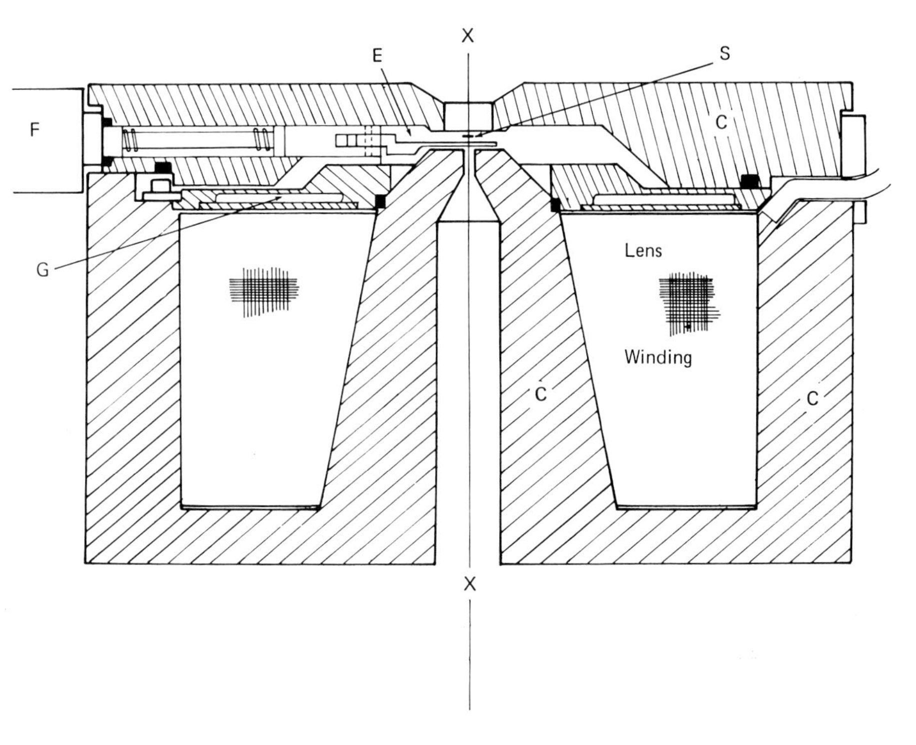

Fig. 2.11. Cross section of a typical objective lens. The specimen is at S and in this lens is in a top-entry cartridge. The objective apertures are carried on the rod E, and adjusted by control knob F. The water cooling channels G are above the lens coil. Note the heavy iron circuit C. XX is the lens axis.

water cooling channels G are placed above the lens windings so as to minimise heat transference to the specimen stage. A very small temperature change can cause an observable specimen drift due to differential thermal expansion of the elements of the stage with respect to the lens itself. The heavy iron cross section C is required to minimise stray magnetic flux from the lower part of the iron circuit. This field can cause misalignment and undesirable image shifts if it becomes excessive.

2.6 The specimen stage

An electron microscope specimen may be in the form of a very thin disc (Goodhew 1972) or, more generally, be mounted upon an electroformed mesh (usually copper) of 3 mm diameter. This is known as the specimen grid.

A variety of designs are available, and grids of 2.3 mm diameter may be used in some applications. A typical grid is shown in Fig. 2.12. The specimen grid must be supported in the correct plane in the lens and is usually clamped into a rod (Fig. 2.13) entering from the side of the lens into the pole piece gap or in a cartridge entering from above the lens so that the specimen is suspended

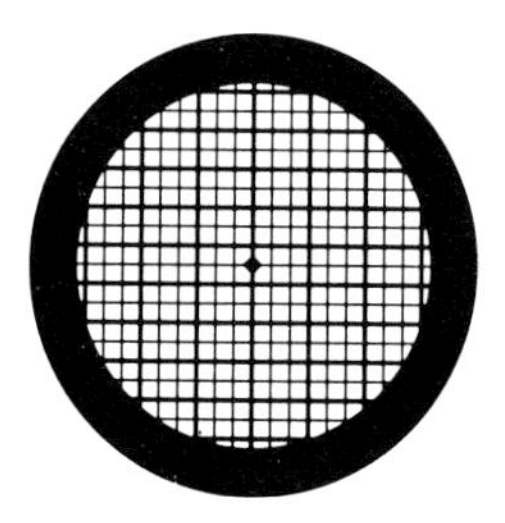

Fig. 2.12. Electron microscope specimen grid. This example has 200 mesh/inch (type New 200) and successive thick and thin bars to aid direction finding. (Courtesy Smethurst High-Light Ltd.)

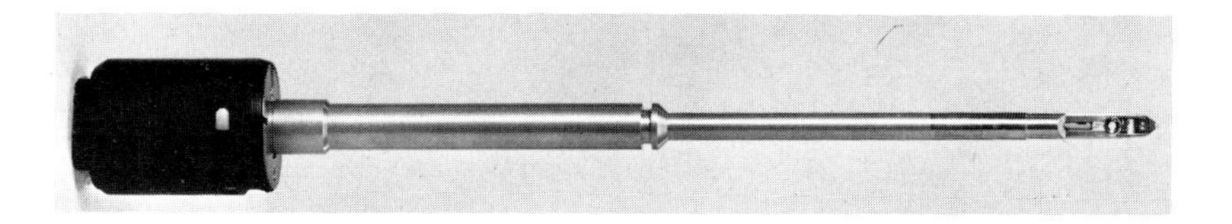

Fig. 2.13. Specimen mounted in side-entry rod. (Courtesy Philips.)

in the gap. Fig. 2.14 shows the specimen cartridge of the Zeiss EM 10 electron microscope in its insertion mechanism. (Of course, when inside the microscope, the cartridge hangs vertically.) The specimen grid is held in position by the end cap which is removed by the special tool shown. The cartridge or rod must provide good thermal contact with the specimen so that any heat dissipated in the specimen is quickly conducted away. The cartridge or rod is carried in the specimen stage proper which provides the translational movement which permits the survey of the specimen and selection of the area to be examined. Fig. 2.15 illustrates how the stage motion is transmitted from the stage control rods via micrometer screw threads to a long reduction lever. This is pivoted either on a crossed spring support or on a very accurate pivot bearing so as to provide a drive to the stage push rod which passes through a vacuum seal in the wall of the microscope column, to bear on the specimen stage carriage. Two orthogonal drives are introduced to bear on the

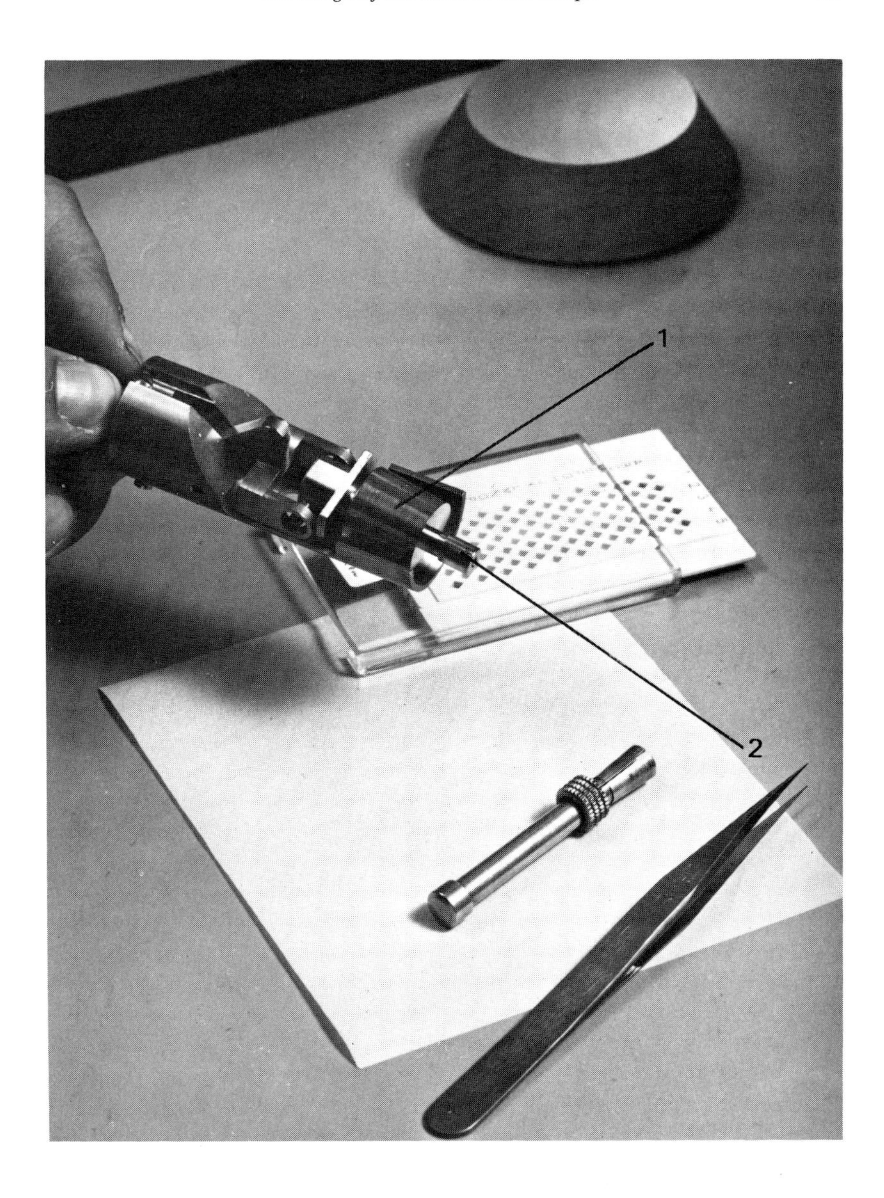

Fig. 2.14. Specimen mounted in a top-entry cartridge for a Zeiss EM 10 electron microscope.

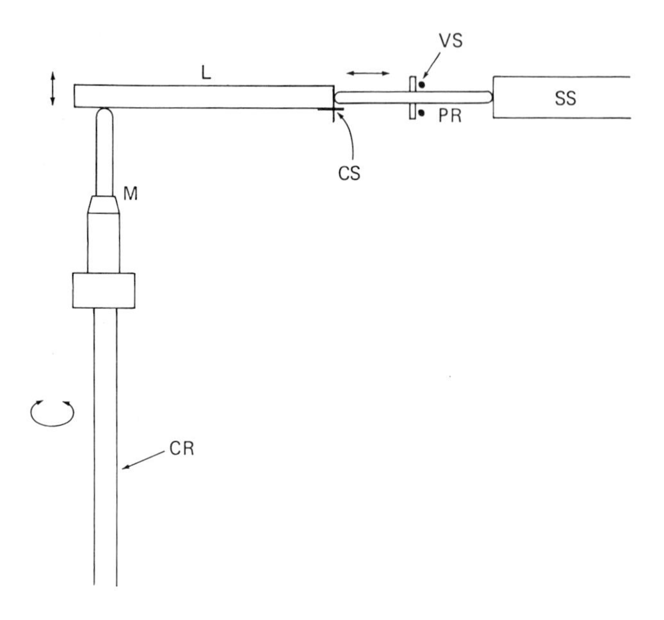

Fig. 2.15. Simplified specimen stage mechanism. The control rod CR drives the micrometer screw M. This operates on the reduction lever L, which is pivoted on the crossed springs CS. The much reduced motion is then transmitted by the push rod PR through the vacuum seals VS onto the specimen stage SS.

specimen stage, and they operate against return springs to avoid any backlash in the motion.

The specimen stage requirements are very stringent. Any part of the specimen grid should be available for survey at high magnification. If this movement is to be controllable, the motions have to be orthogonal and without significant backlash even at the top magnification of the microscope. An allowable backlash might be 5 mm on the viewing screen at a magnification of 200,000 ×, that is 25 nm at the object. Next, the stage must permit the positioning of a feature of the object at the centre of the screen to a similar sort of accuracy, with only two or three movements of the controls. When the positioning has been achieved, there must be no mechanical relaxation or thermal drift which would move the specimen by an appreciable fraction of the ultimate resolution of the microscope during the photographic exposure. This requires a stage stability of better than one atom diameter for a few seconds. These are formidable specifications to meet and are achieved through a judicious combination of fine machining tolerances to produce smooth motion, and frictional forces to reduce drift.

The specimen stage is also required to make good thermal contact with the specimen holder, in order to carry away the heat generated by the electron beam. The specimen holder has to be designed for entry through a vacuum lock and to have a fool-proof way of fitting into the specimen stage itself. Top-entry cartridges normally have a taper which provides both a location guide and a large area of contact with the stage.

2.6.1 TILT STAGES

Probably the most useful addition to any electron microscope is a special stage which permits tilting of the specimen. There are three principal types of study requiring tilting of the specimen.

1. Tilt for stereo viewing, to yield three-dimensional information about the specimen.
2. Tilt into selected crystallographic directions in order to reveal structure.
3. Tilt for interpretation by tilting the specimen through a large enough angle to obtain a different view of a complex structure – particularly useful for sections of biological specimens.

In principle, one kind of tilt stage will provide the facilities for carrying out all of these functions, which are discussed in § 6.11. The tilt stage has to provide a means of inclining the specimen in any desired orientation with respect to the electron beam. This can obviously be achieved either by rotating the specimen and then tilting it, or by providing two orthogonal tilt directions, which used in combination, will provide the same end result. In nearly all stages the tilt drives are brought into the movable stage block with flexible drives and then linked with the specimen cartridge. In such an arrangement, the tilting, or tilt and rotation, is carried out about the centre of the specimen; the point of observation may be at some distance from this, and consequently the tilting will be accompanied by a translation, S, of the image (Fig. 2.16). This will require movement of the specimen translation controls to restore the field of view. The effect will be much more apparent when working at high magnifications and very accurate controls become necessary if comfortable operation is to be retained.

It will readily be seen from Fig. 2.16 that the normal specimen tilt also causes the image to move out of focus. When the tilt angle, θ, is large, this distance may be important in affecting the magnification calibration. For example, if one examines a region near the edge of the field of view, say 1 mm from the specimen centre, then, for a tilt angle of 30°, the change in height, PP′, is 0.58 mm, which for an objective focal length of 4 mm represents about a 15% change in focal length. Of course, for areas near the centre of the grid,

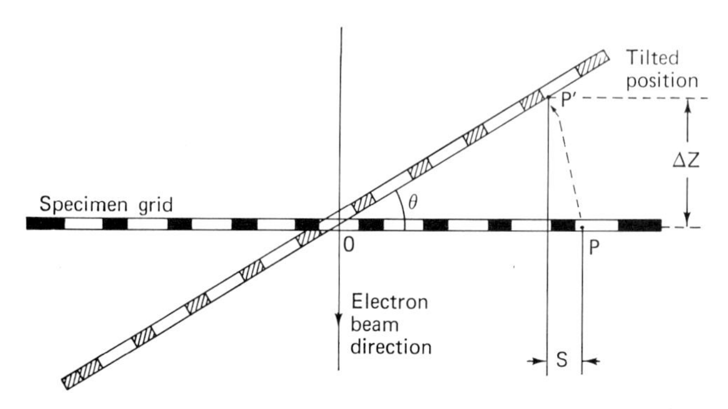

Fig. 2.16. Effect of tilt of the specimen on the point of observation and the focal plane. For a tilt θ the point P moves laterally a distance S and the focal plane moves ΔZ. Note that the effect increases with the distance OP.

the effect becomes negligible at any angle of tilt. It would obviously be desirable to tilt the specimen in any direction about the point of observation. This requires that the specimen translation controls are within the tilt motion, and it poses severe mechanical design problems. The problem has been solved for one axis of tilt by the designers of Philips microscopes (the arrangement is shown in Fig. 2.17).

The figure shows a horizontal cross-section of the microscope column W, in the plane of the eucentric specimen stage GS which has a system of tubes attached to its right hand side. The innermost tube T_1 (lightly shaded), has a hemispherical left-hand end bearing against the spherical surface B_1 and pivoting at A. This tube carries the specimen rod H_s whose inner end is seen at E. The specimen rod may be moved axially by a lever motion from the micrometer M via the push rod R. A lateral motion is transmitted by the tube T_1 which is moved via the lever L_1 which is driven by the tube T_2 when the nut N is turned. When the tube T_3 resting in the ball bearing B_2 is rotated about its axis, the tube T_1 and the specimen holder follow it, so that the angle between the specimen and the horizontal axis changes. This tilt is about the axis of the instrument, (and hence the point of observation), and is quite independent of the position of the specimen. The lever L_2 permits the adjustment of the specimen in a vertical plane, so that it lies accurately on the axis of tube T_3, and hence avoids translation of the image as the specimen is tilted. The objective aperture rod and centring control are shown at H_D.

In principle, the specimen may be tilted through any angle, but in practice it is limited to 60° since at large angles the electrons will no longer be able to pass through the grid bars of the specimen support. The initial setting of the

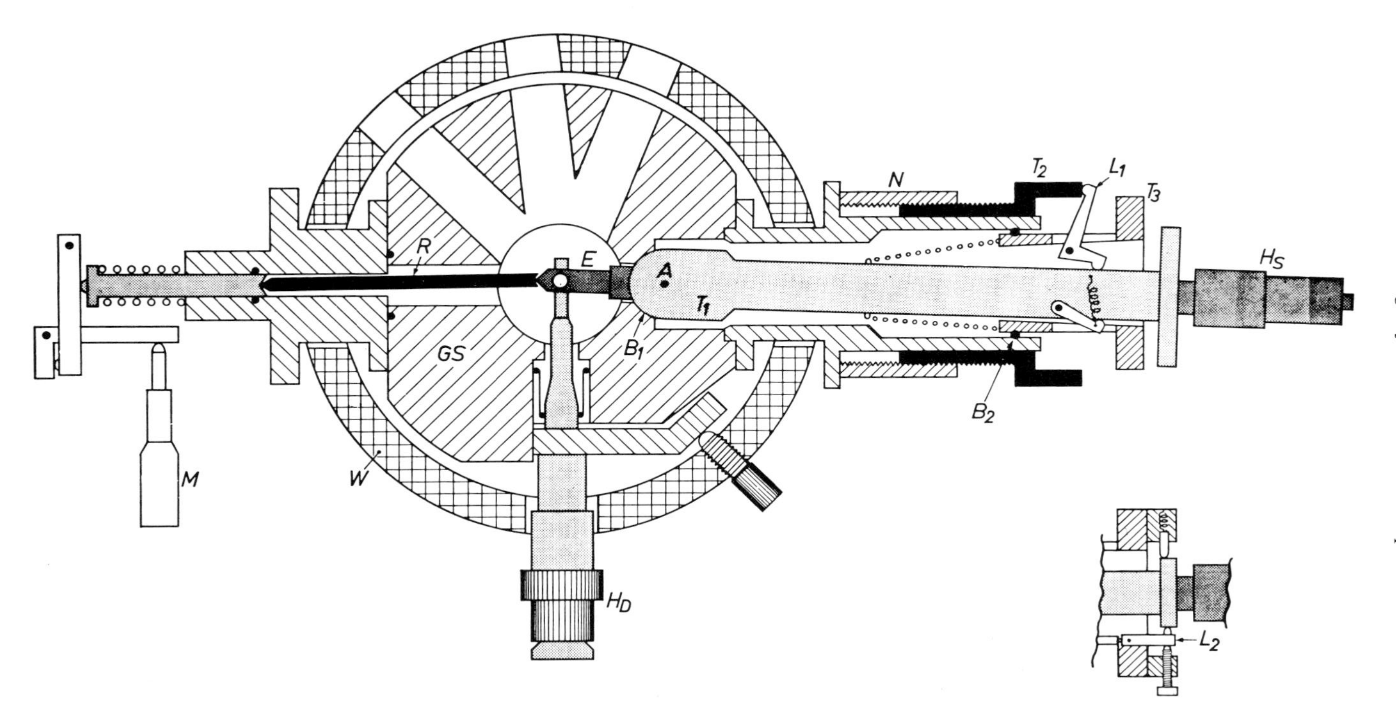

Fig. 2.17. Cross-sectional diagram of the eucentric stage designed for Philips electron microscopes. For a detailed explanation of the operation, see text.

adjustments must be done with care if the full tilt is to be obtained without change of focus setting, or loss of the field of view.

Unfortunately, the great convenience of this stage (known as an axis-centred or eucentric stage) has only been attained in a relatively bulky mechanism which requires an objective lens with a rather large space between the pole pieces. This results in larger aberration coefficients and some loss in resolution. This sort of dilemma is a common problem for designers but is often not fully appreciated by the user. For many applications, the great convenience of an axis-centred stage will make it attractive, in spite of the slight loss in performance. For other problems, loss of resolution may not be tolerable, and limitations of image movement, and indeed of the maximum tilt angle available, may have to be accepted. The requirement for very high resolution leads to a very strong lens field and hence rather small dimensions for lens pole pieces. These then restrict the space available to accommodate the specimen (in normal and tilted positions), the objective rod and the anticontaminator.

The inconvenience of a non-axis-centred stage can be partly overcome by arranging for the tilt motions to be electrically driven and controlled by foot pedals, while the hands control specimen traverse to retain the field of view.

Only in very high voltage electron microscopes (§ 9.2) where all dimensions are considerably increased, do the mechanical requirements look reasonably easy of fulfilment in a very high resolution lens.

2.6.2 MULTIPLE SPECIMEN STAGES

Rod-type specimen holders (for side-entry stages) lend themselves to multiple specimen mounts, allowing quick consecutive examination or comparison of different specimens. This is a great convenience for routine operation. The

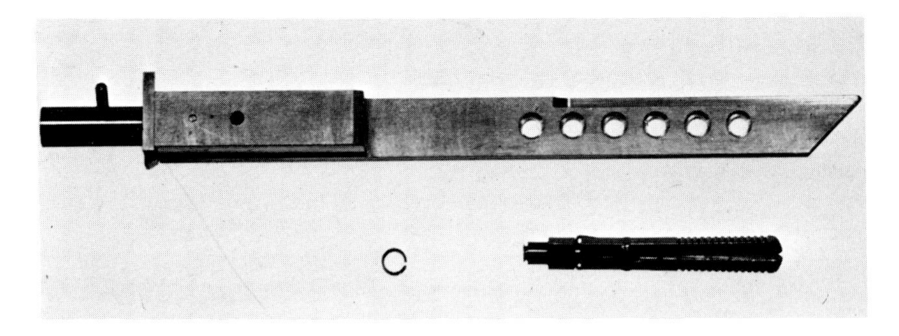

Fig. 2.18. Multiple specimen holder for side entry stage (after Lucas 1968a). Specimens are retained in countersunk holes by means of circlips inserted with the special tool shown.

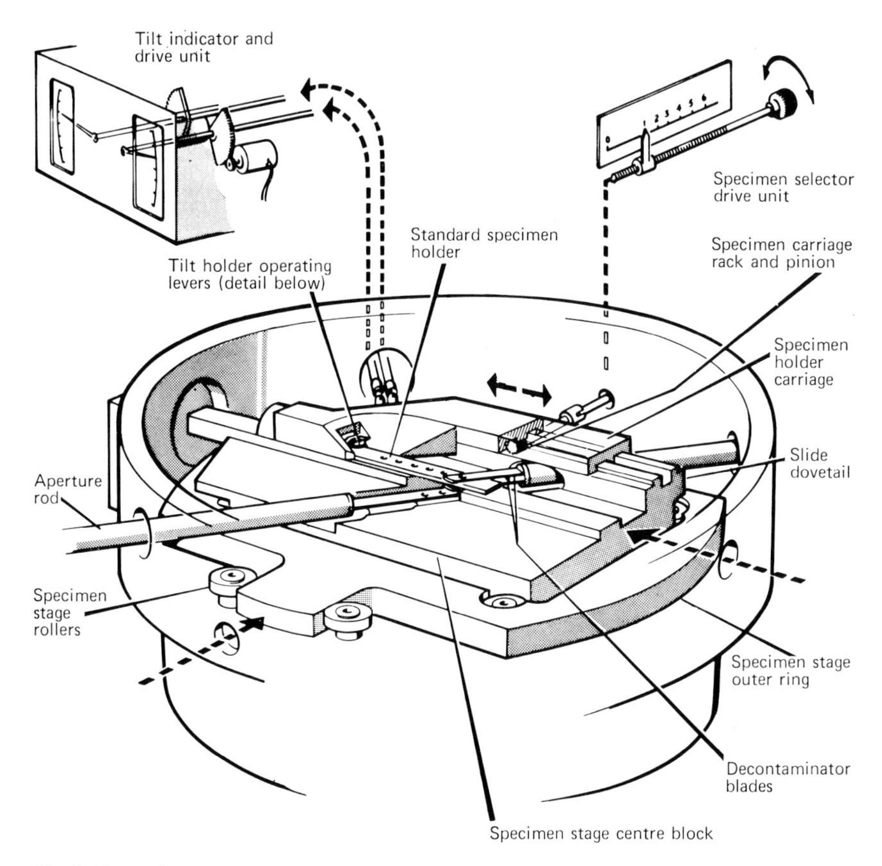

Fig. 2.19. Schematic diagram of Lucas type stage mechanism for accommodating a multiple-specimen holder or a serial section stage. The main specimen movements are indicated. The specimen holder is introduced on the upper specimen carriage and its position is indicated on the external scale.

idea was first suggested by Liebmann and realised in a commercial instrument by Page (1956). An elegant modern design for the AEI EM 801 is due to Lucas (1968a) and is illustrated in Fig. 2.18. Each grid is placed in a recess in the specimen holder and held in position by a circlip, which is inserted or removed with the special tool shown alongside the holder. The specimen stage to carry this rod is of a special design (Fig. 2.19) and has an extra substage which carries the rod and which moves along one of the stage traverse directions. Thus, while the main stage movement of ± 1.5 mm is given by the normal stage controls, the special substage can be moved 2.5 cm in a continuous motion, motor driven if required. The position of the rod is recorded on a scale on the outside of the instrument. The six specimens can very quickly be interchanged by movement of the substage.

This stage is also used to give the movement for the serial section stage which is a development from the multiple specimen holder. In order to support the specimen, a long (25 mm) narrow (100 μm) slit is provided and on this a long series of serial sections may be mounted so that they may be

Fig. 2.20. Serial section stage. Designed by Lucas (1968b) for F. S. Sjöstrand. The long slit is 2.5 cm long and may be 100 or 300 μm wide. The full length can be continuously traversed.

examined without interference from support grid bars (Fig. 2.20). This facilitates the examination of corresponding areas from successive sections. The design was evolved by Lucas (1968b) following a suggestion by F. S. Sjöstrand.

2.6.3 HIGH CONTRAST SPECIMEN HOLDER

This is simply a specimen cartridge for a top entry stage, shorter than usual, which holds the specimen further away from the centre of the objective lens. The objective lens therefore has to be operated at a longer focal length in order to focus the image. This results in a higher contrast image (§ 3.3). The high contrast specimen holder has two other uses: it can be used to provide a lower range of magnifications than those normally provided; and it provides a longer effective diffraction camera length.

2.7 Special stages for specimen treatment

Apart from the standard specimen holder for the specimen, and those providing tilt of the specimen, a whole range of other special stages have been designed for examining specimens under different physical conditions, and for performing experiments within the electron microscope.

2.7.1 FURNACE HEATING STAGES

A small furnace constructed of bi-filar windings surrounds the specimen cartridge and the whole region around the specimen is raised to a high temperature (most stages achieve 650 °C; some achieve 800 °C). Fig. 2.21

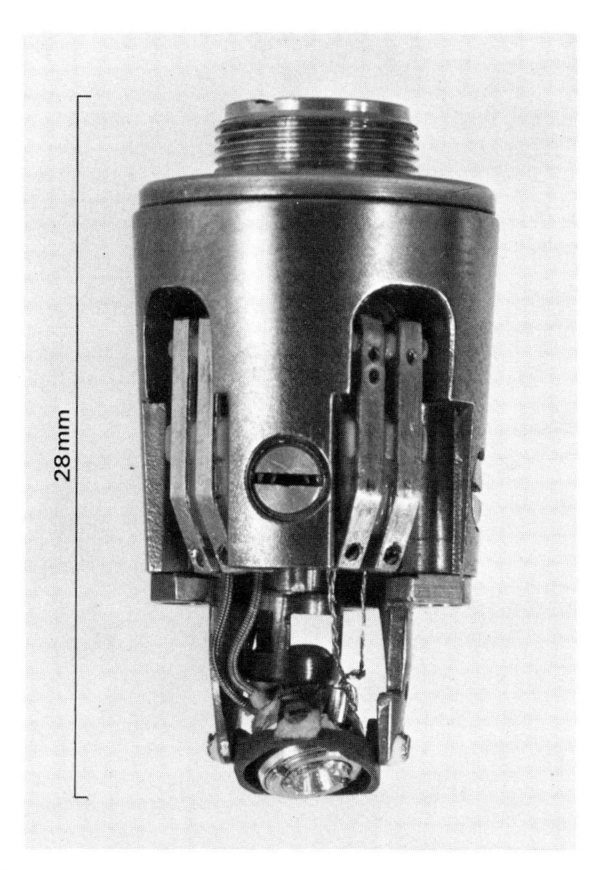

Fig. 2.21. Top-entry furnace heating cartridge with double tilt motion for Siemens microscopes. The contact strips for the entry of the heater current can be seen, with the leads to the tilting gimbal.

shows a specimen cartridge (top-entry) double-tilt heater stage designed for Siemens microscopes. The whole tip of the cartridge can be tilted by applying a drive at the centre of the cartridge. The heater winding in the tip can be seen. The photograph shows how the electrical supplies are introduced through spring contacts which are made as the cartridge enters the stage. A close-up of the specimen region in a rod-type specimen holder for a Philips microscope is illustrated in Fig. 2.22, and shows the heater leads to the small furnace heater. This type of heater gives a rather stable temperature of the specimen which can be measured by a thermocouple but the disadvantage is the relatively high thermal input. This may make it necessary to provide a thermal

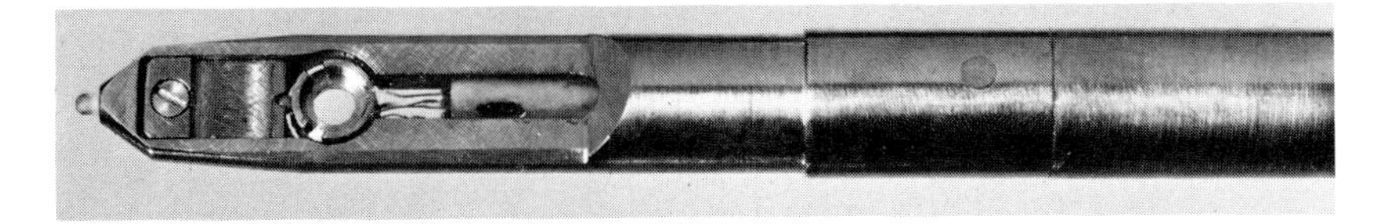

Fig. 2.22. Close-up of the specimen region in the rod furnace heater for the Philips electron microscopes.

screen to protect the objective lens pole pieces and it generally also results in a rather long period before stage stability is achieved – times of 30–45 minutes may elapse before thermal drifts are reduced to the level where photography may be attempted. Typical furnaces are described by Itoh et al. (1954) and Valdrè (1965).

2.7.2 GRID HEATER STAGE

In this stage the heating current passes through the grid mesh supporting the specimen; only the centre portion of the grid reaches a high temperature and the problem of ensuring good thermal contact between the heater mesh and the specimen arises. Fig. 2.23 shows the heater cartridge for the AEI microscopes on its loading stand. The end cap which clamps the heater mesh is shown in the raised position with a mesh clinging to it. It is not normally possible to measure the temperature directly, but the heater current can be calibrated to give the temperature of the centre of the grid to fair accuracy. This type of heater can easily achieve 1000 °C and with a carbon mesh, temperatures as high as 2000 °C have been reported. The total heat input is small, and thermal equilibrium is achieved within a few seconds so this type of heater stage has applications where the furnace heater is unsuitable (Agar and Lucas 1962). It is important that a DC supply is used, and that the supply leads are twisted to reduce the field which would otherwise cause a deflexion of the electron beam.

2.7.3 COLD STAGES

In these stages the specimen cartridge is cooled by thermal contact with a reservoir of liquid nitrogen (Leisegang 1954; Schott and Leisegang 1956) or by a constant flow of cold nitrogen gas. Temperatures between −130 °C and −170 °C are achieved, depending on the thermal efficiency of the stage. In general, the gas flow type of stage yields higher performance because it avoids vibration caused by boiling of the liquid nitrogen. It also lends itself to better temperature control.

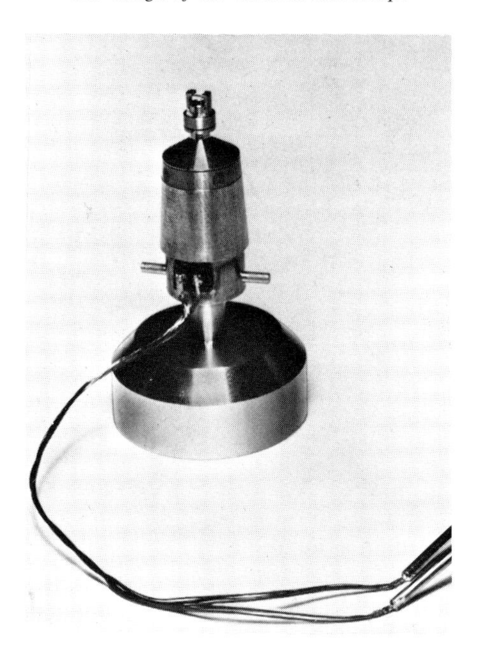

Fig. 2.23. Grid heater cartridge for AEI microscopes. The heater leads are twisted within the body of the cartridge to reduce magnetic fields. The cartridge is shown on its stand, with the clamp jaws in the open position ready for exchange of the heater mesh.

Fig. 2.24. Cooling stage for an AEI microscope (after Swann and Swann 1970). The specimen movement drives are applied to the stage body A. The central part of the stage is an annulus C cooled by nitrogen gas brought in by the flexible bellows on the right.

A top entry cooling stage is shown in Fig. 2.24, and is the design by Swann and Swann (1970) for an AEI microscope. The cooling annulus C is shown attached to the transfer tubes; this cools the centre part of the stage which is thermally insulated from the outside by being supported on three point supports. The stage enables simple tilt cartridges to be used, and is thermally very efficient, so that a temperature as low as $-180\,^{\circ}\mathrm{C}$ can be attained. A heater enables the stage to be warmed up quickly when specimen changes are to be made.

Liquid helium stages have been designed which enable temperatures lower than $10\,^{\circ}\mathrm{K}$ to be achieved. These stages are rather complex and are normally built by experimenters for their own purposes (e.g. Valdrè 1964; Cotterill 1964; Watanabe and Ishikawa 1968; Valdrè and Goringe 1970).

2.7.4 COMBINED HEATING, COOLING AND TILTING STAGE

An ingenious stage has been evolved by Mills and Moodie (1968) which achieves two high tilt motions while allowing specimen cooling to $-140\,^{\circ}\mathrm{C}$ or heating to $1200\,^{\circ}\mathrm{C}$ in a gas atmosphere. This is accommodated in a 100 kV microscope, in spite of space limitation. Most other such complex stages have been designed for high voltage microscopes where much more space is available.

2.7.5 STRAINING STAGES

These stages provide means for straining the specimen while it is being observed (Imura et al. 1970). The stress may be applied by a mechanical drive, or the extension may be achieved using electrically-heated bimetallic strips. Separate designs are needed for metal specimens, where extensions of less than 10% are required, and for rubber and plastics where extensions of more than 100% may be called for. A desirable feature is that the extension should be symmetrical about the centre of the specimen. This minimises the movement of the field of view during the extension.

A side entry stage lends itself to a very simple direct drive for the stress, and

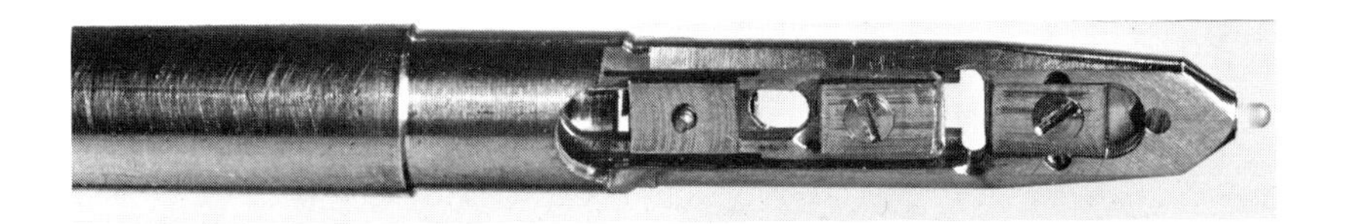

Fig. 2.25. Straining stage within a rod entry holder for Philips microscopes. The jaws may be drawn apart by a screw thread drive.

a typical rod stage for the Philips instruments is shown in Fig. 2.25. The specimen is clamped between the two jaws. A cartridge stage for a Siemens instrument is shown in Fig. 2.26 in which the force is applied by a transversely-moving wedge which causes the supports for the specimen to be driven apart symmetrically.

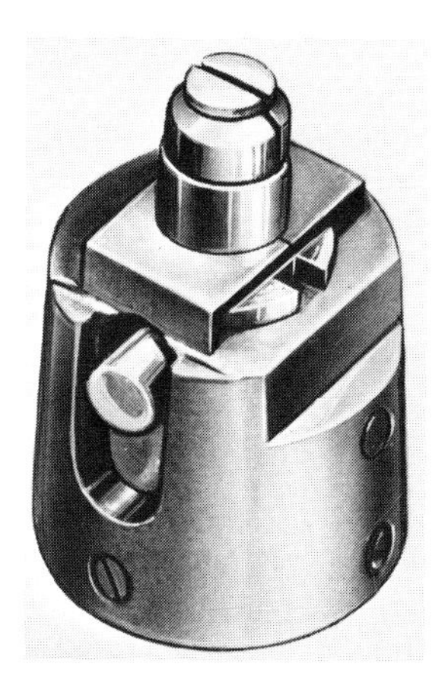

Fig. 2.26. Straining cartridge for Siemens microscopes. The jaws are driven apart by a wedge operated from the pin in the side of the cartridge.

A tilting straining stage has been described by Lehtinen et al. (1967). The coming of the high voltage microscope has greatly extended the value of dynamic experiments and the refinement of the straining devices required.

2.7.6 GAS REACTION STAGES

It is possible to enclose the specimen either in a cell sealed with two thin windows, or with small apertures above and below the specimen to restrict gas flow into the microscope vacuum. The gas pressure around the specimen may then be raised to atmospheric pressure or a good fraction of it, in order to examine gas reactions in the electron microscope. A number of such stages have been described (Heide 1962; Hashimoto et al. 1966; Hashimoto et al. 1968).

2.7.7 STAGES FOR MAGNETIC STUDIES

There are a variety of such stages but in essence the specimen holder is required to place the specimen in a field free space (or nearly so) so that a controlled applied field may be introduced. Normally, therefore, a magnetic

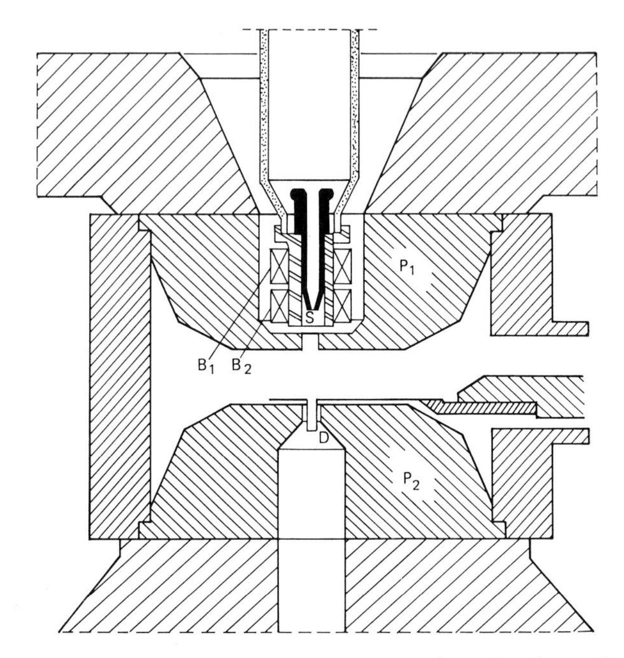

Fig. 2.27. Stage for magnetic studies (after Dupouy et al. 1968). The specimen S is well above the lens gap between the pole pieces P_1 and P_2 and therefore in minimal field. Experimentally applied fields are produced by coils B_1 with compensatory coils B_2 to return the beam to the axis. The long focal length of the lens demands the positioning of the objective aperture at D.

stage involves using long focal length operation of the objective lens, so that the stray field at the specimen from the objective lens gap is minimal. Special coils for forming the experimental applied magnetic field may also be provided. Fig. 2.27 shows the experimental arrangement of Dupouy et al. (1968). The objective lens pole pieces have a small bore so that very little field penetrates up to the specimen region (less than 1 oersted). The experimentally applied field is produced by coils B_1, and compensating coils B_2 bring the illumination back onto the axis. It is necessary to have a special design for the objective aperture in order to place it in the back focal plane, which is down in the bore of the lower pole piece. Other designs have been described by Fuchs and Liesk (1962) and Bowman and Meyer (1970).

2.8 Objective apertures

As described in §1.9, the objective apertures are placed in the back focal plane of the objective lens. They are introduced into the lens gap in a rod which usually carries three or four different aperture discs; or a strip drilled with different hole sizes (Fig. 2.28). The end of the rod is accurately centrable on the objective lens axis by two micrometer drives outside the microscope column and the controls also provide a click-stop mechanism for selecting the different apertures. The apertures themselves are made from a refractory metal, usually platinum or molybdenum, since considerable energy is

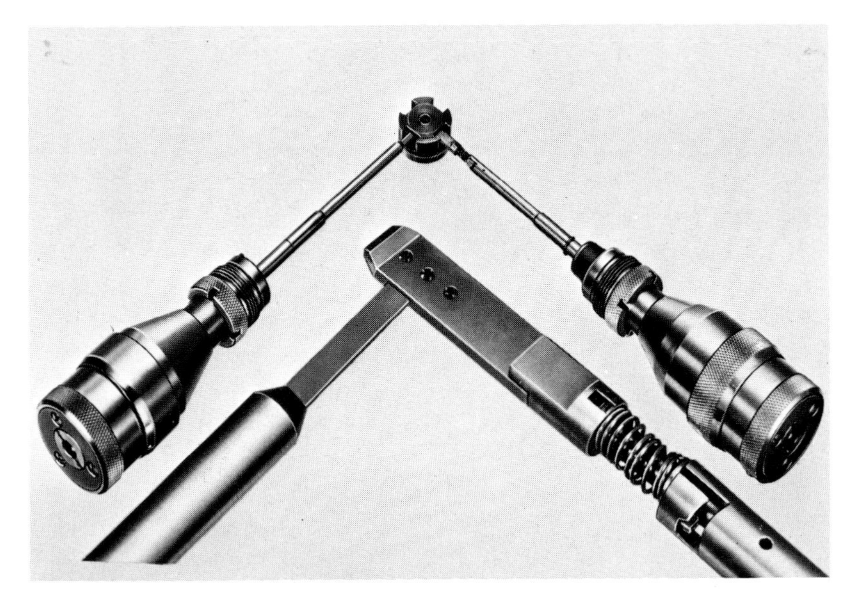

Fig. 2.28. Adjustable objective aperture mechanism, showing the rods entering the pole piece spacer. Outer knurled knobs give orthogonal micrometer movements. Aperture selection is by the inner knurled ring on the right-hand control. Enlarged picture shows how the movement of the three apertures is controlled.

dissipated in them by the scattered electrons. The apertures have to be readily interchangeable during operation of the microscope, and for some purposes they must be withdrawn from the beam altogether. The mechanism has to be accurate enough to return the apertures to their centred position around the axis when they are replaced.

A recent development in objective apertures has been the production of very thin apertures. They heat in the beam and so do not collect a layer of contamination, as do thicker apertures (contamination is discussed in § 6.14).

2.9 Objective astigmatism corrector

This is an arrangement of magnetic poles or electrostatic electrodes which can set up a cylindrical lens field in any desired orientation and strength so as to oppose the residual astigmatism of the objective lens. The usual corrector (Fig. 2.29) has eight poles, linked together in pairs, and fed by some convenient network of current to excite fields designed to provide the cylindrical

Fig. 2.29. Objective lens astigmatism corrector showing octupole arrangement of coils.

lens. The whole system can be aligned with the objective lens axis by feeding individual poles with small excitations, so as to avoid any movement of the image as the corrector is varied. The whole assembly may be in the objective lens gap, some distance from the lens axis, or it may be lower down the bore of the objective lens.

2.10 The projector system

The design of the projector lens system has been discussed at some length in § 1.10. The projector lenses themselves are rather simple, as they do not have to accommodate any special components within them. There must be a translational movement for each projector lens to permit correct alignment of the instrument (discussed in § 4.6). The lenses are usually equipped with limiting apertures to cut out scattered electrons which would cause loss of contrast in the final image. Immediately above the first projector lens is another adjustable aperture, similar in construction to the adjustable condenser aperture. It is used in diffraction operation. The other main feature of the projector system is the inclusion of magnetic shielding to reduce the effects of any ambient stray fields on the image.

2.11 Viewing system

Below the projector system is the viewing chamber, which accommodates the fluorescent screen on which the image is rendered visible to the observer, and which has from one to three ports through which the screen may be observed by the operator and additional observers. The fluorescent screen is needed both to enable the operator to view the image for selection of an appropriate area, and for focusing the image before recording it on a photo-

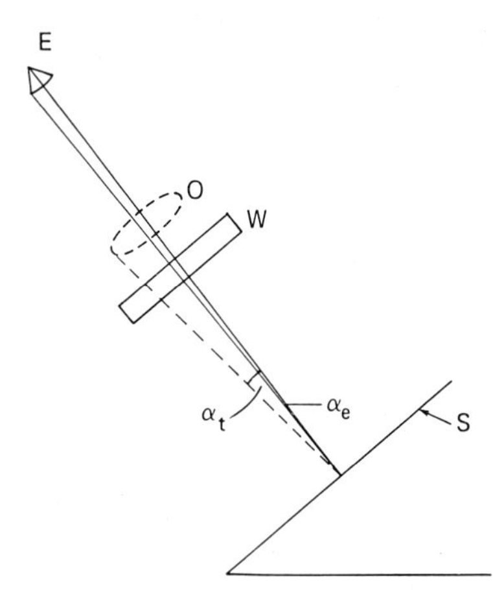

Fig. 2.30. Diagram to illustrate how the use of a viewing telescope (object lens O) increases the angular aperture of viewing the image formed on the screen S from α_e (by the unaided eye E) to α_t. W is the lead glass viewing window.

graphic plate. It is necessary to provide additional optical magnification for this to be done accurately enough, and all microscopes are now provided with a binocular telescope for viewing the fluorescent screen. The objective lens aperture of such a telescope should be very large so that it collects light from a wide angular aperture from the screen. If the aperture is large enough, the increased collection angle can balance the loss in illumination intensity due to the magnification of the system and a magnification of 10 × may generally be obtained without loss of light (Fig. 2.30). This is very important indeed, as this extra magnification could only be obtained electron optically by increasing the electron beam intensity on the specimen plane by a factor of 100. Even were this available, it would greatly increase the damage to the specimen.

In some microscopes (e.g. the Philips series), a small extra screen is provided for focusing. This is about halfway down the viewing chamber (it is swung into the electron beam as required). The image magnification on this screen is only half as great as that on the main screen, so, although the image is brighter, the resolution is more limited. When a 35 mm camera is placed in an electron microscope, it is placed even higher in the viewing chamber as described in § 2.1, so that most of the image on the final screen is contained within its small format. The advantage of such an arrangement is that a large number of photographs may be recorded with a small expenditure in photographic material. On the other hand, the electron optical magnification is low, and hence relatively large optical magnifications are needed to obtain prints of a useful magnification. The electron noise conditions may be unfavourable unless a very slow film is used (see § 7.2.3).

2.12 The camera

In order to obtain a permanent record of the image, photographic material has to be introduced into the microscope for direct exposure by the beam. In the past, photographic plates were used almost exclusively, but sheet film and roll film are now coming into greater use. Whichever is used, however, the photographic material has to be enclosed in a light tight cassette, and must be transported into the path of the electron beam for exposure. Because the photographic emulsion takes some time to degas in the vacuum, there must be a considerable number of exposures available before the work must be stopped to reload the camera. Most instruments carry 18 or 24 photographic plates or sheets of film and sometimes roll film as an addition or alternative.

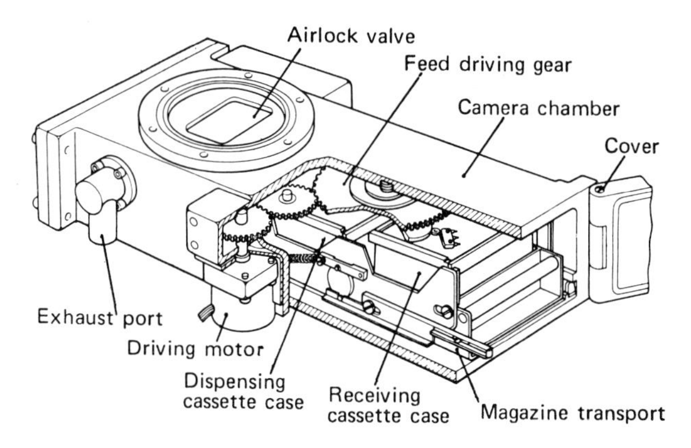

Fig. 2.31. Diagram of a plate camera for an electron microscope. (Courtesy JEOL Ltd.)

A typical plate camera is shown in Fig. 2.31. The unexposed plates are held in a dispensing cassette case. When the camera is actuated, a motor drives one plate forward into the path of the electron beam (the beam having been stopped by the interposition of a shutter blade just below the projector lens). When the plate is in position below the camera airlock valve, which is of course open at this time, the shutter is opened for a time determined by the exposure measuring device. This may be a photocell looking at the fluorescent screen, or the collected current on the screen itself, which is insulated from the rest of the instrument. When the plate has been exposed and the shutter is again closed, the transport mechanism carries the plate to the receiving cassette case, where it is stored until the operator wishes to open the camera chamber to remove all the exposed plates. In most modern instruments, a serial number is automatically recorded on the plate to assist later identification and the magnification of the image and the kilovoltage of operation are often included as well. A 70 mm roll film is generally loaded as an alternative camera to the plate camera. It has the advantage of allowing a considerably larger number of exposures (up to 50) before re-loading and has a reasonably large picture format (about $2\frac{1}{2}$ inches square). Since tank development is used, it is essential to use accurate exposure measurement so as to obtain correct development of all the exposed frames. Alternatively a 35 mm camera is used as described above (§ 2.11). Photography is discussed in detail in Chapter 7.

It is possible to record the image with an external camera by using a transmission phosphor at the bottom of the electron microscope column (Anderson and Kenway 1967). A polaroid camera is very convenient in this position since it gives immediate prints for examination. The photographic grain size need not limit the performance if the electron optical magnifications are correctly chosen. The exposure times required are comparable with those for an internal camera.

More recently, the transmission phosphor has been coupled to closed circuit television chains and to image intensifier systems, to enable low intensity images to be viewed in comfort. The applications and limitations of image intensifiers are discussed in § 7.8.

One feature of an electron microscope which occasions some surprise at first is the fact that the photographic material is often in a plane several centimetres away from the viewing screen, where the image is focused. This is possible because of the very great depth of field of the microscope, arising from the very small aperture used for the objective lens. The effect at the object plane has already been discussed in section § 1.4.

For an image resolution of d, and an objective aperture α, the depth of field in the object plane is D_0 (derived in § 1.4)

$$D_0 = \frac{d}{\alpha}.$$

When this depth of field is referred to the image plane it is known as the depth of focus. For a total instrumental magnification M, the depth of focus D_i becomes very large indeed:

$$D_i = \frac{M^2 d}{\alpha}.$$

For $M = 10^4$; $d = 1.5$ nm; $\alpha = 10^{-2}$; $D_i = 50$ metres. It is therefore immaterial that the fluorescent screen and the camera are in different planes, as their separation is negligible compared with the depth of focus.

It is important, however, to know what the physical distances between the projector lens, screen and camera are, because of the resulting significant changes in magnification in the different positions. In particular, it should be noted that most microscope magnification calibrations are in terms of the magnification of the image on the photographic plate. Since in general this is below the viewing screen, the viewing screen magnification may be as much as 20% lower than the indicated magnification. When a small 'focusing screen' is used, the difference is much greater, and the focusing screen magnification may be only half of that on the final plate. This is important when estimating the accuracy of astigmatism correction (see § 4.9).

2.13 The vacuum system

In order to allow passage of the electron beam through the microscope without interference from gas molecules, the pressure within the instrument has to be reduced to the point where there is a very small probability that an electron will encounter a gas molecule. This condition is attained when the pressure within the microscope column is lower than about 10^{-3} Torr (atmospheric pressure = 760 Torr), but a pressure of 10^{-5} Torr or less is desirable.

2.13.1 GENERAL FUNCTION OF VACUUM SYSTEMS

A good mechanical vacuum pump can just achieve a pressure of 10^{-3} Torr but its pumping speed is very low at this pressure, and it is in fact necessary to employ a diffusion pump as well in order to increase the pumping speed and to obtain a pressure significantly better than the threshold value required.

It is important to realise that the mode of operation of a diffusion pump, and the mechanics of 'pumping' gas at pressures of 10^{-3}–10^{-6} Torr, are quite different from those often imagined. There is no question of the pump pulling molecules along the pumping tubes towards the pump. The molecules comprising the residual air in the column move randomly in straight lines unless they encounter another molecule, or until they strike a surface (of the pumping tube, for instance), when they may effectively 'bounce' off the surface or they may be adsorbed on to the surface, later to be ejected when some random quantum of energy, sufficient to break their bond to the surface, is received. In a vessel at constant pressure, there is a random distribution of speed and direction of molecules in motion, and the net transfer of gas is zero. If, however, one end of the vessel is connected to a pump of some kind, there is one 'wall' of the vessel towards which gas molecules travel and from which no molecules return. Thus there is a steady loss of molecules from the vessel, and the pressure is steadily decreased, since there is a net *diffusion* of gas towards the pump.

A diffusion pump operates by having one or more directed jets of oil (or mercury) molecules between a centre column and an outer cooled wall (where condensation takes place). A gas molecule entering the jet from above the pump is trapped and forced downward into the body of the pump, and eventually out through its exhaust. Gas molecules within the body of the pump are not able to penetrate back into the pumped space because of the dense jet (stream) of oil molecules. The pump continues to operate so long as the pumped gas pressure in the exhaust does rise to too high a value. If this *critical backing pressure* is reached, there are enough gas molecules present to break through the oil jet and to stop the pumping action. Most oil diffusion pumps have a critical backing pressure of the order of 2×10^{-1} Torr, and hence require a mechanical backing pump operating all the time to remove the pumped gas, unless they exhaust into a vacuum reservoir of large volume whose pressure rises only slowly.

It would seem as if a diffusion pump should reduce the gas pressure in a system indefinitely. In practice this does not happen, and an equilibrium pressure is fairly soon attained in any vacuum system. The first cause for this is the presence of small leaks of air into the vacuum system, past one of the many seals that are required in a demountable system. These leaks eventually just balance the rate of removal of molecules by the pump. Secondly, in a metal system particularly, there are a considerable number of molecules trapped on the metal surfaces and within the metal itself, and these take a long time to de-sorb. The process of establishing a good vacuum is greatly speeded up if the vacuum vessel can be baked to a few hundred degrees centigrade, but this is often impractical, particularly in an electron microscope! There may also be sources of relatively large numbers of organic molecules from the use of grease on gaskets, or from careless handling of parts of the vacuum system or the specimen holder.

Finally, there is the question of *pumping speed* between the column and the pump. If there were an infinitely large connection, there would be no limitation on the transfer of molecules to the pump except their own numbers and random motion. With an actual pumping connection, in the form of a tube, the surfaces of the tube act as a brake on progress because molecules are trapped on the surface for a greater or less time. The longer the pumping connection, the lower will be the pumping speed and, for a given length, a reduction in internal diameter will quickly reduce the efficiency of the pumping connection. Obviously, if there is only a very long narrow pumping connection, the pressure in the column will be reduced only slowly and no matter how large a capacity of pump is used there will be a serious limitation on the vacuum achieved.

2.13.2 A VACUUM SYSTEM FOR AN ELECTRON MICROSCOPE

It is not easy to provide a good vacuum system for an electron microscope. There are relatively few places along the microscope column where a large diameter pumping pipe can be attached without interfering with the electron optical design or the specimen facilities. Between these main pumping ports, the main vacuum connection is along the pole piece bores which are in general of small diameter and therefore produce very low pumping speeds. Furthermore, there are several small apertures in the instrument which further reduce the pumping speed. Add to this a large surface area of magnetic lens material, which cannot be baked to desorb gas, and one understands why so many microscopes only achieve pressures of the order of 10^{-4} Torr in the centre of the microscope column.

One device which can be used is the cryo-pump. If a surface is cooled below the condensation temperature of a vapour, it acts as an effective pump, since vapour reaching the surface cannot escape again. Thus a liquid nitrogen cooled surface is very effective for organic vapours, though not of course for permanent gases. These cryo-pumps are discussed in the section on contamination (§ 6.14).

It is necessary to have vacuum gauges to record the pressure in various key points in the system. A simple gauge suitable for pressures between atmospheric and about 1 Torr is the thermocouple gauge. This depends on the cooling of a heated wire by the gas molecules striking it. It is quite robust, and is used to monitor the vacuum produced by the rotary pump. It is therefore placed on the backing line to measure the reservoir pressure, and is also used to monitor the pumping out of the air locks. Its operating range is from atmospheric pressure to 10^{-1} Torr. The high vacuum side of the system is normally measured by a Penning-type gauge. This depends on the ionisation of the gas molecules in a high electrostatic field. A permanent magnet around the tube causes the charged particles to execute long spiral paths and hence increases the chance of ionisation. The gauge is effective in the pressure range 10^{-3}–10^{-6} Torr, and is therefore used to monitor the working vacuum in the microscope column. However, the gauge is often placed for convenience rather close to the diffusion pump, and the indicated pressure is usually much better than the actual pressure achieved in the centre of the microscope column or in the electron gun.

A typical vacuum system for an electron microscope is shown in Fig. 2.32. A mechanical rotary pump backs the diffusion pump through valves V_1 and V_7 and simultaneously evacuates the vacuum reservoir. The rotary pump is normally placed some distance from the microscope desk and is connected to it by a vibration-insulating connection. The diffusion pump, which can be sealed off from the microscope column by the baffle valve BV when the column is let up to atmospheric pressure, pumps the column through a number of pumping tubes up the column. The pump is usually mounted immediately beneath the pumping manifold. The initial evacuation of the column is carried out by the rotary pump via the valve V_2 and selector valve

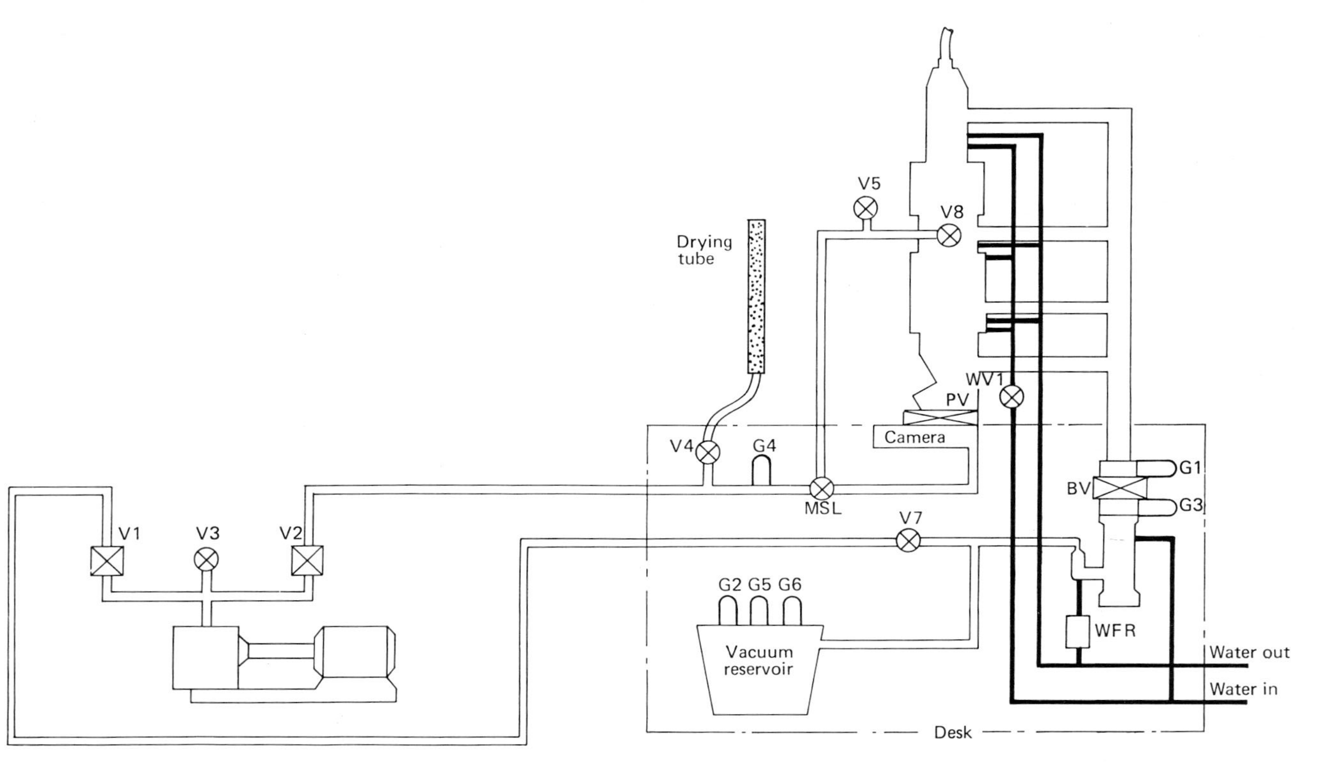

Fig. 2.32. Schematic diagram of the vacuum and water system for an electron microscope. Details of operation are given in the text. The water circuit is indicated in thick black lines. A water flow relay WFR shuts off the electrical supplies to the diffusion pump and closes if the water flow rate is insufficient.

MSL and the camera air-lock, the vacuum sealing plate PV on the camera being left open for this operation. When a low enough pressure has been reached to permit operation of the diffusion pump, the manual selector valve MSL is closed and the baffle valve BV is opened, to allow pumping by the diffusion pump. Once an indicated vacuum of 10^{-5} Torr has been obtained in the column (as indicated by the gauge G1) the rotary pump may be switched off (V_1 and V_2 closing automatically) and the diffusion pump will pump into the vacuum reservoir for long periods. When a specimen is to be inserted into the column, the insertion air-lock V_8 is closed off from the column, and air admitted through V_5. The specimen chamber is re-evacuated via MSL and V_2, and when a suitable vacuum is reached (about 0.01 Torr) MSL is closed and V_8 opened to the column vacuum again.

A similar procedure is used when changing photographic material. In this case, the plate valve (PV) sealing off the camera chamber from the microscope column is closed before admitting air to the camera. After reloading, the camera is pumped out by operating the valve MSL. When the pressure (as measured on the thermocouple gauge, G4) has reached an adequate level, MSL may be closed again and the camera connected to the rest of the column by opening up the plate valve (PV).

When air is to be admitted to the camera or the microscope column, it is drawn in through a drying tube to avoid the admission of water vapour which greatly slows down the pumping, and may affect contamination rate and filament life. Some instruments employ dry nitrogen admission from a cylinder instead of the drying tube. This is especially important in laboratories where the humidity level is high.

It is very important that air at atmospheric pressure should not be admitted to the vacuum system while it is connected with a hot diffusion pump, as this will blow the oil vapour all over the microscope column, and may also decompose the low vapour pressure oil into other substances of high vapour pressure. If this calamity happens, the entire column has to be disassembled and cleaned. The vacuum system accordingly has to be interlocked to prevent incorrect operation of the various valves, and this involves pressure sensing gauges and their associated circuitry. Complete automation of the vacuum system is very complex and expensive but a semi-automated system can be much simpler and is still fairly safe. It is, however, important that an operator should be quite clear about the functioning of the vacuum system and appreciate what the operating sequence must be so as to avoid any accidents. The detail of the operation of the vacuum system varies from one microscope to another, and is described in the Instruction Manuals.

2.14 Electrical supplies

The electrical supplies are required to provide (a) a stabilised source of high voltage to be applied to the electron gun, (b) heater supplies to the filament, (c) bias voltage to the Wehnelt cylinder, (d) stabilised current supplies for the lenses and deflector coils, (e) control circuitry for interlocks and vacuum system.

2.14.1 HIGH VOLTAGE SUPPLY

The high voltage applied to the electron gun is usually generated by a radio-frequency oscillator which provides a 4 kV output. This is fed into the centre of a Cockcroft–Walton multiplier (an array of capacitors and rectifiers), which, with 26 stages can provide 100 kV at the output. A chain of high stability wire-wound resistors is connected right across the high voltage generator to earth (typically a total of 1500 MΩ). A tapping near the earthed end of this resistor chain provides a voltage proportional to the total high voltage which is compared with a reference voltage derived from a standard cell. The difference signal is fed via a feed-back stabilising circuit to the input of the oscillator. The high voltage generated is, of course, negative with respect to earth, since the anode of the microscope must be at earth potential.

The whole of the high voltage generator is enclosed in a tank of insulating oil, carefully filtered free of all particulate matter. Where the stabilisation is required to be particularly good (better than $\sim$ 5 parts per million) the high voltage is transferred to a separate measuring resistor in its own oil or gas pressure tank, which is used for deriving the feed-back signal (this helps to avoid the ‘noise’ generated by thermal movements in the oil, or particles released from the many electrical components in the main oil tank). The high voltage supplies in most modern instruments are stable to only a few parts per million.

It is normally arranged that the microscope can be operated at a series of different values of high voltage, to give a range of contrast and penetration conditions to suit different specimens (§ 6.1). Typically, the operating voltages would be 20, 40, 60, 80, and 100 kV.

The operator should never open a high voltage tank. The chances of introducing some dirt into the oil and causing future instability is too great, without proper facilites for filtering the oil.

2.14.2 FILAMENT AND BIAS SUPPLY

The tungsten filament in the electron microscope gun requires between 2.5

and 3 amps as a heater current (for the normal filament diameter of 0.005 inch). This may be high frequency or DC driven; the present tendency is to have a DC supply to avoid any ripple on the filament supply which would modulate the electron beam. The filament heating circuit has to have a fine control for the use of the operator, so that the filament can be run in the most economic way (see § 1.6). As was also pointed out in § 1.6, a variety of gun operating conditions can be obtained by variation of the value of the self-biasing resistance. This resistance is also housed in the high voltage tank (since the filament and Wehnelt cylinder are at a negative high voltage). It may be a tapped resistor, although many instruments now use the internal resistance of a triode valve since it is infinitely variable and can yield a higher bias value (0–2500 V is a typical range). If the bias voltage is decreased, the beam current is increased and the saturation point is usually achieved at a higher temperature of the filament, yielding of course a higher beam brightness. Thus, control of the bias resistor gives a series of saturation brightness values.

The filament and bias controls must be operated remotely by insulated rods leading out of the high voltage tank. The actual controls are placed on the microscope desk in a position convenient to the operator, since both these parameters need to be altered from time to time during routine operation of the instrument.

A meter on the microscope desk indicates the electron beam current, as a guide to the setting of these controls. In practice, this meter usually records the current measured at the earth end of the bleeder resistance across the high voltage supply. It consequently indicates a standing current due to the small current driven through the very high resistance of the bleeder chain, whenever the high voltage is switched on. There is then an increase in this reading, due to the electron beam current itself, when the filament is heated. If, when the filament heater control is operated, no change is noted in the indicated beam current reading, the filament has burnt out. The standing beam current changes, of course, as the different values of high voltage are applied to the electron gun. If the beam current indication shows kicks or flickering, it is usually an indication of small corona discharges in the gun, due to poor vacuum or some dirt particles.

The supplies to the filament and the Wehnelt cylinder are carried in insulated leads in the core of the high voltage cable to the electron gun.

2.14.3 CURRENT SUPPLIES TO THE ELECTRON LENSES

Individual current supplies are required for each lens, with stabilities of a few parts per million. They must be capable of stable operation over a wide range of current to permit the use of the instrument under widely different operating conditions. Most supplies are now transistorised and therefore operate at relatively low voltages. A schematic diagram of a lens stabiliser is shown in Fig. 2.33. The current for the lens coil L flows through the series transistor T_1. There is a precision variable resistor R in series with the lens coil. The voltage generated across one section of this resistor is balanced against

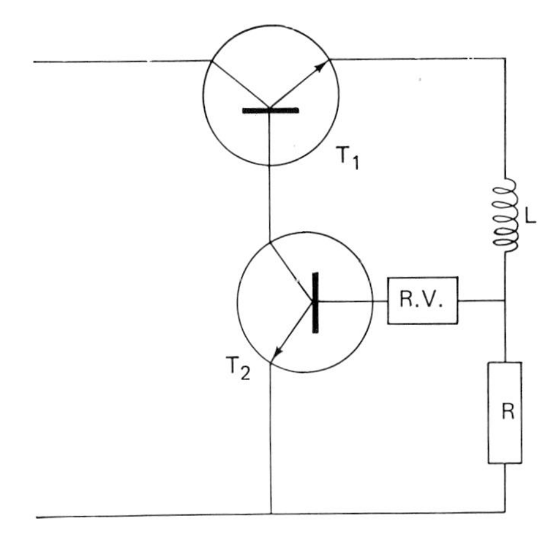

Fig. 2.33. Simplified schematic diagram of a transistor operated stabiliser. T_1 is the series transistor. Small variations in the voltage generated across the resistor R are amplified by transistor T_2 and used to control the flow of current through T_1. RV is a reference voltage.

that from a reference voltage source RV, and fed back through transistor T_2 to control the current through T_1. A new stable value of lens current can be established by varying the series resistor R. This is the lens current control provided on the microscope desk. The older valve driven supplies used exactly the same principle, but operated with relatively high voltages (150–500 V). Since the control switches and potentiometers are in the lens stabilising circuits, it is particularly important that they are of high quality and are kept very clean as otherwise the noise generated will affect the lens stabilisation.

The reference circuit is generally temperature controlled and may conveniently feed both current and high voltage stabiliser circuits. If this is done, a change in high voltage to the gun is accompanied by a corresponding change in the lens currents, which results in automatic retention of focus and magnification with change in kilovoltage.

The operator is not of course provided with a current control to each lens. Much of the programming of lens currents is necessarily built into the instrument so that the correct lens combinations are obtained for minimising distortion, for instance (§1.10). The objective lens current is also adjusted automatically to be near the correct value when the magnification is changed. A further useful coupling is an automatic adjustment of the strength of condenser 2 as the magnification is altered, so as to keep the illumination intensity on the screen approximately constant. Thus, the controls presented to the operator are mainly functional, e.g. magnification, focus, rather than descriptive of the electrical circuits affected (e.g. projector lens current). There is a direct read-out of the magnification setting as a convenience to the operator. The actual lens currents may be monitored by a meter on the control desk.

2.14.4 OTHER CONTROL CIRCUITRY

The astigmatism correctors for condenser and objective lenses involve quite complicated current changes in the multipole correctors, but are reduced in the circuitry to simple orientation and magnitude controls at the microscope desk. There are many other control circuits associated with the vacuum system and the safety interlocks. These are often operated with direct current circuits, since wires carrying alternating current may give rise to fluctuating magnetic fields near the microscope column, which can cause movement of the electron beam and hence loss of performance. In general all power transformers give rise to quite large magnetic fields, and this is why the main electronic cubicle is sited some distance away from the microscope column.

Most instruments are provided with circuit monitoring points which will indicate the correct functioning of the stabilisers and provide a check on the stabilities being obtained. These checks should be noted periodically in order to monitor any slow changes in performance. The check on the main reference circuit is particularly important.

REFERENCES

Agar, A. W. and J. H. Lucas (1962), Use of a new heater stage for the electron microscope, Proc. 5th Int. Congr. Electron Microscopy, Philadelphia *1*, E-2.

Anderson, K. and P. B. Kenway (1967), External photography of the electron microscope image, Proc. 25th Ann. Meeting EMSA, Chicago, p. 244.

Bowman, M. J. and V. H. Meyer (1970), Magnetic phase contrast from thin ferromagnetic films in the transmission electron microscope, J. Physics E: scient. Instrum. *3*, 927.

Bradley, D. E. (1961), Simple methods of preparing pointed filaments for the electron microscope, Nature, Lond. *189*, 298.

Broers, A. L. (1970), The use of Schottky emission lanthanum hexaboride cathodes for high resolution scanning electron microscopy, Proc. 7th Int. Congr. Electron Microscopy, Grenoble *1*, 239.

Cotterill, R. M. J. (1964), A liquid helium stage for the Siemens Elmiskop I, Proc. 3rd Eur. Reg. Conf. Electron Microscopy, Prague *A*, 63.

Crewe, A. V., J. Wall and L. M. Welter (1968), A high resolution scanning electron microscope, Rev. scient. Instrum. *40*, 241.

Dupouy, G., F. Perrier, A. Sequela and R. Segalen (1968), Sur une méthode permettant d'observer individuellement les domaines magnétiques, en microscopie électronique, C.r.Acad. Sci., Paris *266*, 1064.

Fernández Morán, H. (1966), Applications of improved point cathode sources to high resolution electron microscopy, Proc. 6th Int. Congr. Electron Microscopy, Kyoto *1*, 27.

Fuchs, E. and W. Liesk (1962), An arrangement for magnetising objects in the electron microscope, Optik *19*, 307.

Goodhew, P. J. (1972), Specimen preparation in materials science, in: Practical methods in electron microscopy, Vol. 1, A. M. Glauert, ed. (North-Holland, Amsterdam).

Hashimoto, H., T. Naiki, T. Eto, K. Fujiwara, M. Watanabe and Y. Nagahama (1966), Specimen chamber for observing the reaction process with gas at high temperatures, Proc. 6th Int. Congr. Electron Microscopy, Kyoto *1*, 181.

Hashimoto, H., T. Naiki, T. Eto and K. Fujiwara (1968), High temperature gas reaction specimen chamber for an electron microscope, Japan J. appl. Phys. *7*, 946.

Heide, H. G. (1962), Electron microscopic observation of specimens under controlled gas pressure, J. Cell Biol. *13*, 147.

Hibi, T. (1954), Pointed filament and its applications, Proc. 3rd Int. Congr. Electron Microscopy, London, p. 636.

Imura, T., H. Saka, A. Nohara, N. Yukawa and H. Kitagawa (1970), New devices for tensile test in high voltage electron microscopy (simultaneous recording of stress–strain curve and EM images), Proc. 7th Int. Congr. Electron Microscopy, Grenoble *1*, 153.

Itoh, K., T. Itoh and M. Watanabe (1954), The high temperature furnace for the electron microscope, Proc. 3rd Int. Congr. Electron Microscopy, London, p. 685.

Leisegang, S. (1954), Über Versuche in einer stark gekühlten Objektpatron, Proc. 3rd Int. Congr. Electron Microscopy, London, p. 184.

Lehtinen, B., E. Broberg, and L. Dahne (1967), Electron microscope specimen holders for simultaneous straining, tilting and rotation of thin foils, J. scient. Instrum. *44*, 289.

Lucas, J. H. (1968a), A universal specimen stage for a biological electron microscope, 4th Eur. Reg. Conf. Electron Microscopy, Rome *1*, 247.

Lucas, J. H. (1968b), A serial section stage for the EM 801, Personal communication.

Mills, J. C. and A. F. Moodie (1968), Multipurpose high resolution stage for the electron microscope, Rev. scient. Instrum. *39*, 962.

Page, R. S. (1956), Personal communication.

Schott, D. and S. Liesegang (1956), Objektkühlung im Elektronenmikroskop, Proc. Ist Eur. Reg. Conf. Electron Microscopy, Stockholm, p. 27.

Swann, P. R. and G. R. Swann (1970), A tilting cold stage for the AEI EM 802 Proc. 28th Ann. Meeting EMSA, Houston, Texas, p. 372.

Valdrè, U. (1964), A double-tilting liquid-helium cooled object stage for the Siemens Elmiskop I, Proc. 2nd Eur. Reg. Congr. Electron Microscopy, Prague *A*, 61.

Valdrè, U. (1965), A double tilting heating stage for an electron microscope, J. scient. Instrum. *42*, 853.

Valdrè, U. and M. J. Goringe (1970), An improved liquid helium stage for an electron microscope, J. Phys. E: scient. Instrum. *3*, 336.

Watanabe, H. and I. Ishikawa (1968), A liquid helium cooled stage for electron microscope and its application (in Japanese), J. Electron Microscopy *17*, 119.

Chapter 3

Image formation in the electron microscope

Although there are analogies between the electron and light microscopes, the mechanisms of image formation are quite different. Two main processes are at work in the electron microscope, elastic and inelastic scattering.

3.1 Elastic scattering

When electrons in the beam encounter, or pass very close to, an atomic nucleus of one of the atoms in the specimen being examined, they are deflected through relatively large angles without loss of energy – hence the term *elastic scatter*. This is due to the very much greater mass of the nucleus compared to the electron. The number of electrons so deflected increases with the thickness of a given element encountered, and is also proportional to the atomic number of the element. Thus, the scattered intensity in a given solid angle is proportional to the mass thickness of the specimen. If a physical stop of relatively small angular aperture is placed behind the specimen, the widely scattered electrons will be removed from the beam and the image will be less intense in regions corresponding to a greater mass thickness in the object. This correspondence of intensity with mass thickness is known as *amplitude contrast*.

The contrast of a given specimen can be increased or decreased by decreasing or increasing the angular size of the contrast aperture. In the electron microscope, this aperture is placed at the back focal plane of the objective lens, and is known as the *objective aperture*. It is not necessary to have a contrast aperture in the objective lens to obtain some amplitude contrast, because when the widely scattered electrons enter the objective lens they are strongly affected by its spherical aberration, and these scattered electrons are no longer imaged back into the same point in the image corresponding to the object point they started from. Instead they tend to be scattered over a considerable area of the image as a kind of background fog. A strongly scattering part of the specimen nevertheless retains relatively good contrast because a large number of electrons are scattered out of that part of the image and only a small diffuse level of electrons is scattered back. Low contrast parts of the object may, however, be quite difficult to see in the absence of a physical objective aperture, and the presence of the aperture always increases image contrast.

The angle through which electrons are scattered by a given object point is dependent also on the energy of the incident electrons, that is on the accelerating voltage. An increase in electron energy reduces the mean angle of scatter, and vice versa. Thus for a given size of objective aperture, a reduction in voltage will result in a larger number of electrons being stopped by the aperture, and hence a greater contrast results from given image detail.

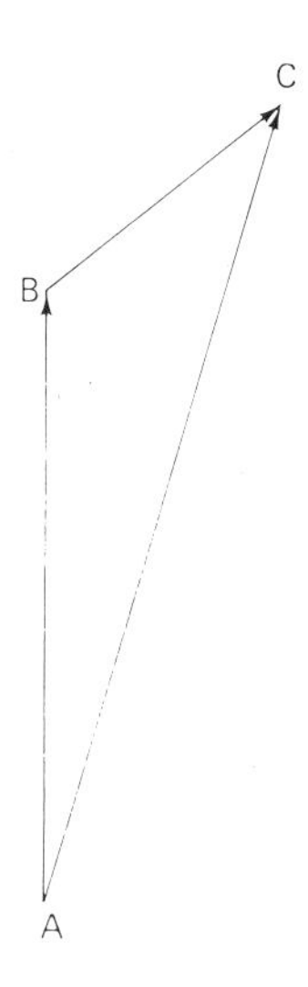

Fig. 3.1. Vectorial representation of the combination of unscattered wave (AB) and scattered wave (BC) to form a resultant wave (AC). Note that the amplitude of AC differs from that of AB and contrast results.

Conversely, a higher accelerating voltage results in a loss of image contrast but also, because more electrons pass through the aperture, the specimen appears more transparent.

The elastically scattered electrons which pass through the objective aperture also contribute to contrast, because (regarding them now as electron waves) they may interfere constructively with the unscattered electrons. The effect may be represented vectorially (Fig. 3.1). The vector BC (this scattered wave) may be in any direction relative to AB, depending on its *phase* relative to the unscattered wave. Haine (1961) showed that a small feature in the object would cause a change in one part of the incident wave front, which could be resolved into a wave in the normal (unscattered) direction; plus a vector roughly in quadrature, when the objective was in focus (Fig. 3.2a). Thus, in this case the amplitude of the resultant wave AC

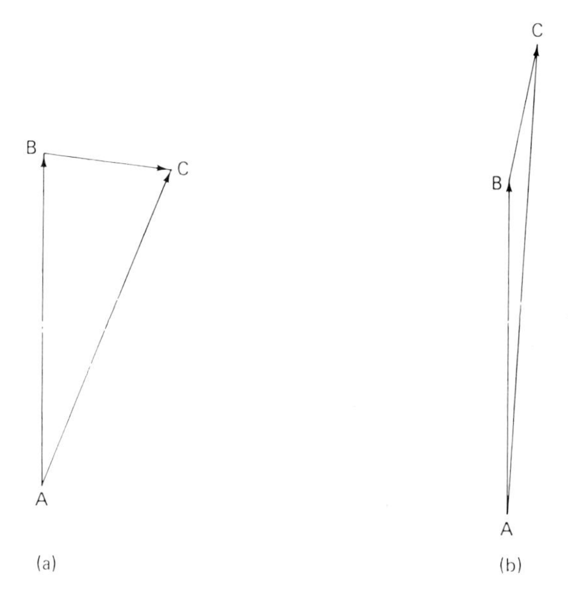

Fig. 3.2. Vectorial representation of background wave (AB), scattered wave (BC) and resultant wave (AC), for a thin specimen. (a) With objective lens focused. (b) With objective focus changed to alter the phase of the scattered wave, yielding amplitude contrast (AC > AB).

is effectively the same as that of the incident wave. Since the eye on the photographic plate can only detect changes in amplitude, no contrast will be perceptible in this condition. However, when the plane of focus is changed, there is no change in phase of the background wave, but the phase of the scattered wave varies quite quickly. Depending on the amount of defocus, the scattered wave may now be in phase or antiphase with the background wave, and a perceptible amplitude difference will appear in the image (Fig. 3.2b). This is known as *phase contrast*, and it may be appreciable, without the defocus causing a significant loss of resolution. The sign of the phase changes as one passes through focus, leading to the well-known contrast reversal of fine image detail at focus, a fact which must be borne in mind on image interpretation (Chapter 8).

The phase contrast observed is independent of the presence of a physical objective aperture; it is also dependent only slightly on the accelerating voltage and on the atomic number of the scattering atoms. Thus, it allows the visualisation of specimen detail from material of low atomic number, which would not give rise to adequate amplitude contrast.

3.2 Inelastic scattering

When an electron in the beam encounters one of the orbital electrons in an atom of a thin specimen, there is an encounter between two bodies of equal mass. This encounter results in energy being imparted to one of the orbital electrons, at the expense of the beam electrons, and the process is known as *inelastic scattering*. The energy loss in the beam electrons in one of these collisions will normally be of the order 10–20 eV; the loss in a single encounter will almost invariably be less than 100 volts. If the specimen under examination is 'thin' (say up to 60 nm) there will generally be no more than one such encounter for any given beam electron, and conditions of *single scattering* hold. The beam electrons suffering such an energy loss will be deflected through very small angles (10^{-4} radians). Inelastically scattered electrons therefore nearly all pass through the objective aperture. The small energy loss corresponds to a change in wavelength of the electrons. Since the inelastically scattered electrons have differing energies, those not close to the beam axis will be distributed as a blur over the image due to the chromatic aberration of the objective lens and they then cause loss of sharpness and contrast in the image. This is not normally critical if the specimen is thin, since only a small proportion of electrons is affected. If the specimen inserted into the microscope is relatively thick, each beam electron will on average suffer a number of encounters with atoms of the specimen, and will suffer multiple energy losses. In such a case, the energy spread in the beam emerging from the specimen may cover some hundreds of electronvolts, and there may be relatively few unscattered electrons, and those electrons which were elastically scattered may also have suffered energy-loss collisions. Under these circumstances, there is no possibility of phase contrast from the specimen since there are no longer one or two well-defined electron wavelengths in the beam suitable for interference. A thick specimen of this type will therefore exhibit only amplitude contrast. Since the energy losses in the specimen are large compared to the voltage variations due to the electron microscope itself, the specimen itself becomes the main limitation to the resolution obtainable in the image. It will be appreciated that this is the case for a high proportion of all specimens examined.

3.3 Control of contrast

The contrast of an image in the electron microscope will be compounded of the amplitude contrast due to elastically scattered electrons stopped by the objective aperture, and phase contrast due to those passing through the aperture and interfering with the unscattered beam. In general, the amplitude contrast is dominant for large structures, while the phase contrast increases in importance for small structures and indeed becomes almost the sole source of contrast for very small object points of low atomic number.

There are various ways in which the contrast in the image can be modified. With any specimen giving rise to an appreciable amount of elastic scattering, the amplitude contrast in the image can be controlled by choice of kilovoltage and effective objective aperture. The amplitude contrast is a maximum for the minimum operating voltage of the electron gun (Fig. 3.3). However, unless the specimen is very thin, the higher chromatic aberration at lower accelerating voltages will lead to an unacceptable loss of resolution. There is also a loss of brightness from the gun at low voltages which makes low voltage operation only suitable for lower magnification operation.

An alternative method of altering image contrast is to change the size of the objective aperture. If the objective aperture size is decreased, from the normal size of about 50 μm, a greater number of the scattered electrons are stopped and the contrast of the image is increased. There is usually little loss in resolution for apertures down to about 20 μm, with noticeable improvement in contrast. If smaller apertures are used, say 5 to 10 μm, there will be a

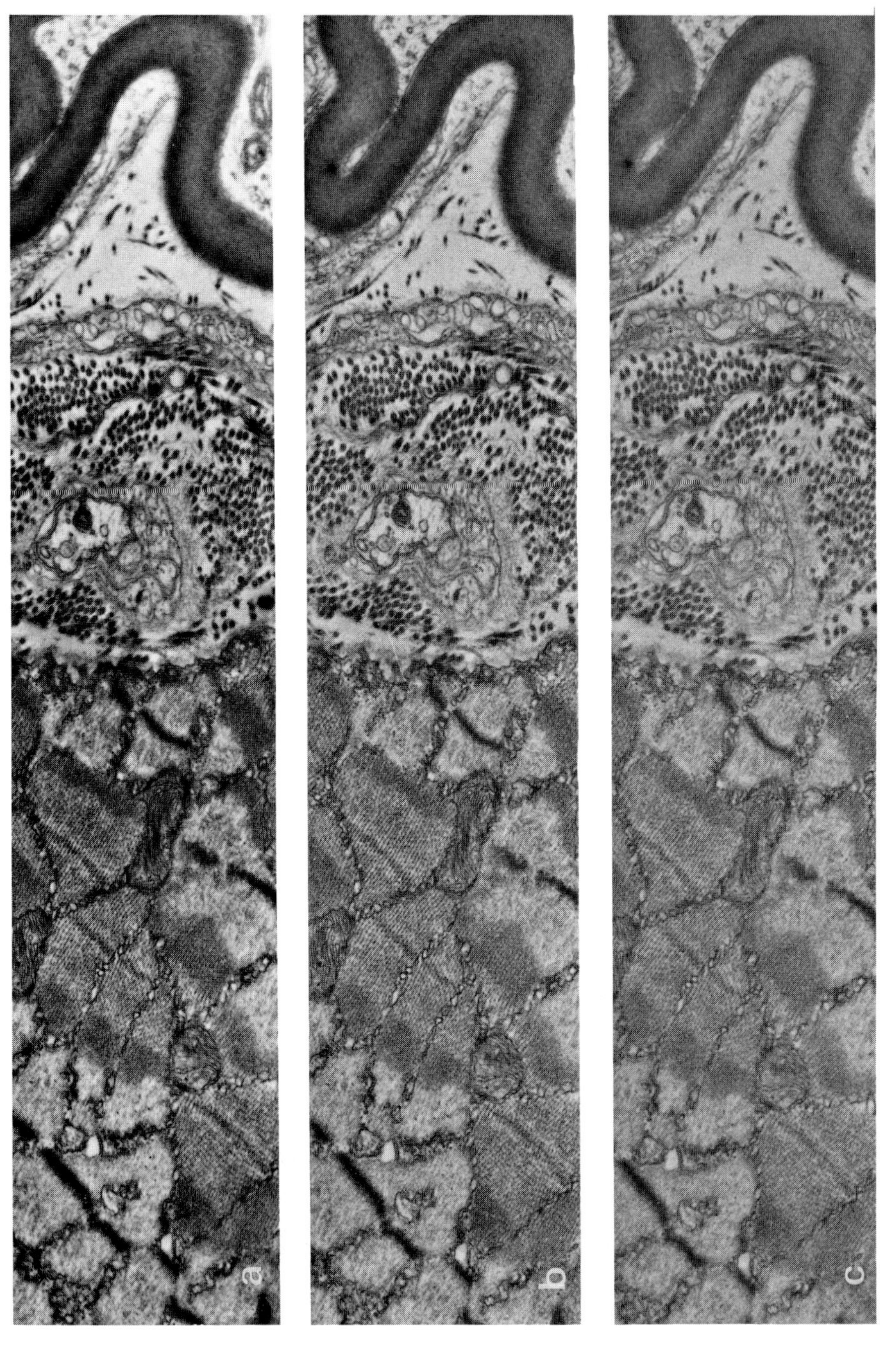

Fig. 3.3.

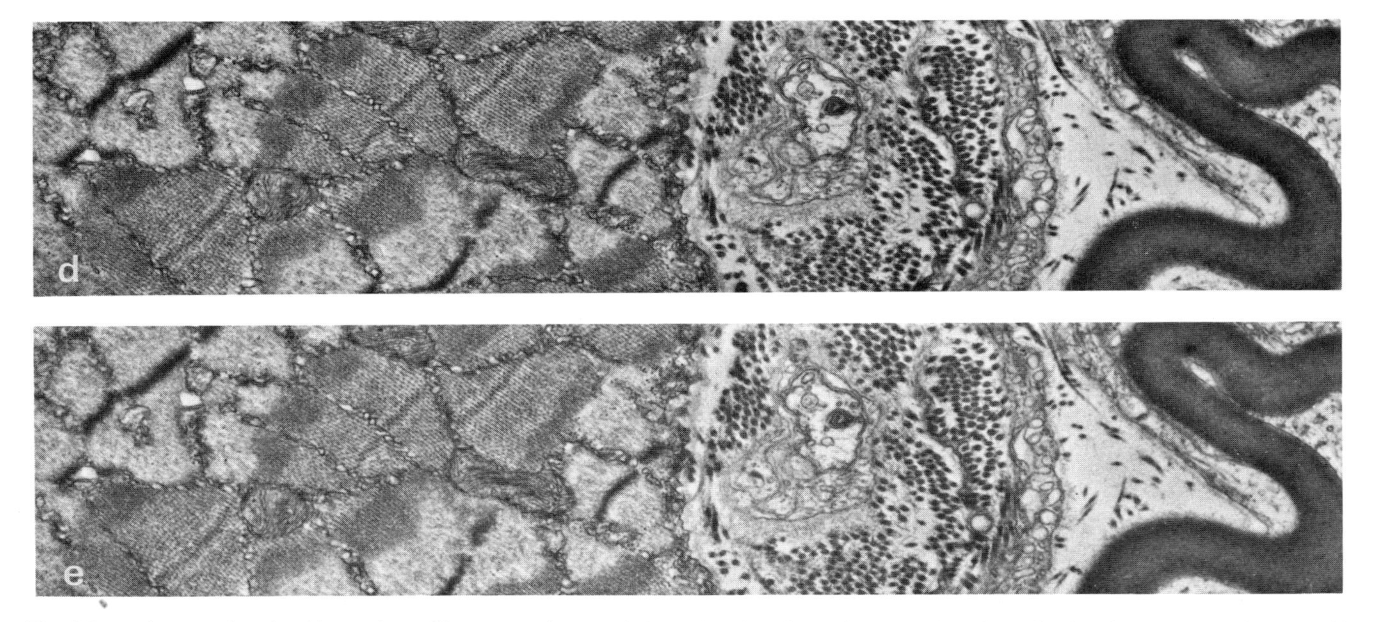

Fig. 3.3. Micrographs of a thin section of heart muscle recorded at (a) 40 kV, (b) 60 kV, and (c) 80 kV, all printed on grade 1 photographic paper; (d) shows the 60 kV micrograph printed on grade 2 paper; (e) shows the 80 kV micrograph printed on grade 3 paper. Magnification 21,000 ×.

deterioration in image quality, but for low magnification work on a very low contrast specimen, the advantage of the increase in contrast may outweigh the disadvantage of image deterioration.

A problem with small apertures is the difficulty in both cleaning them and keeping them clean in the microscope. Because of their small size, they readily contaminate and then adversely affect the image. Whenever possible, thin film apertures should be used for this type of work (see § 6.3).

An alternative method of reducing the objective aperture is to increase the focal length of the objective lens, since the objective semi-aperture angle is given by

$$\alpha = \frac{\text{diam. of aperture}}{2 \times \text{focal length}}.$$

Many manufacturers supply a high contrast holder (§ 2.6.3) which enables the objective lens to work at a longer focal length, hence reducing the effective objective aperture for the same physical aperture diameter. Fig. 3.4 shows

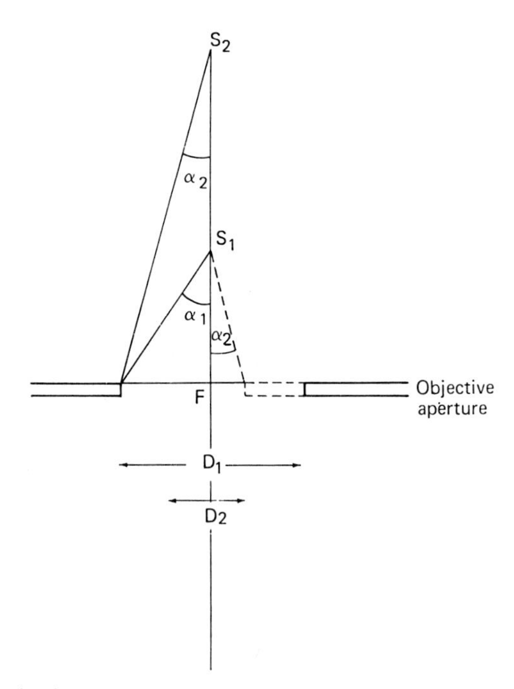

Fig. 3.4. Illustrating how the amplitude contrast is increased by reduction of the effective angular aperture of the objective lens as the specimen is moved away. The dotted rays on the right-hand side of the axis show how a similar objective angular aperture may be obtained by fitting a smaller objective aperture of diameter D_2.

how a lens operating at focal length S_1F and with an objective aperture D_1 allows all scattered electrons out to an angle α_1 to pass through the aperture. If the specimen is moved away so that the focal length is S_2F, then the limiting aperture angle is α_2. On the other side of the axis, the dotted aperture of diameter D_2 is shown limiting the aperture for a specimen at S_1 to α_2. It is seen that the longer focal length increases contrast to the same degree as the appropriate smaller objective aperture. The advantage of this mode of operation is that increased contrast is obtained with larger, more easily cleaned, objective apertures. Since the specimen is now out of the normal plane for high resolution operation there will again be some loss of resolution.

The density of the photographic plate plays an important part in the contrast of the recorded image (see § 7.2.6). Without changing any other condition in the microscope image contrast may be altered by simply increasing or decreasing the exposure time. The micrographs in Fig. 3.5 were taken at exposure times to give an optical density of (a) 0.5, (b) 1.0 and (c) 1.5. In order to print these to the same contrast it was necessary to use a grade difference of paper between each print, i.e. to print (a) on grade 3 paper (Fig. 3.5d) and (b) on grade 2 paper (Fig. 3.5e). The use of the printing process to restore contrast is not, of course, a recommended procedure. It can be seen that Fig. 3.5d is significantly poorer than Fig. 3.5c which is from a well-exposed plate.

3.4 Diffraction contrast

It has been shown that the total contrast in the electron images arises from both amplitude and phase contributions. Diffraction contrast arises from the elastic scatter of electrons from planes of atoms in crystalline material and is thus a special case of amplitude contrast. Where the electron beam encounters crystalline material, there is a strong preferential scattering in certain well-defined directions; these are governed by the relation

$$2d \sin \theta = n\lambda$$

where d is the crystal lattice spacing, θ the angle the incident beam makes with the lattice plane, λ the wavelength of the electrons, and n an integer.

When the crystal is so oriented that a strong beam of scattered electrons is directed along the axis through the objective aperture, it is said to be in the Bragg condition (for a more detailed discussion see Beeston et al. 1972). Since most of the electron energy is concentrated in these Bragg directions from a crystalline material, a crystal grain will appear bright if the scattered

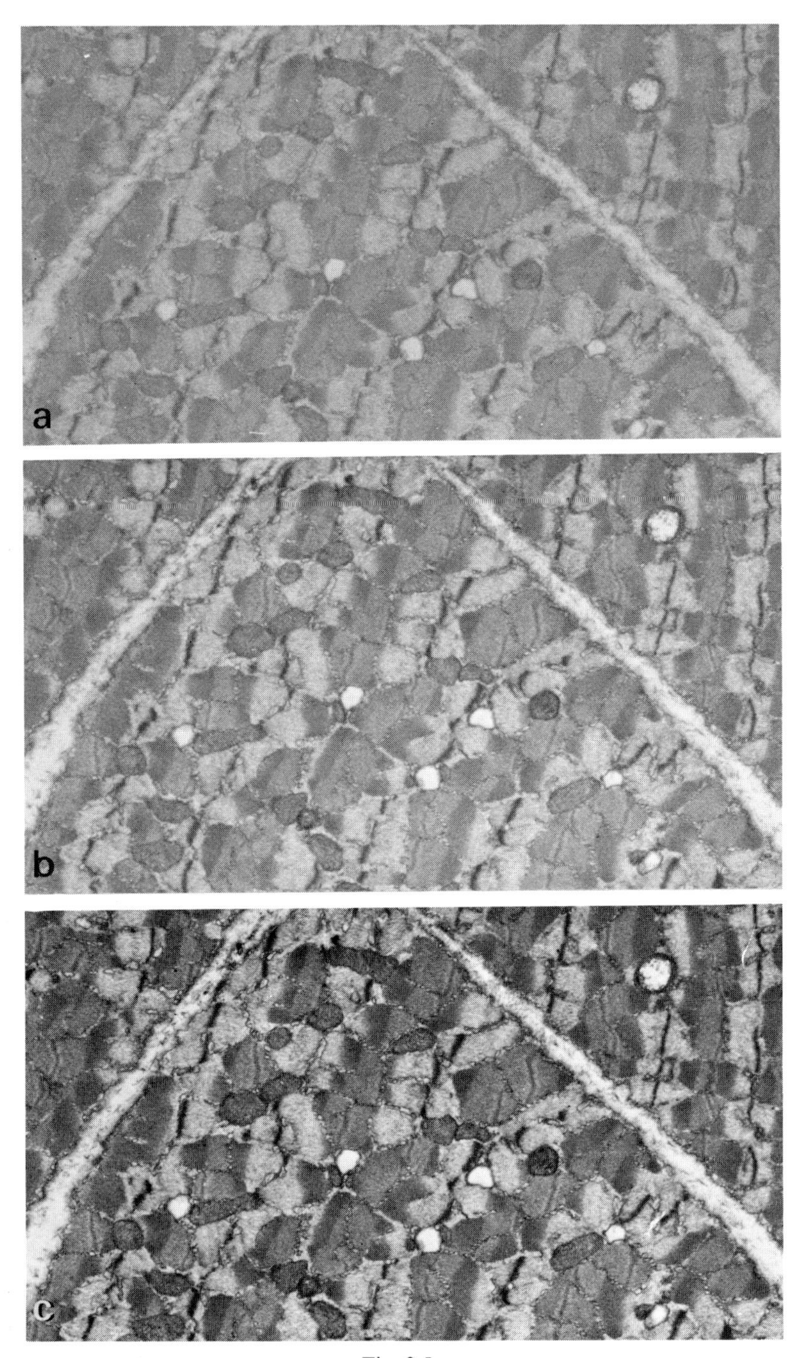

Fig. 3.5.

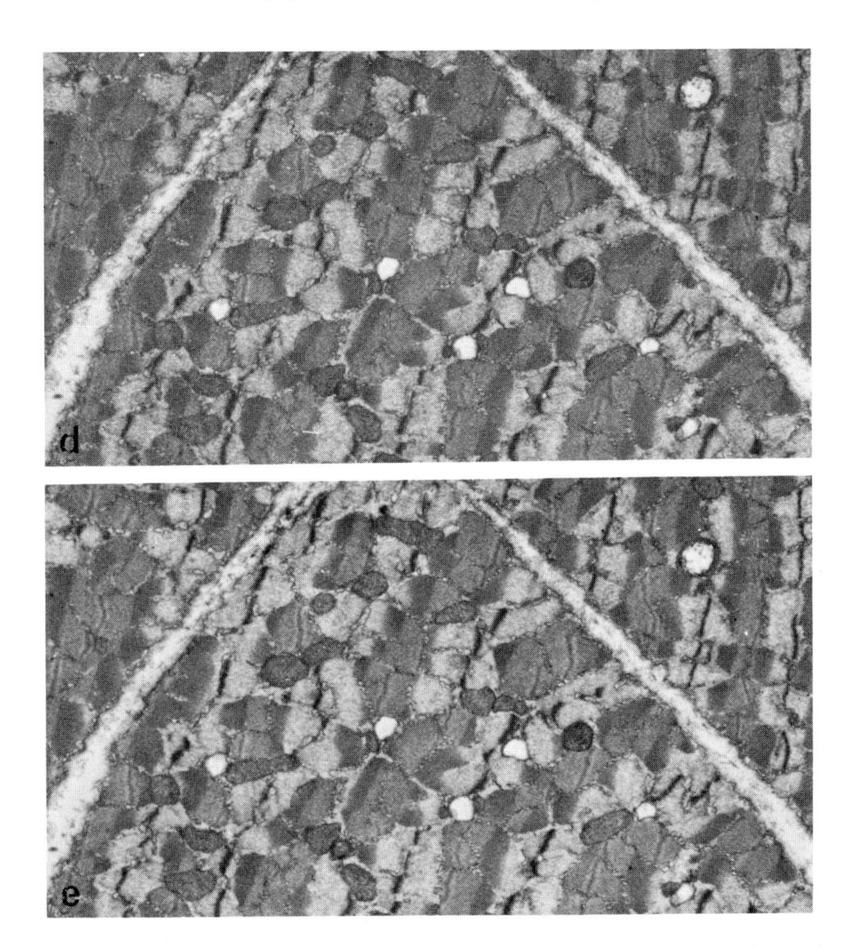

Fig. 3.5. Micrograph of a thin section of heart muscle: (a) exposed to a density 0.5 on the plate; (b) exposed to a density 1.0 on the plate; (c) exposed to a density 1.5 on the plate. Figs. (a), (b) and (c) were all printed on grade 1 paper. (d) Fig. 3.5(a) printed on grade 3 paper; (e) Fig. 3.5(b) printed on grade 2 paper. The increased density of exposure has increased contrast equivalent to two grades of hardness of the printing paper. Magnification 8400 ×.

beams pass through the objective aperture and dark if stopped by the objective aperture. Thus, in a polycrystalline material, such as aluminium (Fig. 3.6), some grains will appear bright, and some dark, depending on their orientation with respect to the electron beam. Because of the small angle subtended by the objective aperture, the image of a crystal grain may be changed from bright to dark contrast by a very small inclination of the specimen. Indeed, if the specimen (usually a metal foil) should bend slightly, due to thermal stress, it may be sufficient to change the image contrast drastically.

For a similar reason, defects in the crystal structure, such as dislocations and stacking faults can be observed in high contrast.

Examples of both amplitude and diffraction contrast are illustrated in Fig. 3.7, which is a micrograph of a carbon extraction replica from a steel specimen. Most of the detail in the image is visible due to amplitude contrast; the large extracted precipitates show this strongly. Along the grain boundary a thin crystal, which has been extracted, exhibits dark bands or fringes. These fringes are due to diffraction contrast and do not represent electron-dense regions of the crystal. They arise from the bending of the

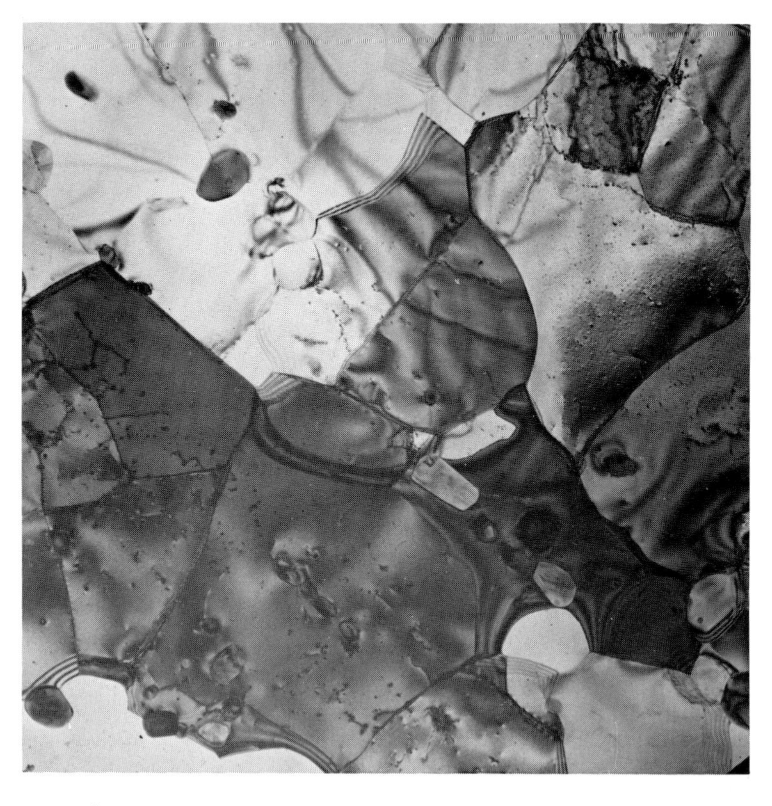

Fig. 3.6. Polycrystalline aluminium film, of almost uniform thickness. The different crystal grains appear light or dark depending on the crystal orientation with respect to the electron beam. Magnification 15,000 ×.

Fig. 3.7. Extraction replica from a steel specimen. The thick black precipitates are opaque to the beam and yield strong amplitude contrast. The thin flaky precipitate at the top of the picture shows black contrast bands which arise from diffraction contrast and are not related to the mass thickness of the specimen. Magnification 10,000 ×.

crystal, which causes some crystal planes to be so oriented that the scattered electrons all hit the objective aperture and are lost from the image. In a perfectly flat crystal, fringes may also appear due to the intensity of the diffracted beam varying sinusoidally within the depth of the crystal. Fringes due to diffraction effects will often be seen to move under the influence of the electron beam, due to the heating and consequent bending of the crystals.

The above explanations only sketch the simplest outline of diffraction contrast. It cannot adequately be described so briefly and for any studies of solid state specimens it must be treated at length. The reader is referred to the main source book (Hirsch et al. 1965). The effects will be discussed in detail in another book in this series (Howie 1974).

3.5 Other methods of contrast improvement

The electron optical means of affecting contrast in bright field are described above and the modes of dark-field operation (§§ 3.10.3 and 3.10.4) are also important. In addition, there are a number of ways in which a specimen may be treated before examination in the electron microscope which will affect the contrast obtained. As these methods are dealt with in detail in other books in this series, the methods will only be listed here.

(a) Shadowing of the specimen surface by a heavy metal *in vacuo*.

(b) Positive staining of biological material by attachment of heavy atoms to specific chemical groups in the specimen.

(c) Negative staining of biological specimens by surrounding them by a layer of dense material.

3.6 The formation of a Fresnel fringe

When the edge of a hole in a support film, or a sharp discontinuity in an object, is viewed at high magnification, some very characteristic changes are observed as the focus of the objective lens is altered. When the specimen is in focus, the edge has low contrast, but when the objective lens is underfocused, the edge is outlined with a bright fringe, and when overfocused, with a dark fringe. These are known as *Fresnel fringes*, and they arise from the interference between scattered and unscattered electron beams.

Following Haine and Mulvey (1954), consider a coherent (parallel) beam of electrons incident on the specimen S (Fig. 3.8a). At some point, behind the specimen, an electron beam BA scattered in the specimen interacts with the unscattered electrons CA.

If the point A is a distance Z behind the specimen, the path difference Δ between the two beams is

$$\Delta = \frac{Z}{\cos\theta} - Z + \psi,$$

where ψ is the effective path difference introduced by passage through the specimen.

When this path difference Δ is equal to a whole number of electron wavelengths, there will be a maximum of intensity, and for an odd number of half-wavelengths, an intensity minimum.

This may perhaps be better visualised with the aid of Fig. 3.8b. A plane

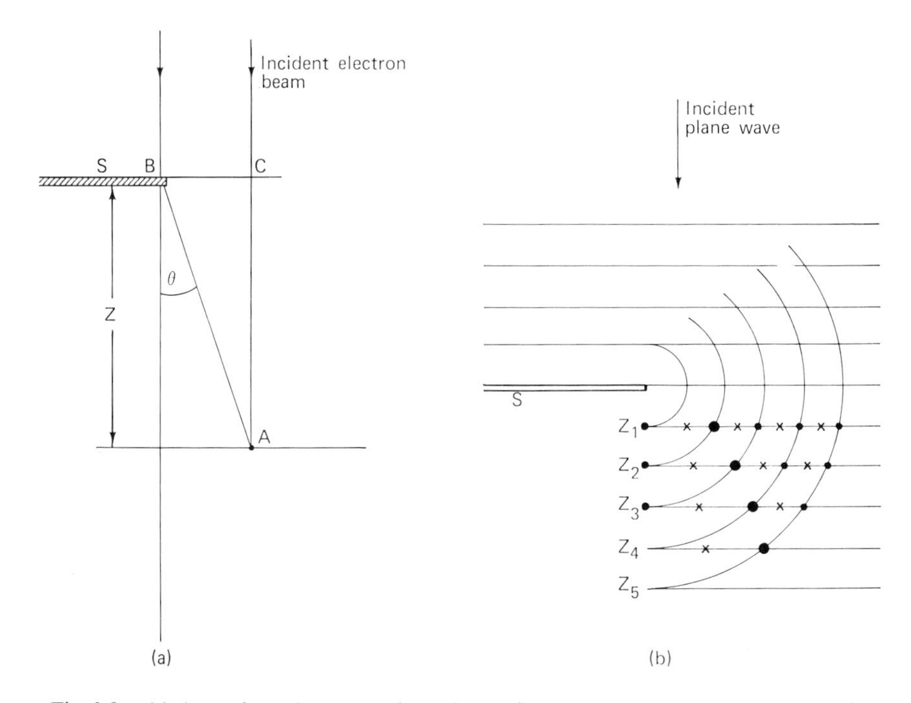

Fig. 3.8. (a) A specimen S scatters the incident electron beam which interacts with the unscattered beam at a point A, distance Z behind the specimen. If this plane is viewed with the objective lens (i.e. the lens is overfocused) a Fresnel diffraction fringe will be observed. (b) A plane wave meets a specimen S and scattered spherical waves interfere with the plane waves. Maxima of intensity are denoted by dots, and minima by crosses.

wave is shown approaching a specimen S. Due to interaction with the specimen, spherical waves are generated at the specimen edge, and these can interfere with the plane wave. Where the maxima of the two waves coincide, there is a maximum of intensity (bright fringe). These maxima are indicated by dots in the planes $Z_1, Z_2 \cdots$ behind the specimen. Minima are shown by crosses.

It is seen that in each plane there are a series of maxima and minima of intensity, but the distance of the first maximum from the shadow of the edge of the specimen increases as Z increases. The distance Z represents the distance out of focus of the objective lens.

In principle, there should be a series of maxima for each value of the integer n. However, the fringe will only be formed if the illumination is coherent and the degree of coherence required increases sharply for higher order fringes.

3.7 Coherence

The term coherence defines the condition of an illuminating wave as it approaches an object. If the wave emanates from a very small source, all points on the wave front are said to be coherent with one another and if they recombine after striking an object, are capable of interference with one another. The resultant intensity in an image plane will be the square of the vectorial sum of all the amplitudes due to the different Airy disc distributions (§1.1) resulting from the scattering points in the object. If, on the other hand, the illumination comes from an extended source, there will be a range of phase differences between the waves arriving at any one object point from the different parts of the illumination source. The larger the source size, the smaller the region of the object over which the illumination appears coherent. In order to obtain a coherent source in the electron microscope, it is therefore necessary to ensure that the effective source is small, i.e. that the illuminating aperture angle is small, and this is achieved by using a condenser system

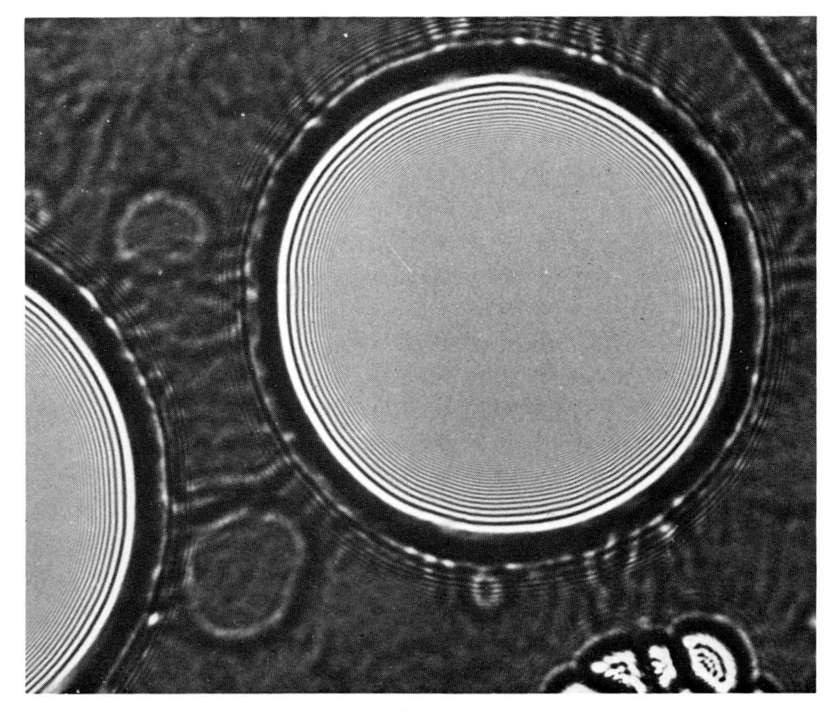

Fig. 3.9. Micrograph of holes in a carbon film photographed with the objective lens defocused. Exposure time 3 min. Magnification 43,000 ×.

which can be adjusted to produce a very small effective illumination aperture (see § 1.7.1).

It is clear that when imaging regular objects, which will give rise to strong interference effects, the optimum results will be achieved with coherent illumination. It is therefore particularly important for resolving lattice images and Fresnel fringes, and for observing other phase-contrast effects in the electron microscope. This is discussed in more detail in § 3.8.

Normally it is only possible to image the first order ($n = 1$) Fresnel fringes since these fringes are viewed at high magnification, necessitating nearly focused illumination to provide adequate intensity for focusing. The illumination aperture has therefore to be relatively large. Trains of fringes are present, but due to the size of the source, they will overlap. To enable these trains of fringes to be seen, a lower magnification must be used and the illumination well defocused to produce a very small illumination aperture. Such a train of fringes is shown in Fig. 3.9. This micrograph was taken at 10,000 × with the condenser strongly defocused. Under such conditions, the intensity of illumination is very low, and for this micrograph a three minute exposure was required.

A smaller source is obtained with pointed filaments (§ 2.2.1) and these enable the illumination to be defocused more than if a standard filament were used, thus reducing the illumination aperture angle, and increasing the coherence of the illumination.

3.8 The imaging of fine structure – optical transfer theory

All that has been said previously concerning image formation can describe quite adequately the performance of an electron microscope in imaging structures down to about 1 nm. Over recent years, however, it began to be realised that there were subtle changes in the appearance of the image of very small structures which were not easily explicable. A series of theoretical papers by Hanssen, summarised in a review article (Hanssen 1971) and experimental work by Thon (1966) have done much to clarify the situation.

Hanssen has discussed the problem in terms of optical transfer theory, which is expressed in terms, not of structure size, but of spatial frequencies. Spatial frequencies which may be measured in lines/mm are related to structure size in a way somewhat analogous to that in which the more familiar electrical frequencies, in cycles/sec are related to wavelength. The microscope system is seen as a more or less imperfect filter for these spatial frequencies whose properties may be varied by such parameters as the objective lens focus. At first sight this approach may seem to make the subject very complex, and difficult to relate to practical operation. However, one can make some simplifying statements which will give a general understanding of the effects in familiar terms.

The reason for using spatial frequencies to describe what happens to the image as it passes through the microscope is that, in general, a feature of a certain size in the specimen is described only poorly by a single spatial frequency. Fig. 3.10 shows the 'top hat' intensity plot across an object point

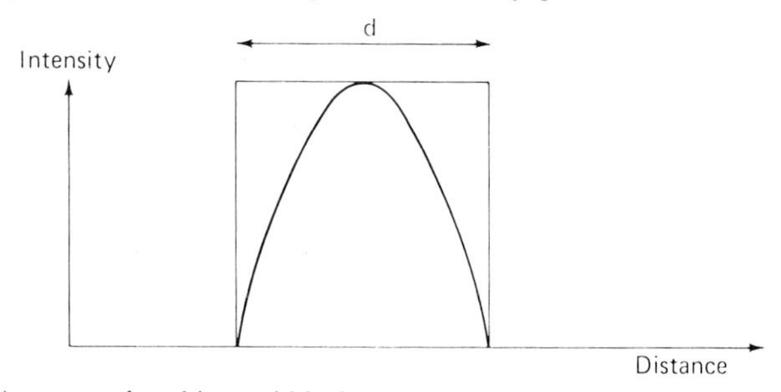

Fig. 3.10. A rectangular object, width d, and a single sinusoidal wave, of wavelength $2d$, showing the poor shape correspondence.

in the specimen. This cannot be represented at all by a simple sine wave of any particular frequency a portion of which wave is shown on 3.10. If the intensity profile in the image is to resemble that in the object, it is necessary to have a vast spectrum of waves of different frequency transmitted through the system; the highest frequency required is defined by the steepness of the inverse slope of the side of the 'top hat'. Where enough frequencies are present, the intensity outside the area of the object point cancels out to zero, and one is left with a good representation of the object area. Fig. 3.11 shows the beginning of the process of improvement in definition – the addition of a few higher frequencies improves the match of the resultant wave front to the true shape of the object area.

Another way of expressing the above statement is to say that in order to obtain a better edge resolution of a small particle, a higher resolution is required than that required to detect the particle itself (since a higher spatial frequency corresponds to a small figure of resolution). In the special case of a long set of crystal lattice planes a very close approximation may be obtained in the image by a single sinusoidal wave of wavelength equal to twice the lattice spacing (Fig. 3.12).

In terms of spatial frequencies, therefore, a lattice image can be produced by a single spatial frequency where the spatial frequency $\Lambda = 1/d$, d being

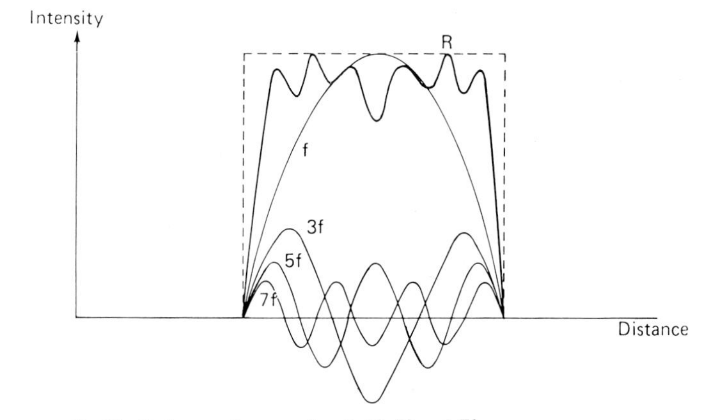

Profile R due to frequencies f, 3f, 5f and 7f

Fig. 3.11. Addition of higher spatial frequencies enable an improved shape profile of the object to be obtained.

the lattice spacing. If an attempt is made to image a small particle with a single spatial frequency, the basic information will normally be carried, but the lack of higher harmonics will affect its intensity profile in the image, perhaps seriously.

In order to preserve familiar terms in the following explanation, reference will mainly be made to structure size, but it must be remembered that, for most objects, the statements made in these terms are only approximations.

Following Haine (1961) and Thon (1966), and a survey by Hilditch (1969), phase contrast will be considered here in terms of electron interference effects. If the electron source is assumed to be coherent, the electrons transmitted through an object without interaction can interfere with those scattered by it.

For a structure spacing d, the Bragg scatter angle will be θ where

$$\theta = \frac{\lambda}{d}$$

where λ is the electron wavelength.

Constructive interference with the unscattered electrons will occur when the path difference between the scattered and unscattered beam is an odd number of half-wavelengths. This path difference is made up of terms representing the effect of spherical aberration C_s of the objective lens, and the amount Δf_0 that the lens is defocused. Following a simple derivation, given in the appendix to this chapter, the amount of objective defocus leading

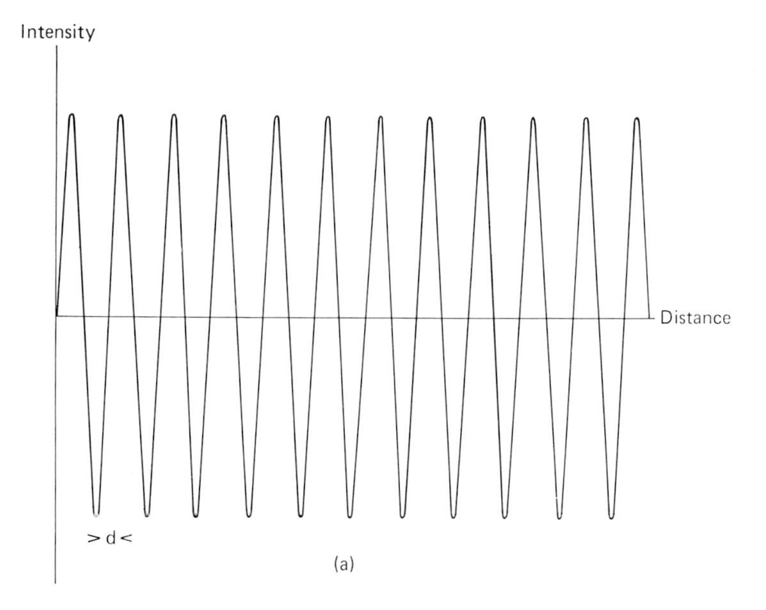

Fig. 3.12. (a) Sinusoidal wave of wavelength $2d$ and (b) a corresponding image of a set of crystal lattice planes of spacing d.

to maximum phase contrast is given by

$$\Delta f_0 = \tfrac{1}{2}C_s\frac{\lambda^2}{d^2} - (2n-1)\frac{d^2}{2\lambda}.$$

The value for Δf_0 was plotted by Thon (1966) for a series of values of n, and Fig. 3.13 shows the family of curves obtained for different values of n for the particular example of an objective lens with $C_s = 1.6$ mm, and an accelerating voltage of 100 kV ($\lambda = 0.0038$ nm). The full lines in the family of curves are lines of maximum phase contrast.

It will be seen that for any particular value of defocus Δf_0 there are a series of values of spatial frequency which will exhibit maximum contrast. Successive curves represent positive and negative contrast and the dotted lines between the solid curves are lines of zero phase contrast. For coherent illumination (very small effective condenser aperture) and for any particular setting of objective lens defocus there is therefore a very complex pattern of contrast arising from a specimen containing a complete range of structures – some are rendered in high positive contrast, some in negative contrast and others, which may be considerably bigger than the nominal resolving power of the microscope, will have zero contrast and not be perceptible in the image.

It will be seen that for an in-focus picture, the phase contrast is near zero for all large structures (note curve X in Fig. 3.13), in accordance with the well-known contrast 'dip' at focus. Maxima of contrast are only obtained at focus for very small structure sizes (which may, however, not be visible for other reasons).

This simple picture is of course modified for practical operating conditions by a number of other factors.

3.8.1 STABILITY OF SUPPLIES

Variation of the objective lens current or accelerating voltage will cause variation in objective lens focal length

$$\Delta f_0 = C_c\left(\frac{\Delta V_r}{V_r} - \frac{2\Delta I}{I}\right)$$

and hence a variation of the effective defocus Δf_0 of the objective lens. If this variation swings the operating point between curves representing maximum positive and maximum negative contrast (e.g. points A and B in Fig. 3.13) the net contrast of the image will be zero. *Any* variation in effective focus will have some effect in reducing the contrast of the image. It will also be noted

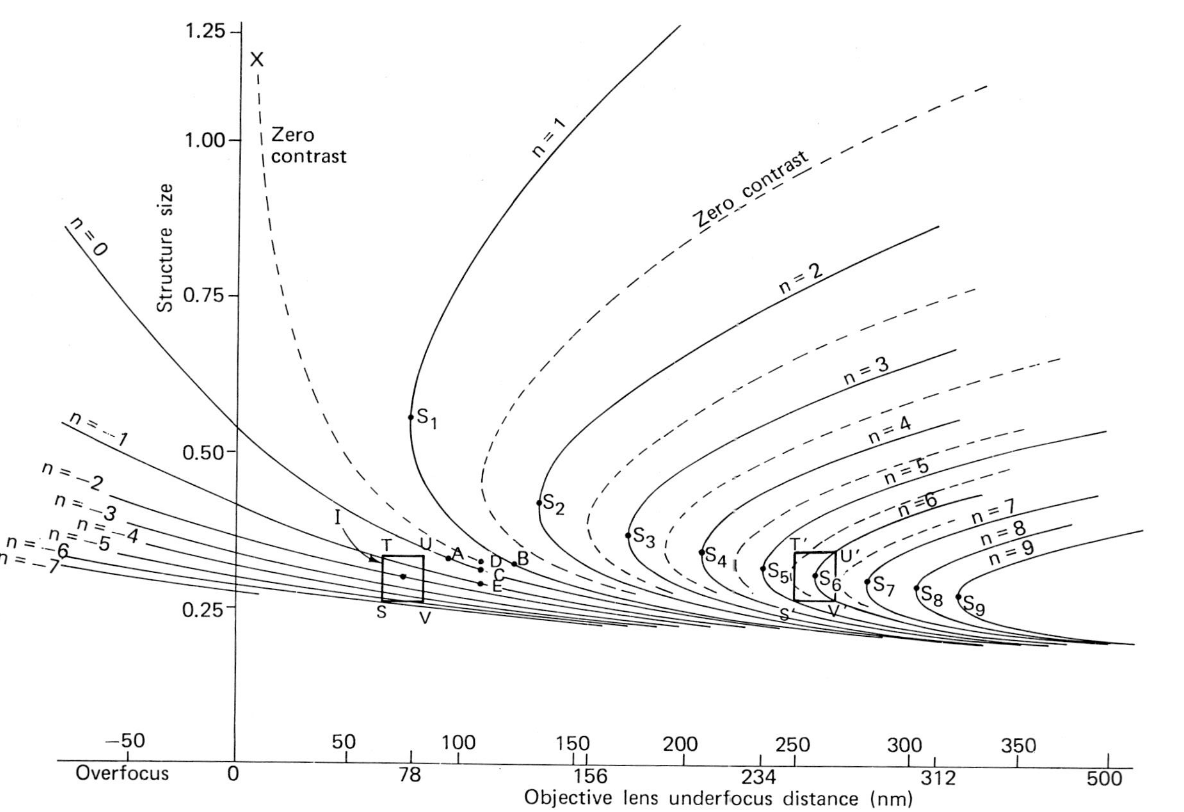

Fig. 3.13. Plot of objective lens defocus f against structure size d (after Thon 1966).

that the effect becomes more important in regions where the curves lie closer together, since the amount of supply instability needed to blur out the contrast is reduced.

3.8.2 NON-AXIAL ILLUMINATION

If the incident illumination makes an angle α with the objective lens axis, then regular structure of spacing x, giving a diffraction maximum at an angle θ, will now give maxima at $\theta + \alpha$ and $\theta - \alpha$, which if α is small, may be written $\theta \pm d\theta$. For a coherent source, the diffraction angles $\theta + d\theta$ and $\theta - d\theta$ would correspond to structures of spacings $x + dx$ and $x - dx$, where

$$\frac{dx}{x} = -\frac{d\theta}{\theta}$$

Thus, the effect of a tilted beam is to give two values of x for each setting of focus, and the resultant contrast will be divided between these values. If the electron source is not completely coherent, but the illumination angle is of finite size $d\theta$, then the energy which would have been concentrated in a single value of structure size will be spread out over a range $x + dx$ to $x - dx$, which may include structure sizes which are in low contrast. Thus the high contrast point C in Fig. 3.13 might be spread over the range DCE, the extreme points lying on curves of zero contrast. In the limit, a sufficiently large illumination aperture will spread the operating range so far that the resultant contrast is too low to be perceived.

3.8.3 THE TURNING POINTS

For certain values of objective lens defocus, the electron beam path difference introduced by the defocus will just balance the effective path difference caused by the spherical aberration of the objective lens. These values occur at the turning points S_1, S_2-------- on the family of curves (Fig. 3.13). At these points, the effects of non-axial illumination, or finite illumination aperture are minimised because there is a considerable range of structure size for which the contrast remains near maximum.

It is possible to represent the joint effects of supply instabilities and finite illumination aperture by a rectangle such as STUV which is drawn for an illumination angle of 1×10^{-3} radians (focused illumination using a physical aperture in condenser 2 of about 250 μm) and for instabilities of $\Delta V/V = \pm 5 \times 10^{-6}$ and $\Delta I/I = \pm 2.5 \times 10^{-6}$ and $C_c = 1.3$ mm.

It can be seen that if this rectangle is placed in position around curve

$n = -2$ for structure of 0.3 nm, it will encompass several regions of positive and negative contrast, and no net contrast will be observable. In position S′T′U′V′, however, most of the contrast is high and of the same sign, and the structure is observable.

Thus, the turning points of the family of curves are very important, because at these points the effect of a finite illumination aperture is minimised.

3.8.4. OPERATING CONDITIONS

It will now be appreciated that, when imaging these very small structures, the optical transfer theory must be used to clarify the best operating conditions for a particular specimen. The effect on interpretation of the image is considered in § 8.2.2. It is perhaps easier to visualise the complexities of the contrast by replotting this information from Fig. 3.13 in terms of the contrast against the space frequency (for convenience a structure size scale is also given), for a series of values of underfocus of the objective lens: curves of this type were first shown by Komoda (1964). Plots of contrast for an in-focus picture, and for an objective under-focus of 78 nm and 234 nm are shown in Fig. 3.14.

The in-focus picture (Fig. 3.14a) is seen to yield very little phase contrast until a structure size of 0.55 nm is reached, when maximum negative phase contrast appears. Thereafter there are a series of contrast maxima of alternate sign. The heavy line represents contrast in the absence of any chromatic effects, and with fully coherent illumination. Putting in a value for chromatic aberration coefficient of 1.3 mm and a supply fluctuation of 5×10^{-6}, the dotted curve results. It will be seen that the maxima of contrast are reduced somewhat. If one postulates an illumination aperture of 10^{-3} radians (a quite normal value when the condenser focuses the beam on the specimen), the thin continuous line shows the resultant loss of contrast. It can be seen that structures of 0.43 nm will have only about 25% contrast, and that for smaller structures, there is no perceptible contrast at all.

If the objective lens is defocused by 78 nm, (Fig. 3.14b) there is a wide range of structure sizes for which there is positive phase contrast, and neither supply fluctuations nor aperture angle cause any appreciable loss of contrast. However, there is zero contrast for structure sizes 0.4 nm, 0.33 nm, 0.3 nm, and reversed phase contrast at structure sizes 0.35 nm, 0.295 nm. But for all structures smaller than 0.35 nm, the contrast is almost negligible if the illumination aperture is 10^{-3} radians.

For an even greater degree of defocus (Fig. 3.14c) the effect of the turning point S_5 (Fig. 3.13) can be seen clearly. The phase contrast alternates between

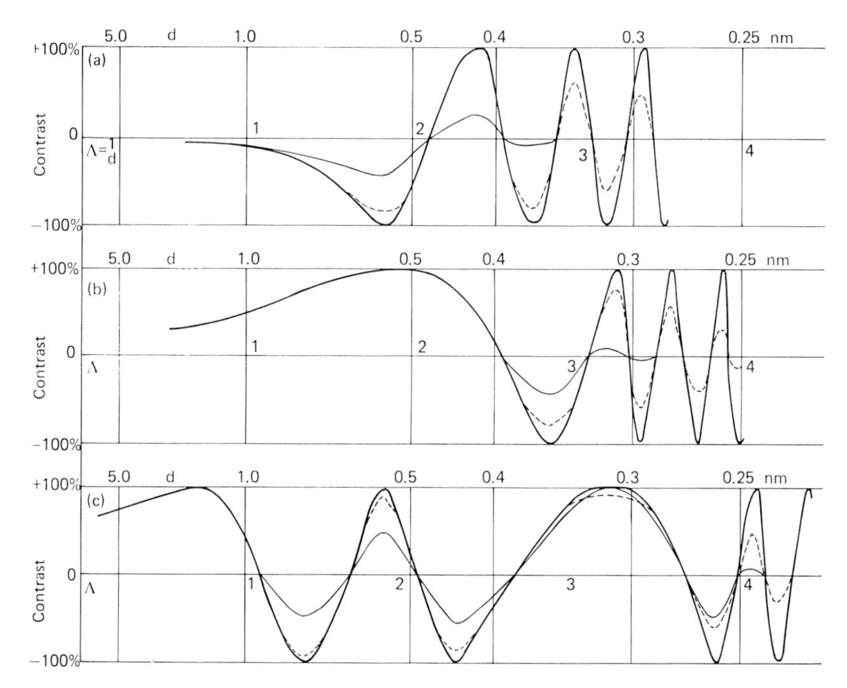

Fig. 3.14. Plots of image phase contrast against spatial frequency Λ and structure size d. (a) Objective lens at focus $\Delta f = 0$; (b) objective lens underfocused $\Delta f = 78$ nm; (c) objective lens underfocused $\Delta f = 234$ nm.

maxima at 1.3 nm, 0.72 nm, 0.54 nm, 0.433 nm, and then enters a broad maximum at about 0.31 nm, where the effect of illumination aperture is minimised. For smaller structures, the contrast is rapidly reduced to zero.

These curves give a clue to desirable operating conditions. For a broad band of phase contrast enhancement of structure, a slight defocus (to about the turning point S_1) is desirable. If a particular crystal lattice spacing is to be imaged, the defocus should coincide with the turning point nearest to this structure size (e.g. S_5 for 0.31 nm). Under such conditions, however, some of the larger structures may have zero or very low phase contrast. There is then an interpretation problem if the crystal lattice is not the only item of interest. It is of course possible to render structure of 0.31 nm spacing in high positive contrast at other degrees of defocus; at 129 nm, 185 nm and 284 nm for example. But in order to achieve the contrast, the illumination aperture would have to be reduced much below 10^{-3} radians at these other operating points, and there could be a problem in having adequate intensity for high resolution focusing and photography.

These curves show that the simple definition of the resolving power of a microscope is no longer really adequate. The spherical aberration of the objective lens changes the scale of the family of curves, but imposes no arbitrary limit on the detectable structure. The detection limit depends on the visibility of the structure; the amount of contrast (and hence visibility) of structure of a certain size varies critically with the setting of objective lens defocus for a given amount of supply instability and for a given illumination aperture. Incidentally, since a misalignment of the objective lens will give rise to a two-valued structure size for maximum contrast and will thus reduce contrast, accurate alignment, by some method such as the fringe test with magnesium oxide crystals (Yada and Hibi 1968) is essential.

It will be appreciated that to maximise contrast in this way the objective lens current control must be provided with very fine incremental steps so that the focus can be found accurately and the correct amount of defocus applied. The technique of maximising contrast is of course difficult because the definition of focus is not easy to determine, and accurate setting will only be achieved with an instrument in excellent condition, and with an experienced operator.

Critical work will demand focal increments at least as fine as 5 nm; steps finer than this can only be justified if the supplies to the objective lens and the high voltage are stabilised to about one part per million.

A further point to be remembered is that the diffraction angle θ corresponding to structure spacing d,

$$\theta = \frac{\lambda}{d}$$

also defines the minimum angular objective aperture which can be used without cutting out the information about this fine structure. As an example, the minimum aperture angle required for the resolution of the carbon lattice, 0.34 nm, would be 1.2×10^{-2} radian at 80 kV, and with an objective lens of 2.5 mm focal length, the physical aperture would have to be larger than 60 μm diameter. In fact, the presence of a physical objective aperture becomes unnecessary for obtaining contrast in very fine structure, since amplitude contrast is no longer important.

It is, of course, essential to correct the residual astigmatism of the objective lens if these effects are to be observed, and to minimise contamination of the specimen.

It will be noted that this examination of the requirements for the achievement of very high resolution bring out the same parameters as those discussed in Chapter 1. However, for the particular requirement to photograph crystal lattice planes which have a precise spacing, a striking improvement in the performance is obtained by a reduction of the illumination aperture (say down to 10^{-4} radians). This enables high contrast micrographs to be obtained from structures quite invisible at normal aperture angles. This procedure places a new emphasis on specimen stage stability and stability of the supplies and suggests the use of very high brightness coherent electron sources for high resolution studies of crystal structures. Most work on the photography of fine lattice spacings has been done with pointed filaments in the gun.

It should perhaps be emphasised that the whole of this discussion refers to the imaging of objects by phase contrast. This implies that the specimen is *very thin* (certainly less than 10 nm). If the specimen is thick, the energy spread in the beam due to multiple scattering of the electron beam in the specimen will blur out the effects described, and there is certainly no point in looking for any of the subtler effects with such a specimen.

3.9 Focusing

Focus can be defined as the condition in which no Fresnel fringe is formed at an image point. This corresponds to the point where the specimen is exactly in the conjugate plane to the image plane of the objective lens, which at any given setting of the intermediate lens is defined by the object plane of that lens. As explained in § 3.2 and § 3.8 above, this condition is also the one in which there is no phase contrast.

In principle, optimum results should be obtained from a focused micrograph, but it is apparent from the discussion in § 3.8 above that there are a number of exceptions to this rule for very small structures, when a certain degree of underfocus is necessary for optimum imaging.

Even when not working at ultimate resolution, there can be some benefits to be obtained from phase-contrast effects, which can supplement the amplitude contrast in a thicker specimen. Reisner (1964) plotted the optimum defocus (underfocus) of the objective lens against the resolution required, and concluded that between 0.3 and 2.0 micrometres of underfocus should be used for structures up to 2.5 nm. The subject was further discussed by Haydon and Taylor (1966) who plotted the Fresnel fringe width against objective defocus to show the optimum contrast enhancement for separating biological membranes and other structures. They suggest quite large amounts of objective lens underfocus may be allowable when examining thick sections. The effect was later studied in greater depth by Johnson (1968) for the particular case of contrast enhancement of the structure of microfibrils in wool. He also showed that a larger degree of underfocus could usefully be used – for example, 6.2 μm to enhance structure of about 6.5 nm spacing. The same figure could be deduced by continuing to plot the first positive contrast curve in the graph in Fig. 3.13 to higher values of Δf_0. The subject has been recently reviewed by Johnson and Crawford (1973).

At very low magnifications, there is a marked loss of contrast at focus, which is best observed with the objective aperture withdrawn. The procedure for employing this focusing method is described in § 6.9.

3.10 Modes of operation of the electron microscope

In the preceding sections, the mechanism of image formation in the normal transmission mode of operation in which the image is formed by combination of the unscattered beam and the scattered electrons, and the imaging lenses are used to focus and magnify the image, has been considered. Considerably more information may be gained from some specimens if other modes of operation are also employed.

3.10.1 ELECTRON DIFFRACTION

Electron diffraction is considered in detail in an earlier volume of this series (Beeston et al. 1972) and will only be briefly mentioned here. A crystalline specimen will diffract the electron beam strongly through certain angles, θ, dependent on electron wavelength and crystal lattice spacing given by Bragg's law,

$$n\lambda = 2d \sin \theta$$

where d is the crystal lattice spacing, λ is the electron wavelength, and n is an integer.

These diffracted electrons are brought to a focus in the back focal plane of the objective lens (Fig. 3.15). In the normal magnification mode of operation, the intermediate image formed by the objective lens is the object plane for the first (intermediate) projector lens. If the strength of this projector lens is so weakened that the object plane of this lens becomes the back focal plane of the objective lens, then the diffraction pattern will be imaged by the first projector and subsequently enlarged by further projector lenses. Both the image plane and the diffraction pattern contain information about the specimen but presented in different ways, and both are always present, to be viewed at will by selection of the strength of the intermediate lens.

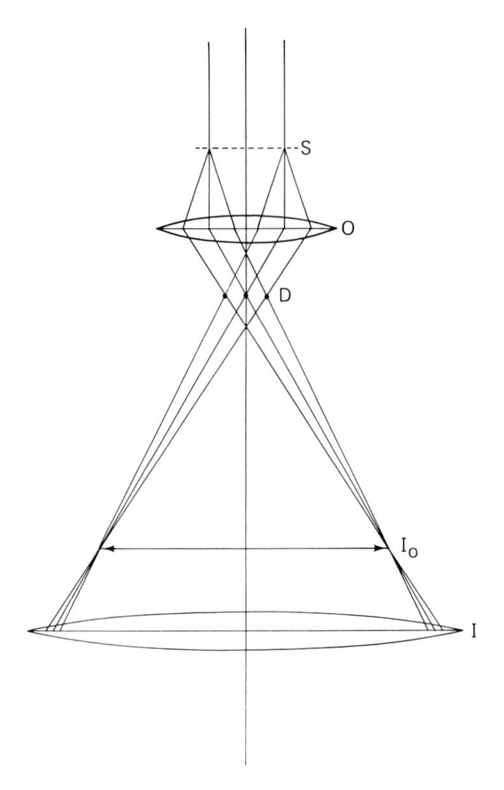

Fig. 3.15. Formation of the diffraction pattern (D) and the intermediate image (I_0) of a crystalline specimen (S) by the objective lens (O). The intermediate lens (I) can be used to image either the intermediate image or the diffraction pattern.

If an aperture is now inserted in the objective lens image plane, it is equivalent to an aperture in the plane of the specimen, but smaller by a factor equal to the magnification of the objective lens (frequently of the order 30–50 times). Thus an aperture of diameter 50 μm in the intermediate aperture plane I_0 can define a 1 μm diameter region at the specimen. This region will then be the one giving rise to any diffraction pattern which is transmitted by the selector aperture. This technique is known as selected area diffraction, and is extremely useful, particularly in materials science investigations.

It is unfortunate that the technique cannot be used to define regions much smaller than 1 μm diameter, because the spherical aberration of the objective lens causes the diffracted beams of higher orders to be imaged into the recorded pattern even when they do not arise from the selected region (Agar 1960; Riecke 1961).

There are other possible errors of the technique; it will be obvious that the objective image plane must correspond with the plane of the intermediate aperture if correspondence between the area imaged and the diffraction pattern recorded is to be maintained. With a three-lens imaging system, there is therefore only one magnification at which the image can be viewed for selected area diffraction, unless the final projector lens strength can be altered. In practice, it is not normally practicable to alter the strength of this lens by more than about a factor of two, without problems from limitation of the field of view. The effective camera length of the system in imaging the diffraction pattern (equivalent to the physical distance through which the diffraction pattern would have to fall in field free space to achieve the same magnification), is

$$L = f_0 M_1 M_2$$

where

L = effective camera length;
f_0 = focal length of objective lens;
M_1 = magnification of first projector lens;
M_2 = magnification of second projector lens.

It will be seen that L cannot be greatly varied in such a system, although in an objective lens where the specimen is inserted from above, a large change in camera length can be achieved by raising the specimen out of the lens so that it works at a considerably longer focal length.

A microscope with a four-lens imaging system has an additional degree of freedom, however. Even though the strength of the objective lens and first intermediate lens (or diffraction lens) are fixed by the restriction that the objective lens must project an image into the selector aperture plane, and the diffraction lens must be focused on the back focal plane of the objective lens, the second intermediate lens can be varied without restriction to give a wide range of camera lengths, a very valuable facility. Furthermore, the 'selected area magnification' – which in a three lens instrument is single valued, because the intermediate lens must image the selector aperture plane, is widely variable when a second intermediate lens is fitted.

3.10.2 HIGH DISPERSION ELECTRON DIFFRACTION

The conventional electron diffraction pattern can yield information on structures up to a few tenths of a nanometre (say 0.4 nm). For larger structure spacings, the diffracted information is scattered through very small angles

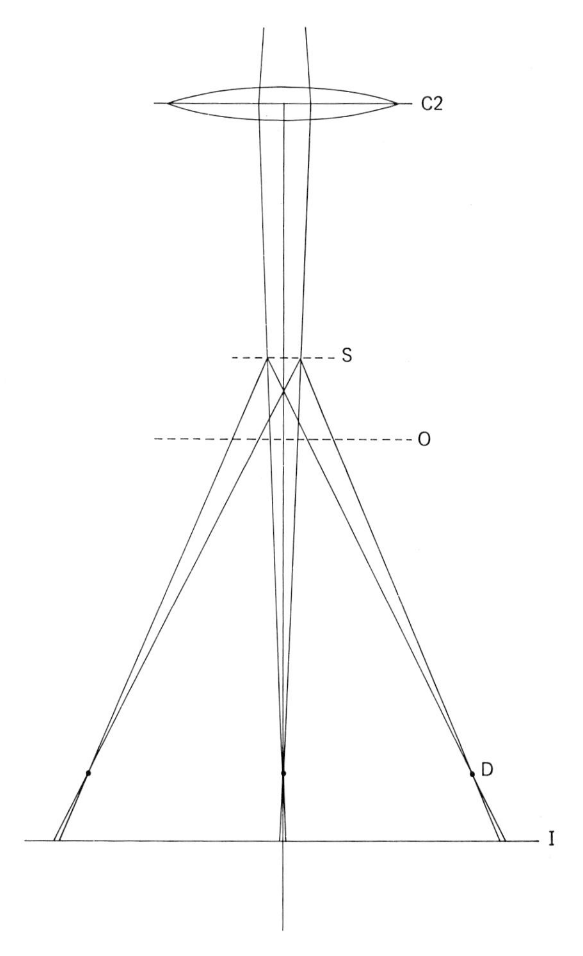

Fig. 3.16. Formation of a high dispersion diffraction pattern. The objective lens (O) is switched off and the pattern (D) is formed, highly magnified, in the image plane of the second condenser lens (C2). It is further magnified by the intermediate (I) and projector lenses.

and tends to be lost in the halation from the very intense central spot of the diffraction pattern.

The technique of high dispersion diffraction, or low angle diffraction, was first described by Bassett and Keller (1964). The objective lens is switched off, and the electron beam is focused by the second condenser lens on the object plane of the intermediate lens. This condenser lens is working at a long focal length, and the pattern has already spread out by the time it reaches this plane

(Fig. 3.16). Thus, very long effective camera lengths are obtained, and diffracted beams due to structure some tens of nanometres in size can be recorded. This is a powerful technique, since it is exceedingly difficult to record X-ray diffraction patterns from such structures, and the exposure times are very long. A good review of the method was given by Ferrier and Murray (1966).

3.10.3 DARK FIELD MICROSCOPY

A bright field image in electron microscopy is formed when the unscattered electrons of the incident beam combine with the scattered electrons as modified by passage through the objective lens and objective aperture. If the unscattered electrons are removed, the image is formed only from the scattered electrons and a *dark field image* is produced. This image is of much greater contrast than a bright field image although, of course, the intensity is much reduced.

The simplest way in which a dark field image can be formed is by displacement of the objective aperture disc so that the aperture in the disc accepts no part of the central unscattered beam (Fig. 3.17a). This image is of poor quality, because the aperture is accepting off-axis electrons subject to larger aberrations than those on the object lens axis. The parts of the image formed by electrons passing through the outer zones of the objective lens are blurred more than those formed by electrons nearer the axis.

Potentially the dark field technique is very powerful when used to image a highly crystalline object, since the objective aperture may be centred around a chosen diffraction spot in the back focal plane and the dark field image will be formed only from the electrons scattered by a chosen crystal plane. A rather simple operational technique can yield dark field images of high quality (Fig. 3.17b). The incident electron beam is tilted over at such an angle that the diffracted beam of interest travels down the objective lens axis and thus is subject only to the minimum aberrations suffered by a bright field image. This technique of *high resolution dark field imaging* is extensively used in conjunction with diffraction studies of materials science specimens.

Most microscopes now include a convenient electrical tilting system for the electron beam which enables the operator to steer the incident beam in any required direction by a simple control. After the area of interest of the object has been studied in bright field, the instrument is switched to the diffraction position and the back focal plane of the objective lens is viewed (the objective aperture is withdrawn). The centre point of the diffraction pattern is either aligned onto a central mark on the viewing screen or its position

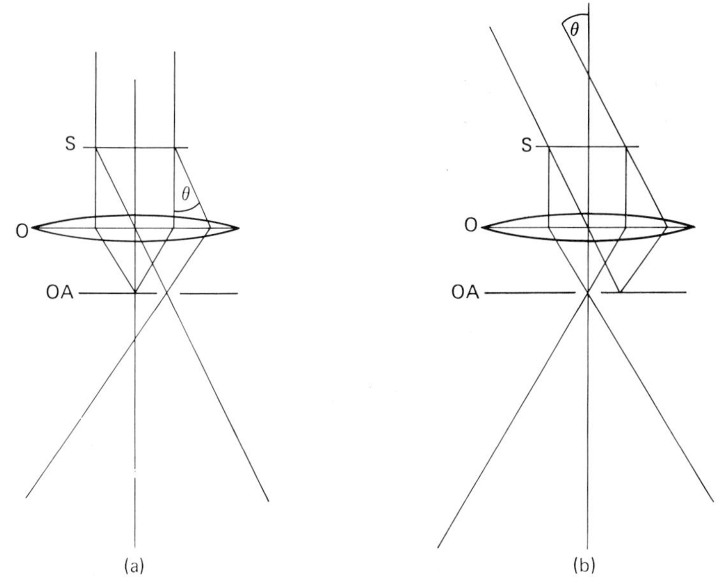

Fig. 3.17. Formation of dark field images. (a) Normal illumination. The objective aperture (OA) is displaced from the axis of the objective lens (O) to accept the beam diffracted by the specimen (S) through the angle θ, and cuts off the unscattered beam. (b) The illuminating beam is tilted through an angle θ so that the diffracted beam passes down the axis and through the objective aperture (OA).

recorded by using the diffraction stop as a marker. The selected reflection in the diffraction pattern is now moved to the central mark by adjustment of the electrical beam tilt controls. The objective aperture is inserted and centred round the spot. When the microscope is switched back to the imaging condition a high resolution dark field image will be obtained. With duplicated tilt and shift controls, it is possible to switch quickly between bright and dark field images without readjustment. In order that this technique should operate smoothly and easily, it is essential that the tilt control should introduce only tilt of the beam without introducing lateral shift of the beam at the object plane. Careful trimming of the control circuits is necessary to set this condition up satisfactorily. This trimming will normally be part of the initial commissioning of the instrument.

One further complication which may ensue is that once the beam tilt angles are large, the deflector system may introduce some astigmatism into the beam, and this will have to be corrected on a special astigmatism corrector if the full beam intensity is to be utilised.

The use of the technique is illustrated in Fig. 3.18; (a) shows a bright

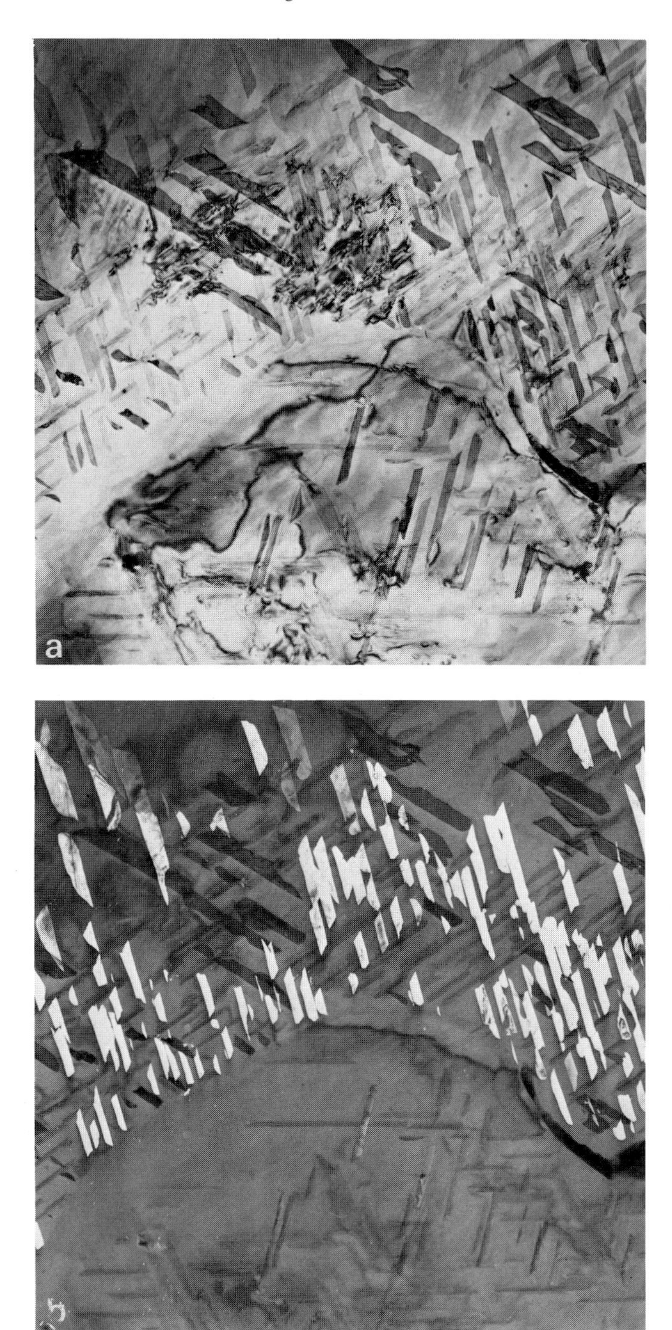

Fig. 3.18. Bright (a) and dark (b) field images of precipitates in an aluminium–copper alloy. Note how the dark field image highlights only the crystals in the appropriate orientation to diffract electrons through the aperture.

field micrograph of an aluminium-copper alloy; when the appropriate crystal reflexion is directed along the objective lens axis, a dark field picture (b) results, with the suitably oriented precipitates in high contrast. Note that the remaining precipitates in different orientations remain in low contrast.

A more recent use has been found for the dark field examination of amorphous materials, and this is described in the following section.

3.10.4 STRIOSCOPY

Most biological materials are stained with heavy metal compounds and fairly readily acquire adequate contrast, even in rather thin sections. In a few instances, however, the contrast is inadequate. Very difficult subjects are the long-chain molecules, such as DNA, which are normally stretched out as long strands on the supporting film. The smaller strands have so little contrast as to be almost invisible, but if viewed by *strioscopic dark field* the contrast is greatly enhanced. With this mode of imaging, the central undeviated beam is stopped by either a physical stop on the objective aperture (Dupouy 1967) or by using an annular aperture (Dupouy et al. 1969; Dubochet et al. 1971). The aperture-stop technique is illustrated in Fig. 3.19. The central beam stop aperture is a normal objective aperture with a fine wire welded across its centre. This stops the undeviated beam and only allows electrons scattered by the specimen to form the image. Dupouy's results were obtained at high accelerating potential (1000 kV), where contrast is particularly low. The normal bright field image is of very low contrast and very little structure can be observed. When the central stop aperture is introduced, the results are striking; detail which was invisible in bright field appears in good contrast and much fine structure can be clearly seen. Unfortunately, due to the asymmetry of the objective aperture, the astigmatism correction for the bright field does not hold in dark field, and since Fresnel fringes are absent in dark field, astigmatism correction in dark field is extremely difficult.

The more elegant solution of the annular aperture, illustrated in Fig. 3.20, overcomes the difficulty of correcting the astigmatism. An annular condenser aperture replaces the normal aperture for the dark field mode. The second condenser lens is brought to a focus on the specimen, in which condition the illumination appears uniform over the specimen (when the condenser lens strength is altered, the hollow beam can be seen, of course). The unscattered electron beam will fall onto the edge of the objective aperture and will be prevented from reaching the image. The diameter of the objective aperture must be carefully chosen to match the diameter of the annulus in the con-

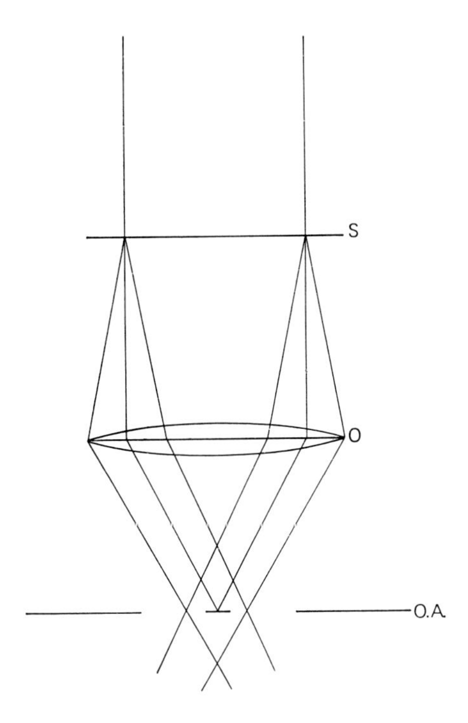

Fig. 3.19. The central aperture stop for dark field microscopy. A small wire on the axis stops unscattered electrons from reaching the final image. Electrons scattered by the specimen (S) are deflected by the objective lens (O) to miss the stop in the aperture (OA).

denser aperture. Any electrons scattered by the specimen will be able to pass through the objective aperture and will form a dark-field image of high quality.

There are several advantages of this mode of operation over other methods of obtaining dark field images. Firstly, the method gives completely symmetrical illumination about the axis, and consequently there is no danger of asymmetric charging of the objective aperture, and hence change of astigmatism. The annular condenser aperture may be replaced by a normal aperture to provide a bright field image for astigmatism correction, which will thereafter remain unchanged when the annular aperture is re-introduced. For amorphous objects, the method is more efficient in electron collection than the method of beam tilting (which is however very efficient for crystalline material).

A beautiful example of the dark field technique is shown in Fig. 3.21 which is a micrograph of the bacterium *Pseudomonas putida* recorded in dark

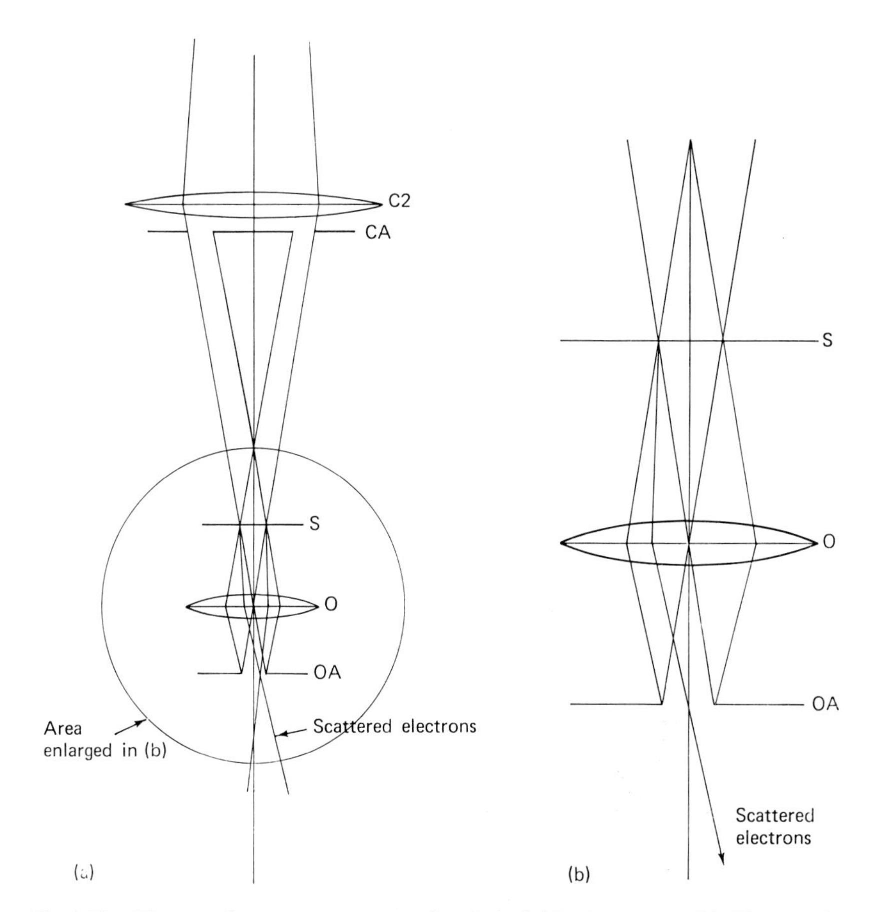

Fig. 3.20. The annular aperture system for dark field microscopy. (a) The annular condenser aperture (CA) forms a hollow illuminating beam which illuminates the specimen (S) evenly when the condenser (C2) is focused. Unscattered electrons are stopped on the edge of the objective aperture (OA), but electrons scattered by the specimen pass through the aperture. (b) Area of the objective lens enlarged to show the rays more clearly.

field illumination at 3 MV in the microscope at Toulouse. The subject of dark field microscopy has recently been reviewed by Dubochet (1973).

APPENDIX

Calculation of path difference introduced by spherical aberration and defocusing.

Path difference between axial and marginal ray due to spherical aberration of objective lens $= \frac{1}{4}C_s\theta^4$

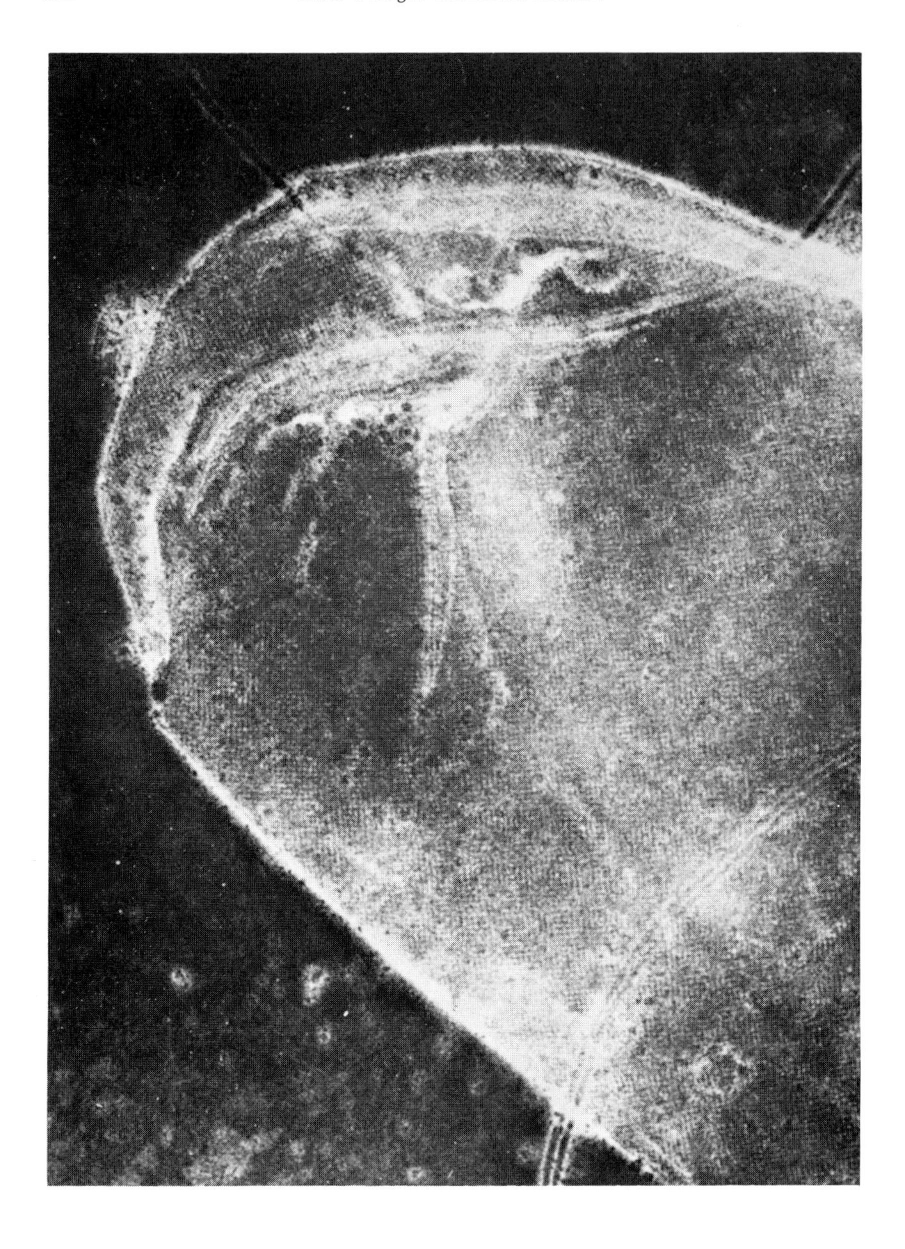

Fig. 3.21. The bacterium, *Pseudomonas putida* in dark field illumination. Accelerating voltage 3 MV. Magnification 25,000 × (Courtesy G. Dupouy).

Path difference due to defocus of objective lens

$$= -\Delta f_0 \theta^2 \qquad \text{(sign introduced)}$$

For maximum signal,

$$(2n-1)\frac{\lambda}{2} = \tfrac{1}{4}C_s\theta^4 - \Delta f_0\theta^2$$

For structure spacing d,

$$\theta = \frac{\lambda}{d} \qquad (\theta \text{ small})$$

$$(2n-1)\frac{\lambda}{2} = \tfrac{1}{4}C_s\frac{\lambda^4}{d^4} - \Delta f_0\frac{\lambda^2}{d^2}$$

or

$$\Delta f_0 = \tfrac{1}{2}C_s\frac{\lambda^2}{d^2} - (2n-1)\frac{d^2}{2\lambda}.$$

REFERENCES

Agar, A. W. (1960), Accuracy of selected area microdiffraction in the electron microscope, Brit. J. appl. Phys. *11*, 185.

Bassett, G. A. and A. Keller (1964), Low angle scattering in the electron microscope: applications to polymers, Phil. Mag. *9*, 817.

Beeston, B. E. P, R. W. Horne and R. Markham (1972), Electron diffraction and optical diffraction techniques, in: Practical methods in electron microscopy, A. M. Glauert, ed. (North-Holland, Amsterdam).

Dubochet, J. (1973), High resolution dark-field electron microscopy, J. Microscopy *98*, 334.

Dubochet, J., M. Ducommun, M. Zollinger and E. Kellenberger (1971), A new preparation method for dark field electron microscopy of biomacromolecules. J. Ultrastructure Res. *35*, 147.

Dupouy, G. (1967), Contrast improvement in electron microscope images of amorphous objects, J. Electron Microscopy *16*, 5.

Dupouy, G. (1973), Performance and applications of the Toulouse 3 million volt electron microscope, J. Microscopy *97*, 3.

Dupouy, G., F. Perrier, L. Enjalbert, L. Lapchine and P. Verdier (1969), Accroissement du contraste des images d'objets amorphes en microscopie électronique, C. r. Acad. Sci. Paris *268*, 1341.

Ferrier, R. P. and R. T. Murray (1966), Low angle electron diffraction, Jl R. microsc. Soc. *85*, 323.

Haine, M. E. (1961), The electron microscope (Spon, London).

Haine, M. E. and T. Mulvey (1954), The regular attainment of very high resolving power in the electron microscope, Proc. 3rd Int. Congr. Electron Microscopy, London, p. 698.

Hanssen, K. H. (1971), The optical transfer theory and the electron microscope: fundamental principles and applications, in: Advances in optical and electron microscopy, Volume 4, R. Barer and V. E. Cosslett, eds. (Academic Press, New York and London), p. 1.

Haydon, G. B. and D. A. Taylor (1966), The optical under-focus enhancement of contrast in electron microscopy, Jl R. microsc. Soc. *85*, 305.

Hilditch, D. H. (1969). Phase contrast in the electron microscope, A.E.I. Engineering Memorandum EM/58 (unpublished).

Hirsch, P. B., A. Howie, R. B. Nicholson, D. W. Pashley and M. J. Whelan (1965), Electron microscopy of thin crystals (Butterworth, London).

Howie, A. (1974), Applications of wave theory in electron microscopy, in: Practical methods in electron microscopy, A. M. Glauert, ed. (North-Holland, Amsterdam), in preparation.

Johnson, D. J. (1968), Amplitude and phase contrast in electron microscope images of molecular structures, Jl R. microsc. Soc. *88*, 39.

Johnson, D. J. and D. Crawford (1973), Defocusing phase contrast effects in electron microscopy, J. Microscopy *98*, 313.

Komoda, T. (1964), Resolution of phase contrast images in electron microscopy, Japan J. appl. Phys. *3*, 122.

Reisner, J. (1964), Quantitative methods for estimating and improving performance with the electron microscope, R.C.A. Scientific News *9*, 1.

Riecke, W. D. (1961), The exactness of agreement between selected and diffracting areas in Le Poole's selected-area diffraction technique, Optik *18*, 278.

Thon, F. (1966), On the defocusing dependence of phase contrast in electron microscope images, Z. Naturf. *21a*, 476.

Yada, K. and T. Hibi (1968), Factors affecting the contrast of lattice images. I. Focusing of objective, J. Electron Microscopy *17*, 97.

Chapter 4

Alignment and adjustment of the electron microscope

Electron microscopes are now relatively easy to operate because much of the alignment can be pre-set, but the operator is still very dependent upon correct alignment if the instrument is to yield its rated performance and if it is to be simple and convenient to operate.

The requirements are fairly easily stated: the operator wishes to be able to change magnification without losing the centre of the field of view; to vary the illumination on the object without the illumination becoming uneven or off-centre; to focus the image without it moving across the screen; and to be able to switch from one mode of operation to another without loss of illumination. These requirements will be met if the microscope has been well designed and if it is correctly aligned – that is if the axes of all the lenses are in optical alignment.

It is normally unnecessary to do more than check the accuracy of alignment of an instrument by confirming that it behaves as outlined above. The full alignment procedures described below would only be necessary if the instrument had been completely dismantled. Because instruments differ considerably in detail, the order in which the alignments are carried out may differ from that given here. The detailed procedure for alignment of an instrument will be found in the instruction manual. As far as possible, however, the general criteria on which misalignment may be identified are given below. Before the instrument is switched on, be sure to note the radiation hazard mentioned in the introduction to Chapter 6 and in § 6.6.

4.1 The electron gun

The electron gun is effectively a triode arrangement: a heated filament, a negatively-biased Wehnelt cylinder, and an anode at high positive potential with respect to the cathode (§ 1.6 and § 2.2). The hole in the Wehnelt cylinder is usually small (of the order of 1 mm) and consequently a small displacement of the filament tip will rapidly generate a highly asymmetric electrostatic field at the filament tip which will distort the emission and throw the centre of the beam off axis (Fig. 4.1). This can be detected very sensitively by viewing the spot pattern from an under-heated filament (Fig. 4.2). This can be seen by refocusing the second condensor lens to image the filament tip on to the screen. This is done with all the lenses switched on and the condenser aperture in position. When the beam current is very low and hence also the bias voltage is low (§ 1.6) a spot pattern is formed. If asymmetric, as in Fig. 4.2, it indicates misalignment.

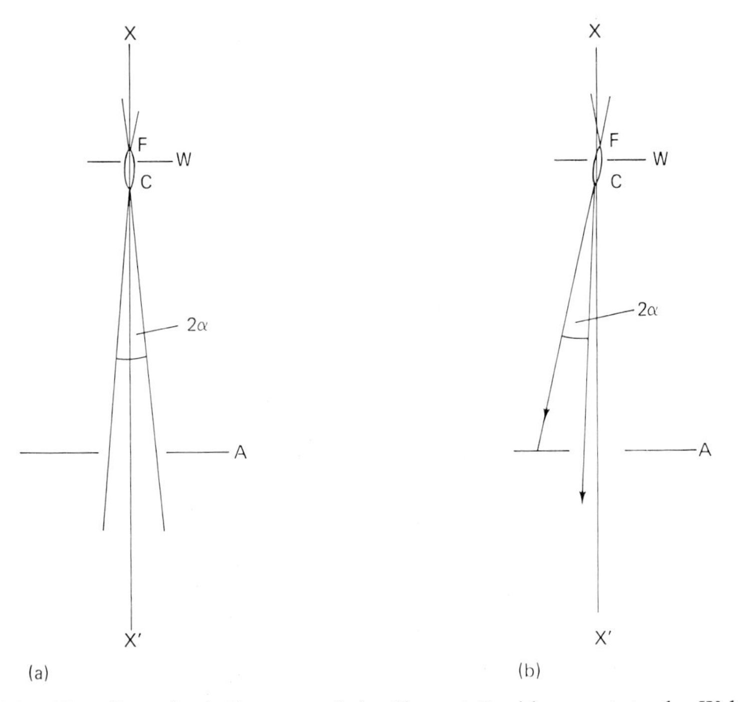

Fig. 4.1. The effect of misalignment of the filament F with respect to the Wehnelt cylinder W. (a) Aligned; the beam passes through the anode A and then symmetrically down the axis XX′ of the condenser lens. (b) Misaligned; part of the beam is lost on hitting the anode A, reducing the total beam angle 2α.

Fig. 4.2. Spot pattern formed by a filament misaligned with respect to the gun axis.

The initial filament centring is provided either by a precentred mount or by a centrable block which the operator adjusts before inserting the filament into the gun (§ 2.2.1). There is no way to recentre the filament during operation, since this would require an insulated control rod which would be difficult to incorporate, since the filament is at a negative high voltage. If the filament distorts slightly during its life, the illumination has to be adjusted by relative motion of cathode and anode. The entire cathode assembly on its

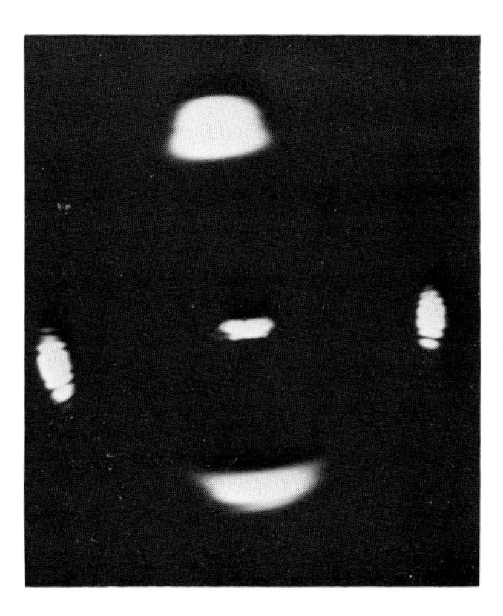

Fig. 4.3. Symmetrical spot pattern formed by a correctly aligned gun.

high voltage insulator can be moved with respect to the microscope column, or in some instruments the anode can be moved, so as to restore the symmetry of the filament spot pattern (Fig. 4.3). This condition gives a near perfect centration. It may sometimes be possible to improve it still further by observing the intensity of the focused spot (of the fully 'saturated' filament) and moving through the alignment point of each gun traverse control in turn to determine the point of maximum intensity. If the filament tip should move an appreciable fraction of the radius of the hole in the Wehnelt cylinder, it will no longer be possible to restore even illumination and the gun will have to be opened to atmosphere and the filament recentred mechanically or replaced.

4.2 The condenser system

Correct adjustment of the electron gun should result in the electron beam entering the first condenser lens along the lens axis or very nearly so. Since this lens is usually used as a strong demagnifier of the source, the demagnified source image is a rather stable point in the electron optical system and may be regarded as the effective electron source for the microscope, assuming there is no intention of altering the strength of the first condenser lens. (See § 4.4 for adjustment of condenser 1.) This image is projected on the specimen by the second condenser lens (§ 1.7) and since this second lens is continually being varied in strength during operation of the instrument, it soon becomes apparent if it is not aligned properly. If it is assumed that the illumination is focused on the specimen plane to say a 2 μm diameter spot, this will just fill a 10 cm viewing screen with illumination at 50,000 × magnification. If the axis of the second condenser lens is displaced by even 0.5 μm the misalignment will be very apparent as a partially off-centre spot and it will only be possible to illuminate the whole screen by defocusing the condenser lens slightly and losing illumination intensity (Fig. 4.4). The condenser alignment is usually carried out with electrical deflectors which can be sensitively adjusted (§ 1.8). The criterion for alignment is very simple – merely that the focused illumination should be symmetrical about the viewing screen centre (since this is on the mechanical axis of the microscope column) and this is achieved by focusing the illumination spot on the screen and then moving this spot to the screen centre, that is from A to 0 in Fig. 4.4. It is assumed here, and in all other alignment instructions, that the screen is in the horizontal position for such adjustments. When the screen is in a tilted position for telescopic viewing, the screen centre no longer coincides with the instrument axis.

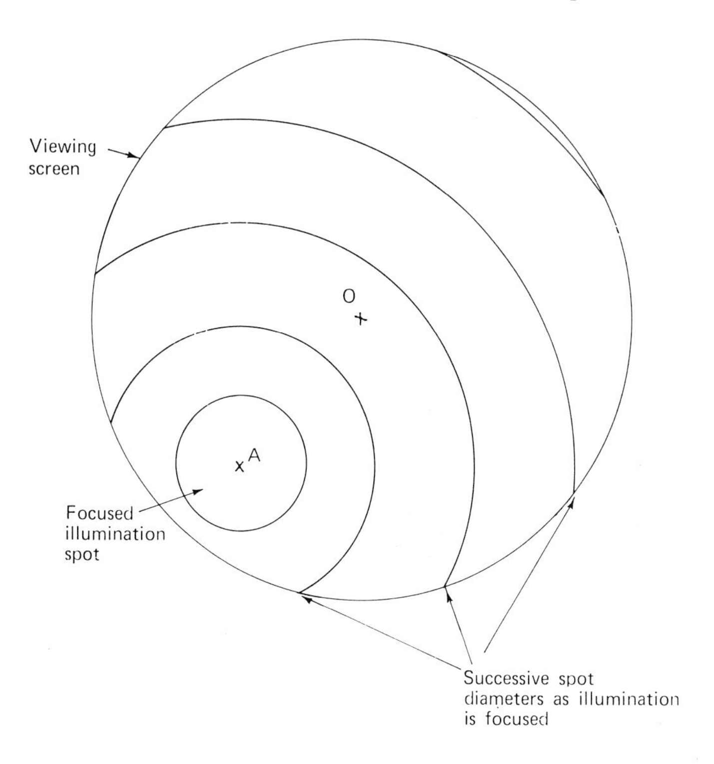

Fig. 4.4. The appearance of the illumination on the viewing screen with a misaligned condenser. The successive circles represent the limit of the illuminated area as the condenser lens is focused. The centre of the focused spot is marked A. The screen centre is at O.

4.3 Centring the condenser aperture

Although the focused illumination be centred, the illuminated area will not expand uniformly about the screen unless the condenser lens aperture is centred also. With the aperture in position, the strength of the lens is varied from one side of the focused spot position to the other. If the condenser aperture is not centred, the spot will appear eccentric to the screen centre on either side of focus (Fig. 4.5). Starting from one such eccentric position, the out-of-focus spot should be centred on the screen by adjustment of the aperture centring controls; further variation of the strength of condenser 2 will then merely result in the spot expanding and contracting about the screen centre.

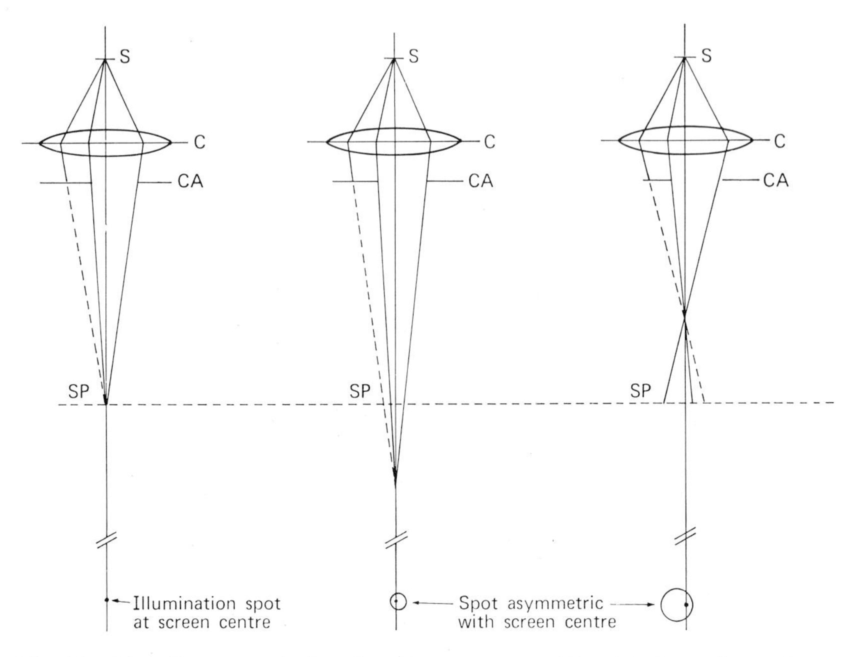

Fig. 4.5. The effect of a misaligned condenser aperture CA with change in condenser focus. (a) With the illumination from the source S focused on the specimen plane SP the spot appears symmetrical at the screen centre. As the illumination is (b) underfocused and (c) overfocused, the illumination appears asymmetric to the screen centre.

4.4 First condenser lens alignment

If it is desired to alter the focused beam size at the specimen by adjustment of the strength of the first condenser (§ 1.7), it will be found that the spot will move off the screen unless the electron source in the gun is accurately on the first condenser lens axis. The alignment between these two units must be adjusted until the strength of the first condenser lens can be varied without movement of the focused illumination beam.

It should perhaps be mentioned here that many strong electron lenses have a certain amount of stray magnetic field at a point along the lens axis, and this field will cause an additional small deflexion of the electron beam. It will usually be most noticeable when the lens is working near its maximum excitation. Thus, while in principle a correctly aligned lens will behave exactly as described, it is not infrequent that practical lenses show some divergence from perfect alignment as they approach maximum excitation. It should be possible to obtain sufficiently good alignment to avoid any serious inconvenience in operation, as a manufacturer will not use lenses which have bad alignment properties.

4.5 Correction of beam astigmatism

When a double condenser system is in use, the spot projected onto the speci-

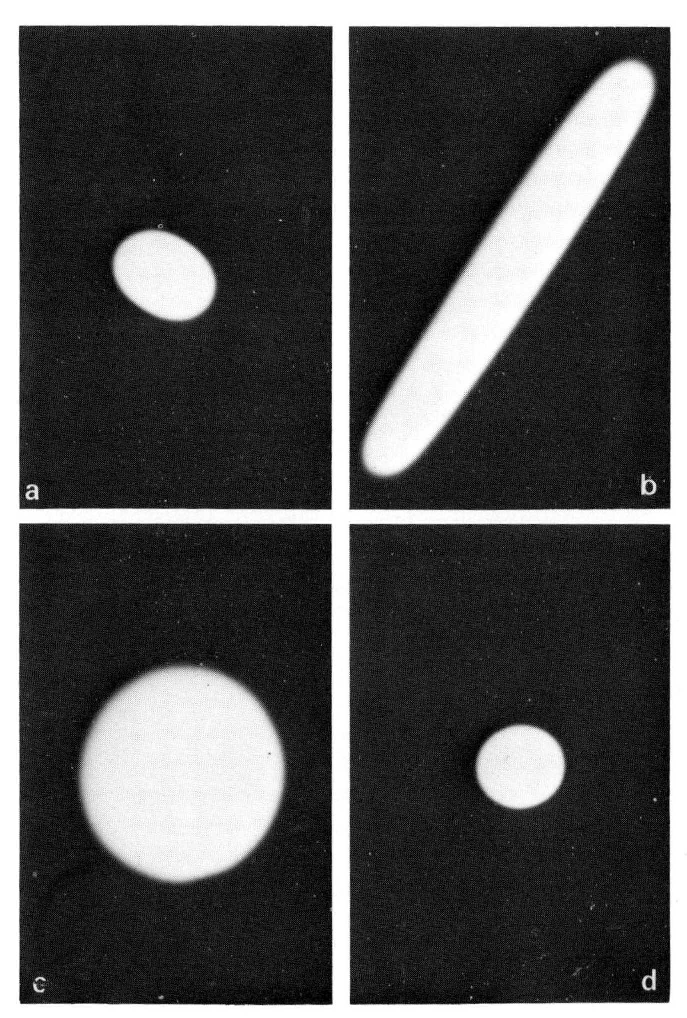

Fig. 4.6. The correction of condenser lens astigmatism: (a) original elongated spot; (b) astigmatism corrector on, maximum strength, orientation perpendicular to lens astigmatism; (c) corrector strength reduced to balance lens astigmatism and (d) lens refocused to minimum spot diameter.

men is usually elongated. This is due to astigmatism in the second condenser lens. This astigmatism must be corrected because it leads to loss of maximum brightness for high magnification work, and also results in an asymmetrical aperture angle which affects the quality of the image.

It is usually sufficient to use a quadrupole or sextupole lens (see § 2.3.1) to correct astigmatism, although octupoles are also used. Correction may be achieved very simply by observing the shape of the focused spot and adjusting the astigmatism corrector controls until the spot is round. Most astigmatism correction controls are of the orientation/amplitude type, which allows a simple logical sequence of correction operations. With the corrector off, the condenser lens is set to give one of the elongated spot images (Fig. 4.6a). The astigmatism corrector is switched on at maximum strength; the orientation control is adjusted until the spot elongation is at right angles to the original direction (Fig. 4.6b). The strength of the corrector is then reduced until the spot is round (Fig. 4.6c). A slight refocusing of condenser 2 to the plane of paraxial focus, rather than the plane of minimum confusion (Fig. 1.7), will produce a smaller spot (Fig. 4.6d).

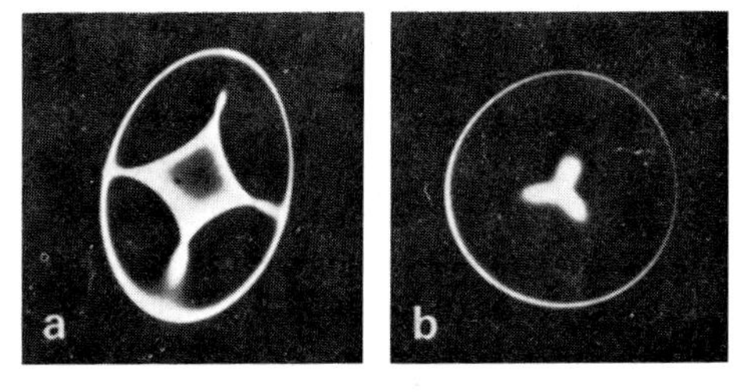

Fig. 4.7. Correction of condenser lens astigmatism by use of the illumination caustic pattern: (a) caustic in the presence of astigmatism; (b) caustic with astigmatism corrected.

Alternatively, the condenser aperture is removed and the caustic pattern formed by the lens is observed (Fig. 4.7). The corrector is adjusted until the caustic forms a three-pointed star. Care must be taken not to focus the spot on the screen when the condenser aperture is removed.

When all these illumination adjustments have been made, the illumination will be sufficiently intense to permit viewing of the image at the highest mag-uification of the instrument and the imaging system can then be aligned.

4.6 Imaging system alignment

The principle of the alignment of the image forming lenses can be appreciated from Fig. 1.5 and §1.2 in which it was shown that the image of an object point on the axis of a lens is also on the axis. Thus the alignment procedure must result in the axes of all the image forming lenses being in line and on the mechanical axis of the instrument. This latter requirement is not very stringent; provided the lenses are mutually aligned the point about which the image expands as the magnification is increased need not be precisely at the screen centre – although the alignment procedure is simpler if it is coincident.

The criterion for alignment for the final projector lens is simple. It will usually be checked with only the objective lens operating and with a specimen in position to produce an image. Increase in strength of the final projector will cause expansion of the image and it will rotate about some point on the screen. This point represents the projection of the axis of the projector lens onto the viewing screen and the lens should be adjusted until this rotation point is at the centre of the screen. This alignment has to be adjusted before the upper imaging lenses are adjusted. In some instruments, no adjustment of this lens is provided since the mechanical tolerances are not difficult to meet in a pre-aligned system.

The objective lens alignment is very important but the procedure will

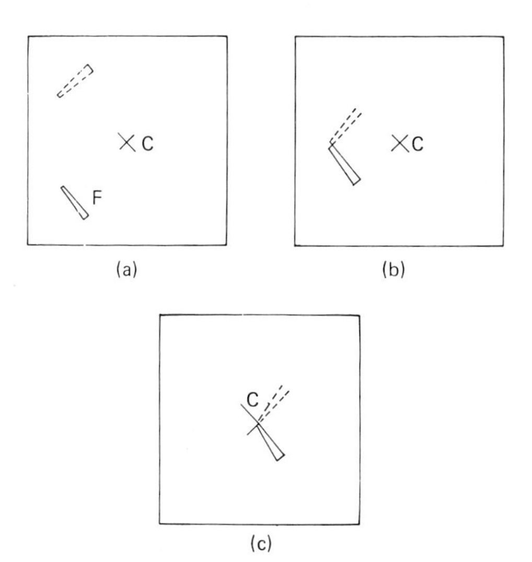

Fig. 4.8. Objective lens alignment by current reversal. (a) Movement of the image of a distinctive specimen feature F on reversal of the objective lens current. (b) Specimen moved to apparent reversal point. Position confirmed by current reversal. (c) Reversal point moved to screen centre C by alignment controls. Position reconfirmed by current reversal.

depend on whether the lens is made adjustable with respect to the viewing chamber or whether all other components are adjusted with respect to the objective lens. The alignment of this lens is complicated by the fact the movement of the image due to changing the objective lens strength is compounded of the image rotation about the objective lens axis, and a shift due to a possible tilt of the illumination with respect to the objective lens axis. (This latter adjustment has to be carried out later see § 4.7.) Haine (1947) showed that the best method of aligning an objective lens unambiguously was to reverse the current in the lens. This does not change the objective lens axis but the rotation of the image changes by twice the angle through which the lens rotates the image. This is normally seen as an image inversion when the current is reversed (Fig. 4.8). By observing the reversal-point, and then moving this point to the screen centre using the objective lens alignment controls, one can optimise the alignment of the objective lens axis with that of the final projector. The procedure has to be done at a relatively high magnification so a final adjustment may be needed after the remaining projector lenses have been aligned. Care must be taken to leave the objective lens current direction in its original sense, i.e. with the image rotation opposing that of the final projector lens.

Once the objective lens axis and projector lens axis have been aligned in this way, the intermediate lenses between them can easily be adjusted. The back focal plane of the objective lens contains a diffraction pattern of the object and a demagnified image of the electron source, which lies on the lens axis. If the intermediate lenses are in turn switched on, and made to project this diffraction pattern onto the object plane of the final projector lens, the criterion for alignment of the intermediate lenses is that the centre of the diffraction pattern is projected onto the screen centre (Fig. 4.9).

Once all these imaging lenses are in line, an object point lying on the axis of the objective lens will be imaged at the centre of the viewing screen whatever the magnification setting, although the image will rotate about the centre and sometimes invert at different points of the magnification range, as explained in § 1.10.2.

4.7 Illumination tilt alignment

When all the rest of the instrument has been aligned as described the illumination tilt has to be adjusted. Although the illumination will have already been adjusted laterally to fall onto the object at the objective lens axis, the illumination may be inclined at a small angle to this axis. If this occurs, the

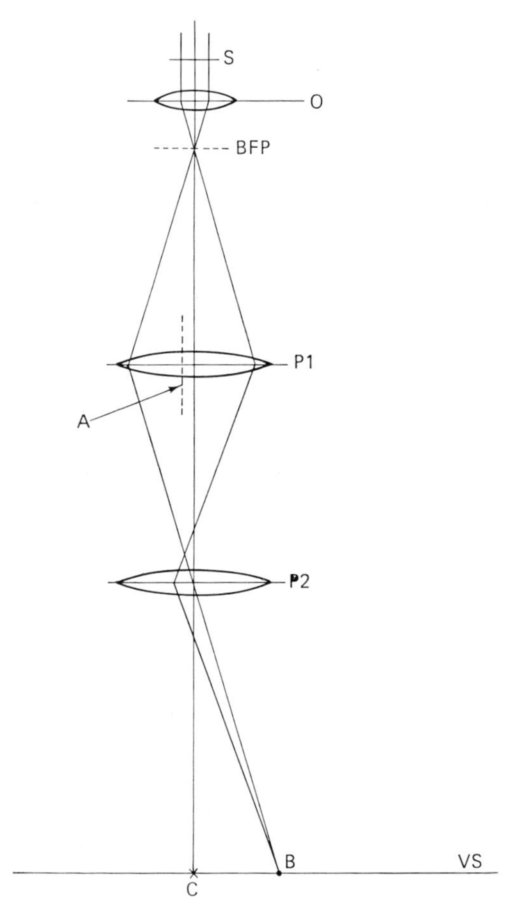

Fig. 4.9. Alignment of the intermediate (or first projector lens) P1. The image of the source is formed on the axis of the objective lens O in the back focal plane BFP. If the axis A of P1 is not on the axis of O and P2 (second projector lens), the image of the crossover is formed at B on the viewing screen VS, away from the screen centre C. When A is aligned with the instrument axis, B moves to C.

image will appear to rotate about some point remote from the screen centre when the objective lens current is changed. Apart from the inconvenience of the image shift at the screen centre that this involves, there is also a decrease in instrument performance which is explained in detail in Chapter 5. The careful correction of this tilt is therefore very important.

Electrical controls, independent of the illumination traverse controls, are provided for correcting the tilt of the illumination. It can be seen from Fig. 4.10 that the image will rotate about the objective lens axis O as the lens strength is changed, but that with a tilted illumination, the change in focus will cause an expansion (or contraction) of the image about the point T. The resultant motion of the image on the screen will result in an apparent rotation centre X. The illumination tilt is corrected by observing the location of the point X, placing a distinctive feature of the image at that point (by movement of the stage controls) and then moving the point to the screen centre by adjustment of the electrical illumination tilt controls. An alternative simple procedure is to place a distinctive image feature at the screen centre with the lens focused, defocus the lens on the medium control, and restore the feature to the screen centre with the illumination tilt controls.

When the instrument has been aligned as described above, it is said to be aligned on the *current centre*, that is the point about which the image revolves for a change in objective lens current. If one imposes a small variation in accelerating voltage, the image may expand about a different point. This is known as the *voltage centre*. There should not be a very large separation be-

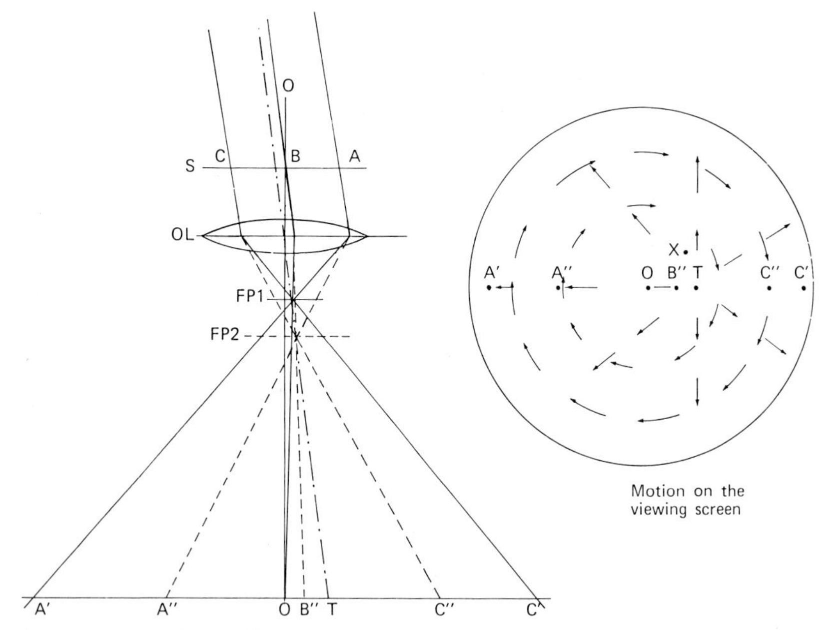

Fig. 4.10. The effect of illumination tilted with respect to the objective lens (OL) axis. OO is the objective lens axis, S is the specimen, T the position of the illumination tilt axis at the final screen. Change of objective lens strength, moves the back focal plane from FP1 to FP2. On the viewing screen point A′ moves to A″, C′ to C″ and O to B″. The expansion of the picture (due to magnification change as the lens strength alters) is centred at T. The rotation centre is at O. The apparent centre of rotation due to combination of these motions is at X.

tween these centres if the overall performance of the instrument is not to be impaired, since either variation can affect the resolution (§ 1.3). The tolerance is of the order of 1 μm of separation (referred to the object plane).

Not many microscopes are equipped with a system permitting small variations of accelerating voltage, so it is often not easy for an operator to use this centring criterion. Most instruments in use therefore will be found to be current centred. For instruments where voltage centring is possible, however, it is to be preferred, as it minimises resolution loss due to electron energy losses in the specimen as well.

4.8 Centring the objective aperture

The contrast aperture in the objective lens is now inserted and centred by viewing the plane of the aperture (by switching to the diffraction mode of operation) in the presence of a scattering object. A bright spot will be seen at the centre of the back focal plane, and scattered electrons will fill the objective aperture which will be clearly outlined. It is centred by using its micrometer adjustments to place it symmetrically around the focused spot.

In some instruments, the objective aperture is not precisely in the back focal plane of the objective lens. It may then be observed at low magnifications as a limitation to the field of view and may be readily centred without switching to the diffraction mode. This technique is useful when using the minimum contrast method of focusing (see § 6.9).

4.9 Correction of objective lens astigmatism

The astigmatism of the objective lens is the main correctable instrumental defect, as described in § 1.3.3. The accuracy with which astigmatism must be corrected is somewhat arbitrary but Haine and Mulvey (1954) consider that it should at worst introduce one quarter of a wavelength path difference in the electron wavefront.

If Z_a is the axial distance between the focal lines and α the objective semi-angular aperture, then for a resolution d (defined by the limit imposed by the spherical aberration and diffraction),

$$\tfrac{1}{2} Z_a \alpha^2 < \frac{\lambda}{4}$$

or

$$Z_a < \frac{1.4 d^2}{\lambda} \qquad \text{for } d = \frac{0.6\lambda}{\alpha}$$

Haine and Mulvey (1954) found experimentally that a Fresnel interference fringe formed around a hole in a thin specimen film had a width $y = \sqrt{\lambda Z}$ where Z is the distance from the specimen film to the object plane of the lens (i.e. the distance out of focus). This becomes a very useful measuring criterion because the fringe width is continuously adjustable by variation of the focus of the objective lens, and its width is a measure of the distance of the focal plane of the lens from the specimen plane in the particular azimuth of the fringe. Thus, an astigmatic lens will give rise to a fringe of varying width around the hole in the specimen film, and this fringe can be used to measure (and correct) the astigmatism of the lens.

If, in an astigmatic image, the maximum and minimum fringe widths are y_{max} and y_{min} respectively

$$Z_a = \frac{1}{\lambda}(y_{max}^2 - y_{min}^2)$$

For astigmatism to be negligible, therefore,

$$y_{max}^2 - y_{min}^2 < 1.4d^2$$

or

$$(y_{max} + y_{min}) \cdot (y_{max} - y_{min}) < 1.4d^2$$

This shows that, if the mean fringe width $y_{max} + y_{min}/2$ becomes large compared with the resolving power, the difference term becomes small. Thus it becomes difficult to measure the asymmetry. For good visibility of the fringe asymmetry, the mean fringe width must be comparable with the resolution being aimed for. In practice, the aim is to make $y_{min} = 0$ while assessing and correcting astigmatism. This is the condition of maximum sensitivity.

The correction procedure is not difficult and is an essential preliminary to any high-quality microscopy, so this is an important adjustment. The first procedure must be to check the alignment of the instrument carefully, as indicated above. Unless the microscope is properly set up, the astigmatism correction will be much more difficult. For one thing, adequate illumination is essential, and for another, if there is image movement as the objective lens is adjusted, the procedure becomes very difficult. The astigmatism of the lens is corrected with the objective aperture withdrawn, and with the anti-contaminator in operation.

The specimen required for astigmatism correction is a thin carbon film perforated with small holes. A hole must be chosen that is nearly round, has no thickened edges, and is small enough for its whole perimeter to be visible under the viewing telescope ($10\times$) at the top magnification of the

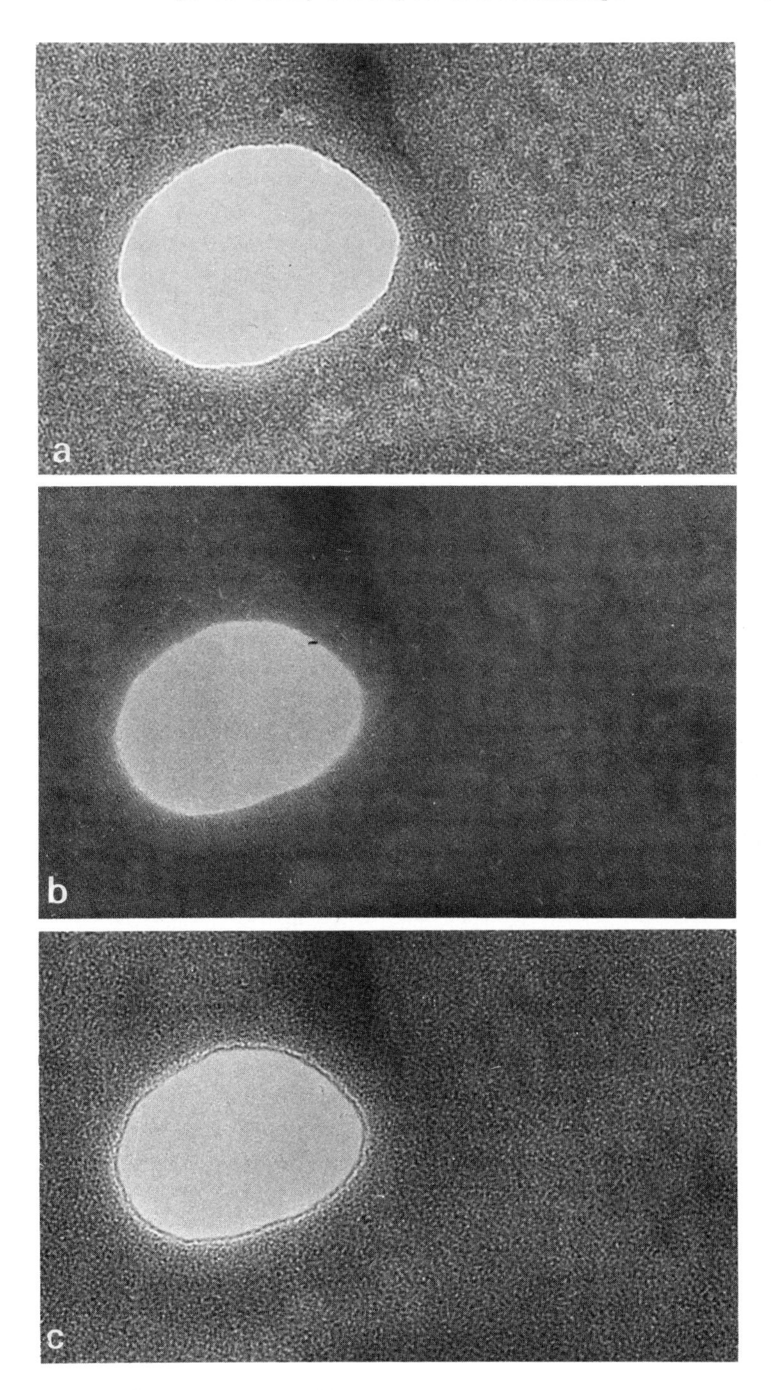

Fig. 4.11. A hole in a thin carbon film: (a) underfocused objective lens (bright fringe); (b) at focus (no fringe) and (c) overfocused objective lens (dark fringe). Note also the change in appearance of the carbon film grain. Magnification 750,000 ×.

microscope (250,000 × need not be exceeded for a 0.5 nm resolution).

The appearance of a hole as the objective lens passes through focus is shown in Fig. 4.11. The overfocused condition is chosen for the correction procedure. The overfocused fringe must be sharp and of good contrast around the hole if the correction procedure is to be successful. If the fringe is not sharp, it is necessary to check the instrument to find the cause before proceeding with the correction (see Chapter 5).

If there is astigmatism in the objective lens, the image will be in focus at one part of the edge of the hole, while some other part is overfocused (dark fringe) (Fig. 4.12a). The aim is to make the fringe uniform in thickness around the hole, even when it is as narrow as the eye can detect (i.e. very near focus). The original direction of the lens astigmatism is determined by removing the objective aperture, switching off the astigmatism corrector and then reducing the strength of the objective lens until only a short arc of fringe remains at two opposite sides of the hole (Fig. 4.12a). Then the astigmatism corrector is switched on at full strength (arbitrary direction). The fringe takes up a new position (Fig. 4.12b). The corrector must now be oriented so that the overfocus fringe is at right angles to the direction it had with the corrector off – that is the corrector is now oriented to oppose the residual cylindrical field of the objective lens (Fig. 4.12c).

There may be rather gross errors in judging a right angle on a tilted screen, so the judgement is made by making the overfocus fringe extend most of the way round the hole leaving only a short segment of focused edge. This should be adjusted to coincide with the position the small overfocused segments first occupied. Then the strength of the corrector is reduced until the fringe width is uniform around the hole, even near focus. Here the correcting cylindrical field is as nearly as possible equal and opposite to the original astigmatism (Fig. 4.12d).

It is not easy to set the orientation control very accurately (better than 10°) and the correction will normally achieve about a ten fold reduction in astigmatism. A more accurate correction is possible with a two-stage corrector, in which a duplicate set of controls can superpose a field on that already set up. The second set of controls enables a new correction to be made on the already reduced astigmatism, and this can be made relatively easily – certainly more easily than a very accurate judgment of orientation of a single corrector. This scheme was first introduced by Page (1963).

The objective aperture is then inserted and the astigmatism rechecked since the astigmatism observed depends not only on the objective lens itself but also on the accuracy of centring and cleanliness of the objective aperture.

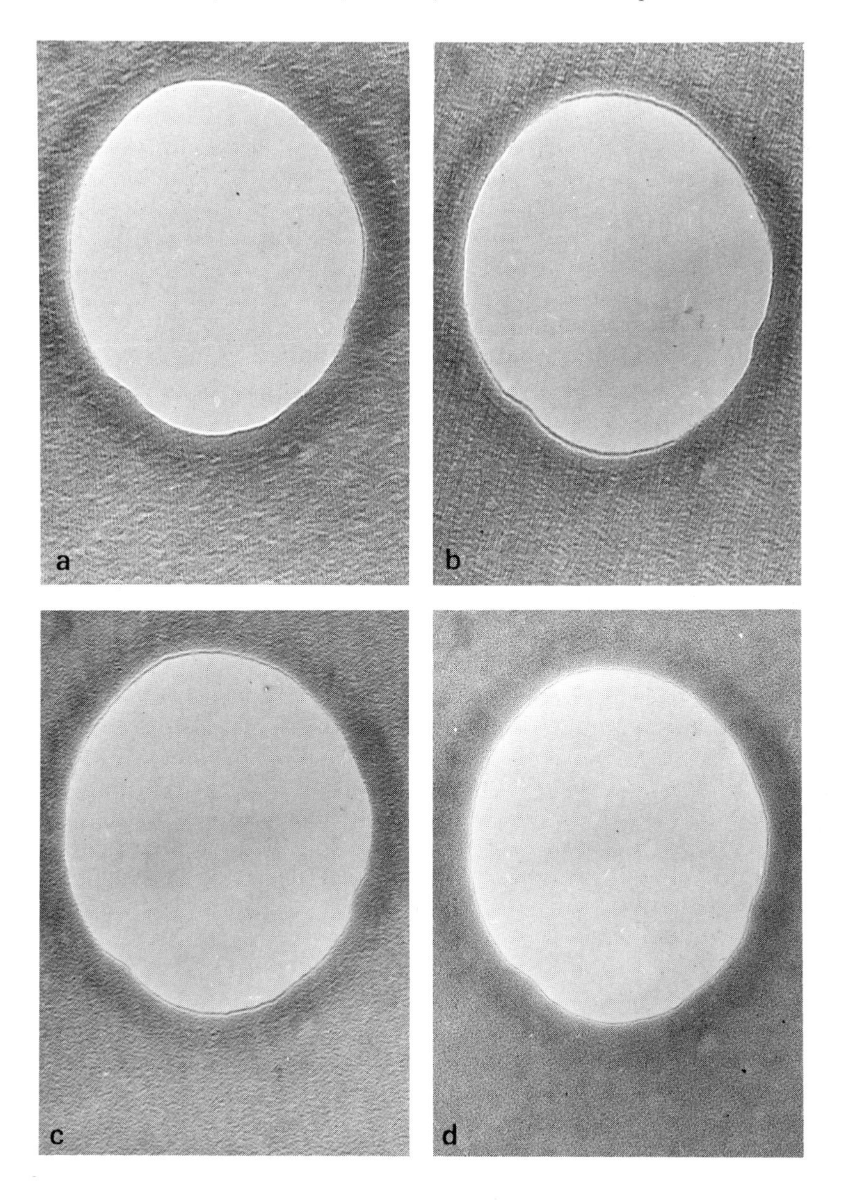

Fig. 4.12. The correction of objective lens astigmatism. (a) A Fresnel fringe showing asymmetry in width due to astigmatism in the objective lens. (b) Astigmatism corrector switched on. Full strength, arbitrary direction. Note increased astigmatism. (c) Corrector oriented to oppose the astigmatism of the objective lens. Note that a short length of underfocus fringe now occupies the place of the overfocused fringe in (a). (d) Corrector strength reduced to obtain a uniform fringe. Magnification 500,000 ×. Fringe width 0.4 nm.

If the aperture is contaminated it will give rise to gross astigmatism and should be replaced by a clean aperture.

One of the effects of astigmatism in the image is to reduce the contrast dip usually observed at focus, which is due to the minimising of phase contrast (§ 3.2). This is because the focal plane for structure in one direction is different from that for structure at right angles to it. If the contrast is plotted against focal position for different amounts of astigmatism (Fig. 4.13), it is seen that there is very little loss of contrast at focus when a large amount of astigmatism is present in the image. This leads to a method for optimising astigmatism correction when it has already been set approximately. After focusing with the finest objective lens control as accurately as possible, the astigmatism corrector controls are used like fine focus controls in order to optimise the contrast dip (some iterative procedure may be necessary) (Page 1964). This method of correction can improve on the astigmatism setting when the fringe width is very fine. It will only be effective under good conditions for phase contrast, i.e. a reduced illumination aperture. This method is very useful because it can be used on any area of a specimen (usually the grain of the specimen support film), without having to search for a suitable small hole around which Fresnel fringes can be observed.

Once the astigmatism correction has been made, it will normally be good for some hours of operation, provided that the contamination rate of the microscope is low. However, there is an important exception to this con-

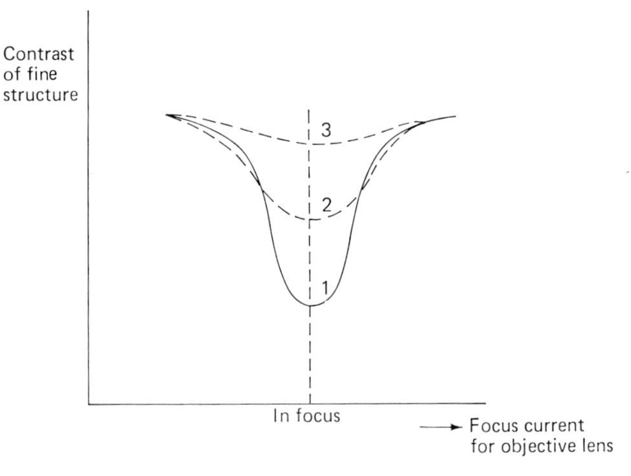

Fig. 4.13. The effect of objective lens astigmatism on the image contrast at focus. Curve (1) no residual astigmatism – sharp contrast dip. Curve (2) slight astigmatism – smaller contrast dip. Curve (3) considerable astigmatism. Imperceptible contrast change.

stancy. When thin foil specimens of ferrous origin are examined, particularly disc type specimens, there may be gross changes of astigmatism with movement of the specimen and a correction will be necessary for every field of view examined. This is because the specimen itself modifies the objective lens field.

The instrument is now in a condition for a critical appraisal of its performance (described in the next chapter). If these tests are passed successfully, it will then be ready for performance to the limits of its capability. For operation at low performance, say for routine examinations of a series of specimens at low magnifications, many of these careful adjustments and checks are not strictly necessary. They will, however, contribute to maximum convenience in operation and a generally higher quality of micrographs than would otherwise be obtained.

4.10 Routine re-checks of alignment during operation

Most of the alignments described are very stable and will not require re-adjustment for long periods; many indeed remain correct until the column is dismantled for a major cleaning and overhaul. Some of the illumination controls do require more frequent adjustment, however. The illumination traverse will be adjusted quite frequently during routine operation. Apart from this, it is worth checking every few hours that the centration of the filament has not altered because of thermal strains, since this can cause gradual loss of peak intensity due to misalignment of the beam.

It is also important to check from time to time (maybe twice per day) that the filament is being operated at the 'saturation' point but not beyond it. This can materially lengthen filament life.

Subsequent to any of these illumination adjustments, the beam tilt adjustment should be checked again for accuracy.

Apertures will require recentring after being interchanged or withdrawn for any reason.

Routine checks also need making of objective lens astigmatism, the frequency depending on the level of performance at which the instrument is being used.

REFERENCES

Haine, M. E. (1947), The design and construction of a new electron microscope. J. IEE *94*, 447.

Haine, M. E. and T. Mulvey (1954), The regular attainment of very high resolving power in the electron microscope. Proc. 3rd Int. Cong. Electron Microscopy, London, p. 698.

Page, R. S. (1963), Personal communication.

Page, R. S. (1964), Personal communication.

Chapter 5

Checking the performance of the electron microscope

Much of the routine work for an electron microscopist consists of the examination of moderately thick specimens at middle or low magnifications (say, below 25,000 ×). For such work, the resolution of the final image is largely determined by the specimen thickness (§ 3.2), and the nominal resolving power of the microscope is not needed. Under such circumstances, there is little point in spending much time in adjusting the instrument to achieve its ultimate performance, although it will be realised that good alignment, for instance, simplifies operation so much that it is worth ensuring that it is maintained.

Nevertheless, every operator should know how to check the performance of an electron microscope and to be aware of the factors affecting it, so that it can be quickly restored to ultimate performance when the investigation demands it. The most important parameter in instrument performance is its *resolving power*, and this is now considered in some detail.

5.1 The resolving power of an electron microscope

The *resolving power* of an electron microscope is the best possible performance as limited by built-in instrumental parameters. The theoretical resolving power may be quoted by the manufacturers. The only practicable way of checking whether it is achieved is by recording a photograph of a suitable object; one can then measure the *resolution* achieved in the image. In order to approach the resolving power, the specimen used should be as near as possible 'ideal', that is, it should be thin enough to avoid limiting the resolution of the image (through chromatic effects for instance).

The actual measurement of the resolution obtained has become a far from simple matter, when one is considering the high performance instruments of today. Some of the imaging conditions affecting resolution are discussed in § 3.8; and in § 8.2, in the consideration of interpretation of the image, the matter has to be examined again. For the present purposes, the commonly used criteria of resolution will be discussed.

5.1.1 THE POINT SEPARATION TEST

This was the first and is still the most generally acceptable criterion of resolution. It is directly comparable with the test for a light microscope – suitable small particles are photographed, and the photograph is searched for two small particles which can just be distinguished as separate. The distance between their centres then defines the resolution achieved on the photograph (Fig. 5.1). The pair of particles do not have to be completely separated, but there should be a sufficient drop in intensity of the profile of each particle to be certain that there are two particles present. It is not enough to record one plate; two successive plates at the same focal setting must be exposed and the same pair of particles must be found on each plate, to ensure that the effect of electron noise (see § 7.2.3) is negligible. Furthermore, several pairs of particles in different orientations must be located, so as to demonstrate that astigmatism or other image defects are not significant.

It is relatively easy to prepare a suitable specimen for detail of 0.5 nm and larger, and for this a thin carbon film, approximately 5 nm thick, is covered by a finely dispersed, evaporated film of platinum–iridium alloy. Although correlation for particles with a separation of 1 nm and greater is excellent, the smaller the separation, the less correlation there is. Dowell et al. (1966) found that on the best pair of consecutive prints obtained during resolution tests, the correspondence in two successive pictures of fine detail decreases with the resolution sought. Their results were summarised in the table given below:

Resolution	*No. of cases*	*Agreement*	*Doubtful*	*Disagreement*
1 nm	20	19	—	1
0.7 nm	20	17	—	3
0.55 nm	28	6	9	13
0.4 nm	29	7	4	18
0.3 nm	19	3	2	14

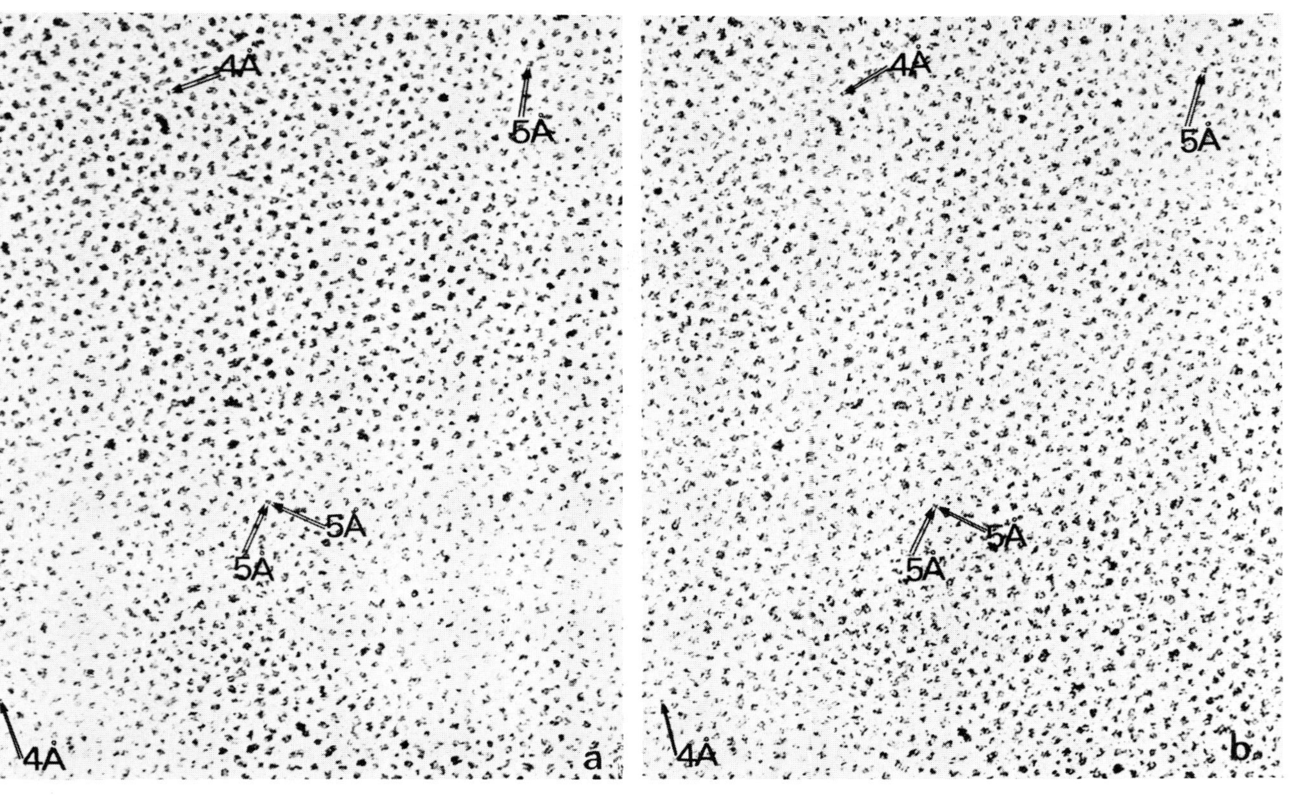

Fig. 5.1. The point-to-point resolution test. Successive micrographs (a) and (b) of evaporated Pt/Ir alloy on a perforated carbon film. Magnification 700,000 ×. Arrows show particles separated by distances marked as a measure of resolution.

The advantage of this type of resolution check is that the test specimen is similar to the type of specimen used in routine work and therefore comparable results can be expected from real (thin) specimens. Dowell's figures show, however, how doubtful it becomes for resolutions below 0.5 nm.

If only a single micrograph is available for a resolution check, it is necessary to ensure that the magnification of the plate is sufficiently high so that specimen detail cannot possibly be confused with the electron noise in the image. With resolutions of 0.5 nm and better, it is difficult to find particles with this separation that can be used as a test object, since this would require particles of 0.2 or 0.3 nm in diameter, i.e. of atomic dimensions. Since the contrast of such structure is due almost entirely to phase effects, a thin carbon film may be used for this type of measurement. An optical diffractometer should be used to check the resolution of such a plate (Beeston et al. 1972). If there is sufficient detail on the plate of a given structure size, a ring will be seen in the optical diffraction pattern and the diameter of this can be easily measured. Individual particles can easily be missed in the general noise of the picture.

5.1.2 THE LATTICE RESOLUTION TEST

The other widely used criterion of instrument performance is *lattice resolution*. If a suitably oriented thin crystal is used as the specimen, then the crystal lattice planes may be imaged as a set of dark lines of high contrast (Fig. 5.2). Since the lattice plane spacing is known accurately from X-ray data, there is a well-defined level of performance for the imaging of each set of lattice planes (Menter 1956) (which must be identified by examination of the diffraction pattern from the crystal). This criterion of performance is undoubtedly aesthetically satisfying and has the appearance of much greater precision than the point separation test, since the specimen itself provides an accurate magnification calibration. There is also no doubt that the clearly defined lines are easily measured to a much higher degree of accuracy than two image points of perhaps poor contrast in a noisy background.

However, the special conditions under which a lattice image is obtained must be remembered. It is only necessary to admit the unscattered electron beam and one diffracted beam, to obtain a lattice plane image, and these two interfering beams yield a sinusoidally varying intensity of the correct period (normally) but this variation does not necessarily correspond to the *position* of the lattice planes in image space. It was shown in § 3.8 that for a lattice plane image, which has a single-valued space frequency, one can choose a particular value of object defocus which yields a sharp maximum of

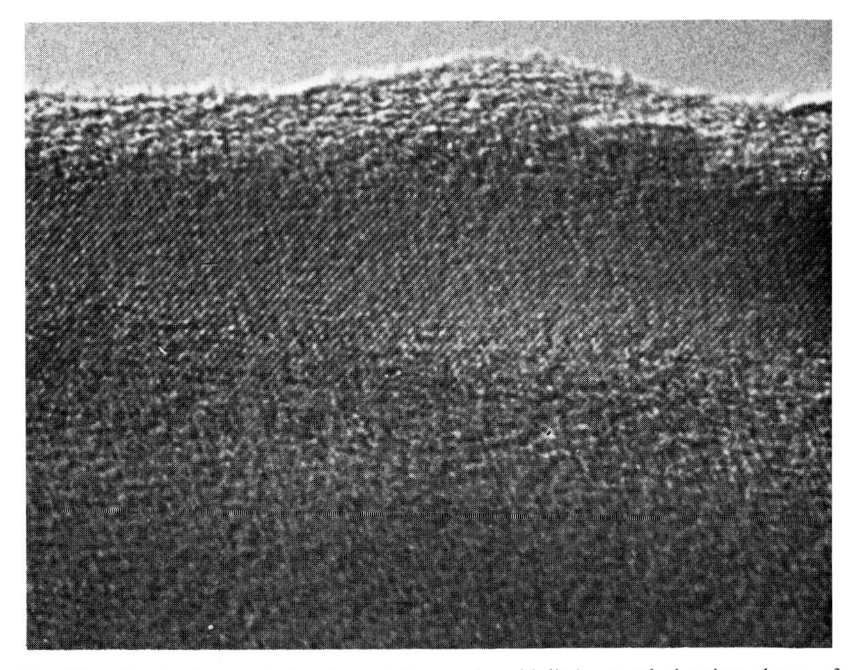

Fig. 5.2. The lattice resolution test: Asbestos (crocidolite) crystal showing planes of spacing 0.448 nm. Magnification 2,250,000 ×.

contrast; the selective filtering of other space frequencies is immaterial for this type of object. Many of the objections to the use of lattice planes as a resolution test are overcome if two orthogonal sets of planes are imaged with equal contrast and resolution, as this eliminates the possibility of unidirectional defects being overlooked. Heidenreich et al. (1968) have suggested that partially graphitised carbon black is an ideal test specimen, since the lattice planes curve in many directions within a small region (Fig. 5.3). The lattice spacing (0.34 nm) is a convenient size for checking the performance of most high quality microscopes today. Some can, of course, be made to yield considerably better lattice resolutions, but usually only with special care.

There are, however, two factors which make lattice plane imaging unreliable as a guide to the results likely to be obtained from an 'ordinary' specimen with a random distribution of structure sizes. Firstly, the lattice imaging requires only two or three narrow-angle electron beams to travel through the objective lens. Hence the full effect of spherical aberration is not imposed on the image. Secondly, because of the information redundancy of a long straight line, the eye will recognise the lattice planes in conditions of

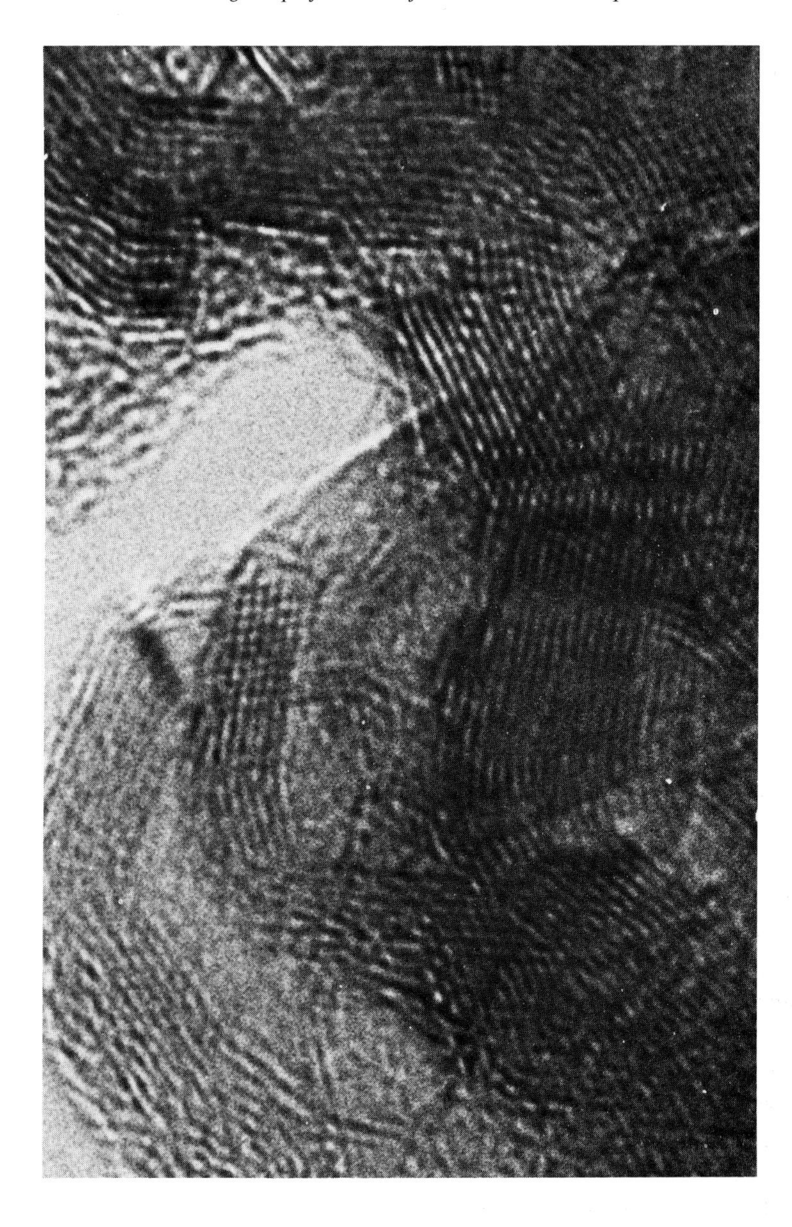

Fig. 5.3. Partially graphitised carbon black, showing lattice planes of spacing 0.34 nm curving around the particles. Magnification 5,400,000 ×.

noise which would make any interpretation of irregular structure only speculative.

The lattice plane test is nevertheless a searching one for an electron microscope. The planes can only be seen if the supplies for the high voltage and lenses are very stable and if the specimen stage is free of drift; these are important pre-conditions for high resolution performance. The 'lattice plane resolution' of an electron microscope should be regarded primarily as a stability test, rather than as a guide to the likely performance with other specimens. It is, however, useful to quote particle separation and lattice plane resolution figures side by side, as they do provide complementary information about the microscope performance.

When reading nominal resolution figures for an electron microscope, it should be remembered that they are performance figures guaranteed by the manufacturers as being reasonably easy of achievement with a fault-free instrument properly aligned, in an acceptable site. Given a good site, and with attention to all the factors which might limit resolution (discussed below), it should be possible with most instruments to achieve a resolution significantly better than the manufacturer's guarantee. However, it must be realised that the maintenance of such a performance will involve significantly more time in maintenance of the instrument and correspondingly less operational time for the examination of specimens. The project must therefore be of major importance if the time is to be spent in pushing an instrument to its ultimate performance.

5.1.3 THE FRESNEL FRINGE TEST

While the accurate determination of the limiting performance of an electron microscope requires the photography of particles or lattices as described above, it is useful to have a visual check on the instrumental performance, and this can be provided by examining the overfocused Fresnel fringe at the edge of a hole in a thin carbon film, as suggested by Haine and Mulvey (1954). The merit of the Fresnel fringe is that it is continuously variable in width, by adjustment of the focus control, and can therefore be used to assess both gross defects and very small ones (Fig. 5.4). It is also of good contrast, and may be observed at high magnification on the viewing screen, with the aid of a telescope.

The operator can therefore watch the effect on the fringe of varying various parameters which might affect instrumental performance. A rough guide to instrument resolution is the width of the finest Fresnel fringe that can still be seen all round the circumference of a hole in the specimen film.

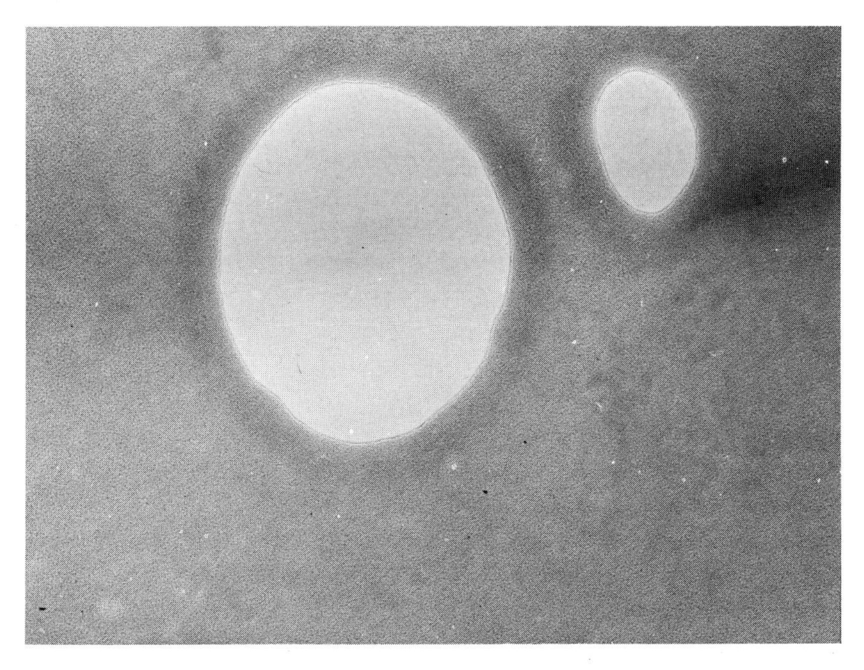

Fig. 5.4. Fresnel fringe at the edge of a hole in a carbon film. Minimum fringe width 0.4 nm. Therefore resolution is at least as good as this figure. Magnification 750,000 ×.

(The use of such a specimen has already been described in the section on correction of objective lens astigmatism, § 4.9).

As explained by Haine and Mulvey (1954), any disturbance to instrumental performance should be visible in its effect on the Fresnel fringe, and this makes possible a systematic check on the various relevant factors. The instrument should be carefully aligned as explained in Chapter 4 before these checks are carried out, as misalignment can give rise to misleading results, and will make sensitive detection of defects more difficult.

The foregoing discussion assumes that the majority of limitations to performance have been overcome, and discusses ultimate performance. In practice, the achieved resolution depends on many other factors, such as:

(a) thickness of the specimen (discussed in § 3.2);
(b) contrast of the image (discussed in § 3.3 and § 3.8);
(c) magnification of the image (§ 7.2.3) (§ 1.10.1);
(d) contamination of the specimen (§ 6.14 and § 5.5);
(e) astigmatism of the objective lens (§ 1.3.3);
(f) focusing of the objective lens;

(g) alignment of the imaging system;
(h) instrumental defects;
(i) site conditions.

Many of these have been discussed elsewhere, in the sections noted; the rest are discussed below.

5.2 Illumination tilt alignment

As mentioned in § 4.7, this is a very important alignment. Fig. 5.5 repeats Fig. 4.10 and shows the effect of an illumination axis tilted with respect to the objective lens. The specimen points ABC are imaged at A′O′C′ in the image plane of the objective lens. If the focal length of the objective lens is altered, the points ABC are now imaged at A″B″C″. Only the image point T does not move with change of focus – the intersection of the tilted illumination axis with the image plane. There are thus two superposed image movements with change of focus – an expansion or contraction about T and a rotation about O. The resultant motion is an apparent rotation about some point X, known as the current centre. In a properly aligned microscope, X is

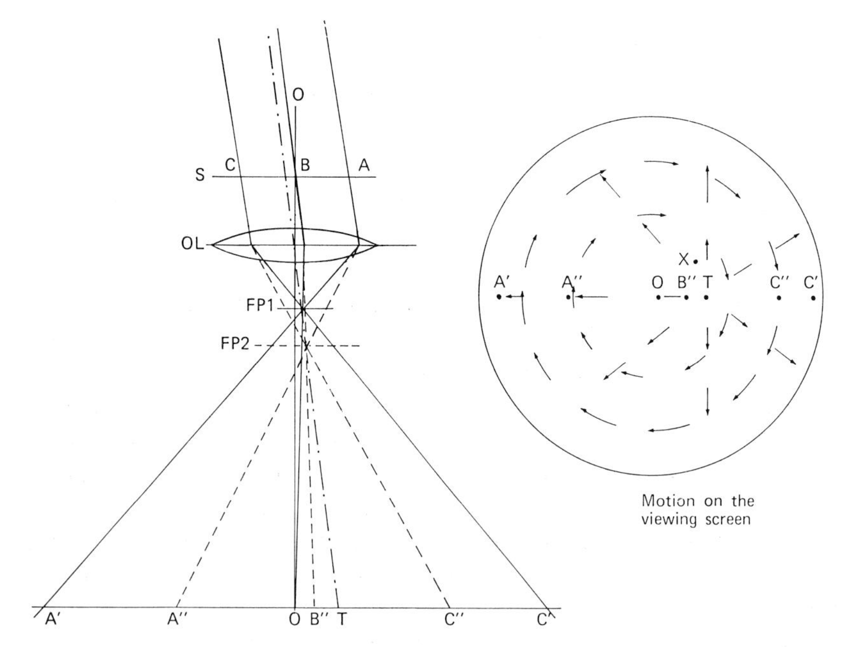

Fig. 5.5. Diagram of tilted illumination showing how it causes rotation of the image about some apparent centre distant from the instrument axis.

moved to be coincident with O, and a small change in objective focal length then causes no movement at the screen centre, but an increasing amount of rotary motion as the distance from O increases, being at a maximum at the edge of the screen.

There is always a small residual ripple in the objective lens current supply corresponding to a continuous small change in focus. The centre of the image will, however, be relatively little affected. Thus, the illumination tilt alignment results in a condition of minimum sensitivity to ripple on the objective lens current supply. It is easy to see that if there is gross misalignment of the illumination tilt, the current centre could be right off the viewing screen and there could be a large image movement in the centre of the visible image due to the small ripple. Correct alignment of illumination tilt is therefore essential to minimise loss of resolution due to unavoidable instabilities in the supplies.

Variation of the high voltage has the effect of varying the strengths of all the lenses simultaneously, and hence the principal effect to be seen is a change of magnification. Thus, a varying high voltage will cause radial expansion and contraction about the 'voltage centre' (normally close to the current centre).

5.3 Stability of the supplies

As described in §1.3.2 a high degree of stability is required both in the high voltage supply and objective lens current supply, if they are not to limit the instrumental performance to a worse figure than that determined by spherical aberration and diffraction. In a correctly aligned instrument, there will be negligible image motion at the screen centre as a result of such instabilities, as explained in § 5.2 above, although there may be changes in objective lens focus which result in a small change in the width of the Fresnel fringe. Thus, small amounts of ripple will cause broadening and loss of contrast in the fringe all around the hole. It will not normally be possible to differentiate between current or voltage instability by viewing a hole at screen centre. By using as specimen a carbon film with many small holes, a number of these can be in the field of view at high magnifications, and one can then examine holes near the edge of the field of view as well. The edges of such holes may well show a unidirectional blur if the instabilities are large and will be radial with respect to the screen centre in the case of high voltage fluctuation, and rotary for objective lens current variation.

Often the ripple levels, although causing some loss in fringe contrast, may not be enough to allow differentiation by viewing a hole at the edge of the

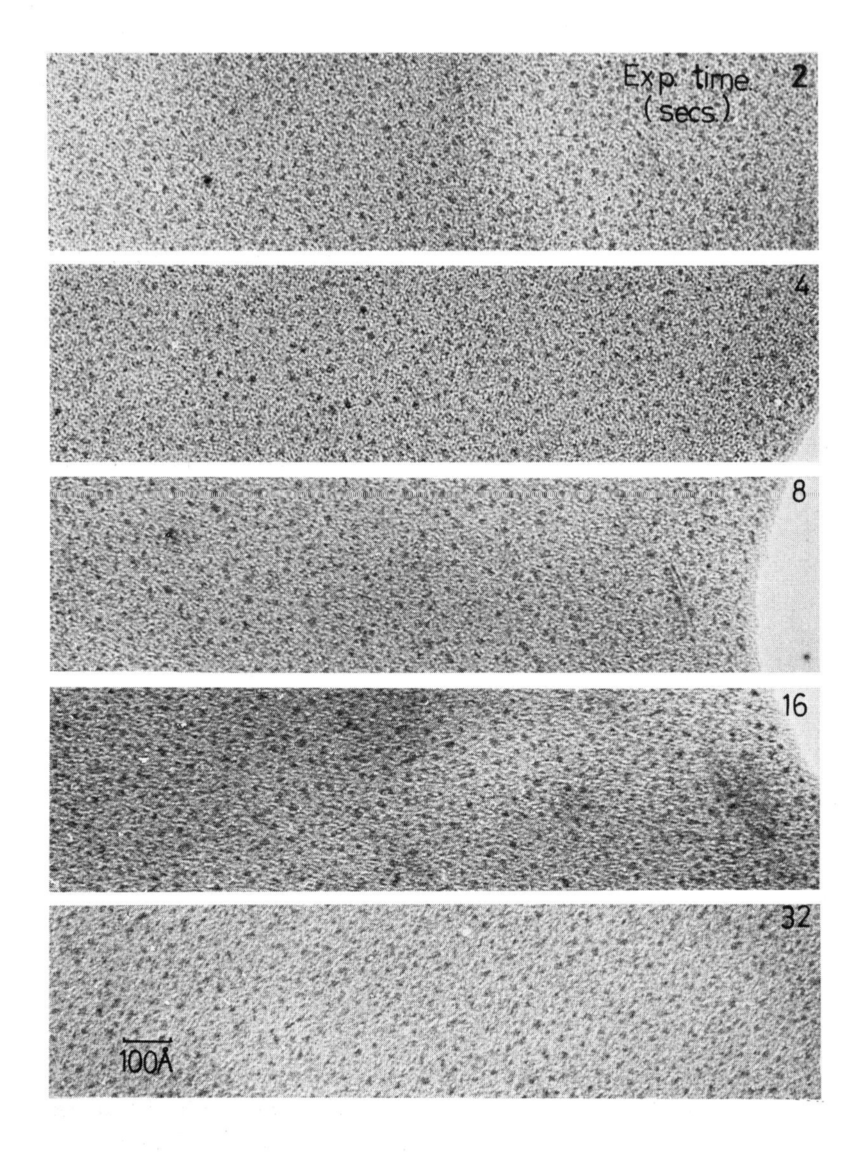

Fig. 5.6. Successive micrographs of a hole in a carbon film, photographed at increasing exposure times. Note how the finest detail is lost in the long exposure period. Magnification 1,100,000 ×.

field of view. In such a case, the effect of the ripple can be greatly magnified by deliberately misaligning the illumination tilt axis, so that the tilt centre lies well off the fluorescent screen. It is usually possible to differentiate the ripples in this way: the objective lens ripple will cause a blur in exactly the direction which the image moves when a small focus change is made. If the blur is in any other direction, and has increased compared with the condition of good alignment, it is probably due to the high voltage supply.

A searching test of supply stability is to record successive micrographs at different exposure times (making appropriate adjustments of the illumination). The pictures are then re-assembled in strips to form a montage as in Fig. 5.6, to show how the increasing exposure time affects the sharpness in the fine detail of the micrographs.

5.4 Specimen drift

As pointed out in § 2.6, the specification for a specimen stage is extremely difficult to meet, and the whole stage mechanism, specimen holder and specimen must be in excellent condition if unwanted movement of the image is to be avoided. The first intimation of trouble may be imperfect specimen motion; backlash in the stage drives, or a rectangular motion when the direction of movement of the stage is reversed, indicate defects which may also cause spontaneous stage motion. Sometimes, some image motion is noticed after the stage drives have been released. Unless this stops very quickly, there may again be mechanical defects in the stage motion. Not all image drift is caused by a defective stage. The specimen support film may be ruptured, and such a film will often move when heated in the electron beam. A buckled specimen grid will often be in poor thermal contact with the specimen and/or the specimen cartridge, and can cause quite serious drift. The specimen holder itself must fit well into the stage carriage, or again the thermal contact will be poor.

Even if all these components fit well, there may be sufficient heat flowing from the lens windings into the stage to cause movement due to differential thermal expansion between the different metallic components. This can be minimised by the manufacturers by good design of the water cooling circuit which should be interposed between the windings and the specimen stage. Nevertheless, even with good cooling, it may take several hours from the time the microscope is switched on until thermal equilibrium has been reached in the microscope column. Some idea of the sensitivity will be gained by considering the expansion of a 3 inch push rod, when heated through 1 °C.

It amounts to 0.825 μm. If this temperature change occurred in 1 hr it would give a specimen drift of 0.23 nm/sec, an unacceptable amount. Thus there must be temperature compensation built into a specimen stage.

The best way to check on stage drift is to place an easily recognisable feature of the image at the centre of the screen and watch it under the viewing telescope. The movement should be imperceptible over a period of 30 sec if image motion is not to affect the photograph. Sometimes the stage can be jolted into a stable position by tapping the column; another useful trick is to reverse the direction of drive of the last specimen movement used by just a fraction of a turn of the control knob. If the drift is persistent, and the specimen grid is flat (check by focusing several different portions of the grid) and the support film is unbroken, some part of the stage must be suspect. Note

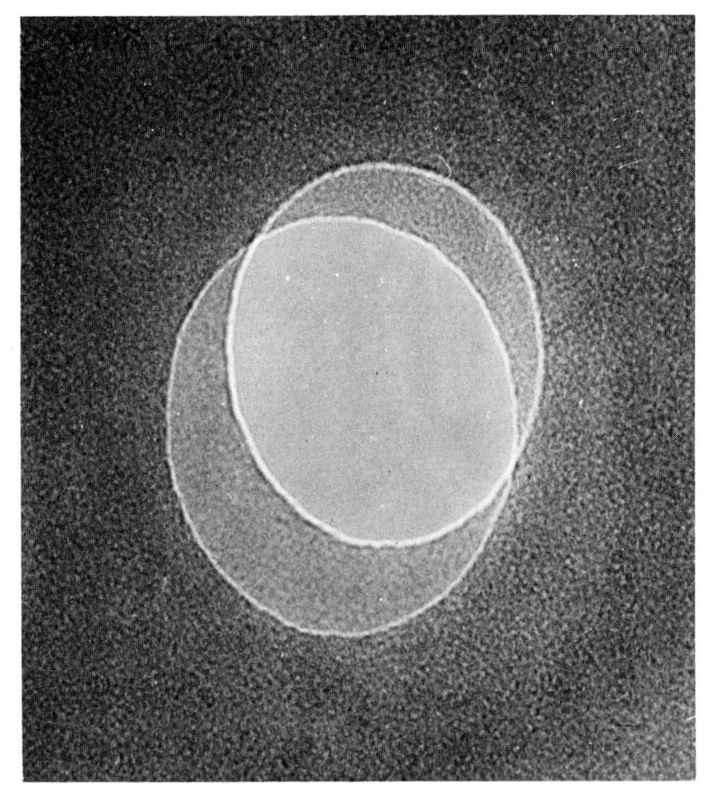

Fig. 5.7. Double exposed micrograph of a hole in a carbon film showing drift of the specimen between exposures, and also the deposit of a layer of contamination, showing up as a reduction in hole diameter. Magnification 300,000 ×.

whether the specimen holder shows any warmth when withdrawn through the airlock (this could indicate bad thermal contact with the stage). Alternatively, if the holder is very cold, it could indicate thermal contact with the anti-contaminator, which is certainly to be avoided. If the drift has not been noticed by observation under the viewing telescope, it would manifest itself on a plate by a unidirectional blur of all the image detail on the plate. Such a blur can be caused by other effects, to be described, but if two successive exposures of the same field have been made, a displacement of the image on successive plates will identify image drift (Fig. 5.7).

If the problem cannot be isolated to the specimen or its mount, the stage may require cleaning. In most instruments it is advisable to leave this to a service engineer, as it is such a delicate mechanism that it is very easily damaged.

5.5 Contamination of the specimen

As will be discussed later (§ 6.14), a layer of contamination may build up on the specimen at a rate of 0.1 nm/sec (6 nm/min) or more. This means that after a short examination, a layer of carbon contamination (usually with a coarse granular structure) will have formed over the specimen region being examined. This layer thickens the specimen locally, causing increased chromatic losses and hence poorer resolution; it imposes an unwanted granular structure and the increased thickness causes attenuation of the electron beam and loss of intensity for viewing and photography. In addition, the contaminant layer may introduce mechanical stresses in the specimen which will modify its original appearance (this happens particularly with thin metal foils). A high contamination rate therefore quickly limits the achievable resolution in any particular region of the specimen. High resolution operation therefore demands operation of an efficient anti-contaminator device (§ 6.14).

A measurement of contamination rate may be made by recording two micrographs with a known time interval between them, and by noting the decrease in diameter of a hole in the specimen support film in the time noted (Fig. 5.7). Then the contamination rate, r is given by

$$r = \frac{d_2 - d_1}{2t}$$

where d_1 is the original diameter of the hole, and d_2 the diameter after time t.

5.6 Focusing of the objective lens

This is an important operation, and as pointed out earlier (§ 3.8), one does not always aim to have the image exactly in focus, since a degree of underfocus may markedly improve the contrast of fine detail. Nevertheless, any gross error in setting the focus will result in enlargement of phase-contrast detail arising from the support film, or the embedding medium in a section. This coarse structure will be related to the degree of defocus, and can completely mask the genuine fine structure of the specimen. Control of focus is therefore important in obtaining the best combination of resolution and contrast in the image.

5.7 Cleanliness of the column

Even when all the previous items have been attended to, an otherwise good microscope may have a poor performance if it has not been kept clean in a few vital places, according to the routine cleaning instructions given by the manufacturer. The cleaning becomes necessary, because of the formation of contamination (§ 6.14). This usually arises from the entry of organic vapour into the microscope column – oil vapours from the vacuum pumps, vapour from grease on vacuum seals or on lubricated surfaces (normally avoided), and grease from handling parts of the microscope without protective gloves. Lesser contributions arise from dirt from the atmosphere, vapours from photographic plates, and adsorbed solvents. All these vapours may interact with the electron beam and remain on inner surfaces of the microscope as insulating layers which can charge up and affect the electron beam.

Particular regions which require attention are:

5.7.1 THE ILLUMINATION SYSTEM

A high degree of cleanliness is essential in the region of the electron gun if a stable image is to be observed on the fluorescent screen. Any dirt, no matter how minute, will lead to micro-discharges in the gun and hence a change in accelerating voltage; this causes a change of focus. When small focus changes are observed in the image, it is far more likely to be due to dirt in the gun than a fault in the electrical supplies. Components which require frequent cleaning are the filament assembly, the anode and gun body.

Contamination in the condenser lenses will not change the focus of the image, but if it is excessive it will reduce the image contrast. Severe astigmatism of the illumination spot is generally due to a contaminated condenser aperture. If excessive, this astigmatism will blur the image in the direction of elongation of the spot. Gross contamination will lead to an insulating region within the condenser system, which can become charged by the electron beam. This can cause a more or less severe deflection of the beam followed by a slow drift back to the original position, which is due to the slow discharge from the insulating area once the beam has been deflected away. The beam eventually returns to its original position and the process then repeats. If this happens very frequently it is normally due to an insulating particle adhering to a surface adjacent to the electron beam.

In instruments where there is a protective screening tube in the condenser lenses (e.g. AEI EM 6, EM 801), it is a simple matter to remove it for routine cleaning and the possibility of these occurrences is much reduced.

5.7.2 THE SPECIMEN HOLDER

If the specimen holder is allowed to become dirty, there may be disturbances to the electron beam close to the very critical image forming region, and this can quickly affect the resolution attainable. One of the most important steps to avoid dirt on the specimen holder is to handle it at all times with a gloved hand, or with a specially designed carrying tool. If the specimen holder is becoming dirty, it will often first be noticed when the edges of the grid are being examined, i.e. when the electron beam passes close to the contaminated part of the holder.

5.7.3 THE OBJECTIVE APERTURES AND CARRIER

If a contaminant layer is allowed to build up on an objective aperture, it can charge up and the resulting electrostatic field can introduce strong astigmatic effects, which vary rather rapidly with time. This prevents an accurate correction of astigmatism, except for very short periods.

The cleanliness of an objective aperture can be checked by correcting the astigmatism with the aperture carefully centred, and then displacing the aperture by half its radius. If the aperture is clean, there should be no detectable change in the astigmatism when the image is viewed at high magnification through the viewing telescope. It should also be possible to withdraw the aperture from the beam, leaving the lens without an aperture, and observe no significant change in astigmatism. The use of thin foil apertures (§ 6.3) avoids most of the problems of aperture contamination because the heating of the aperture by the electron beam raises it to a temperature high enough to prevent the growth of contaminant layers.

5.7.4 THE PROJECTOR LENS

This region of the microscope requires little or no routine maintenance since only a fraction of the electrons emitted from the filament arrive at the projector, the majority having been stopped by the various apertures in the column. Also, slight contamination in this region will not adversely affect the image quality. If gross, it may however cause charging leading to movement of the image.

5.8 Site conditions

Apart from the instrumental limitations discussed previously, the environment in which the electron microscope is placed can prevent the specified performance from being obtained. The site requirements are described in detail by Alderson (1974) in another book of this series. The salient features only will be mentioned here.

The main sources of trouble, which can directly affect the resolution obtained with the instrument, are mechanical vibrations and electric fields. The microscope column is a fairly rigid structure and it will isolate the electron beam from minor sources of vibration. However, once the vibrations exceed some minimum value, they will start to affect image quality. The frequency of the vibration is important, as well as its amplitude, as it is difficult to isolate low frequency vibrations, and the column may have resonant frequencies to which it is particularly susceptible.

The whole site of the laboratory may be subject to ground vibrations, or the local conditions may be inadequate, so that heavy tramping on the floor, or slamming of doors, affect the image. The effect of mechanical vibration is to cause the image to vibrate. It will cause any micrograph to have a directional blur in the direction of the vibration.

5.8.1 SMALL AMPLITUDE MECHANICAL VIBRATIONS

The direction of the blurring may be found to change with the magnification of the image. In this case, the vibration is affecting some upper part of the microscope column. If the vibration is taking place, say, at the join of the projector lens and viewing chamber (in practice, a very unlikely occurrence) the direction of blur will be independent of magnification. It is always worth checking if the blur is in the direction of one of the specimen drives, since they are more prone to be sensitive to vibration than other parts of the column.

Vibration can also arise from the microscope system itself. The first place to suspect is the rotary pump. This normally vibrates quite severely, and the pumping connection is carefully isolated by flexible bellows and clamped sections of pipe. But it can sometimes occur that the bellows are allowed to collapse, and they can then transmit vibration. The boiling of the pump fluid in the diffusion pump can also cause vibration problems, which are more difficult to isolate because the diffusion pump is often rigidly connected to the vacuum manifold at the back of the microscope column. Finally, vibration has been known to be transmitted through the water cooling pipes into the lenses. If the site has previously been trouble-free, these are the places to check first for vibration.

A useful and inexpensive vibration detector is a 4 in Petri dish containing mercury. If this is placed on the microscope desk, it is possible to see vibrations as ripples on the liquid surface. The tips of the fingers lightly applied to a surface are often useful detectors of slightly larger vibrations.

5.8.2 LOW LEVEL AMBIENT MAGNETIC FIELDS

The influence of stray magnetic fields on resolution is as disturbing as mechanical vibration. Again, depending on the magnetic shielding of the microscope and the required performance, the acceptable level of ambient field will vary for different instruments. A typical set of values for an alternating 50 or 60 Hz field is that the maximum horizontal field should not exceed 5 mG with a similar figure for the vertical field (Alderson 1974). For critical work it may be desirable that the field be no greater than 2 mG. Excessive fields will result in a unidirectional blurring of the image in extreme cases, and a loss of image contrast when the field is only slightly in excess of that recommended. A very slowly varying field may cause the illumination beam to move off the viewing screen (due to local electric trains, for instance).

5.9 Magnification

The only aspect of the choice of magnification which may affect instrument performance concerns the exposure time. In order to minimise the effects of image disturbances such as drift, contamination, and instabilities in the electrical supplies, it is usually advisable to limit the exposure time of the photographic plate to the range 1–5 sec. Depending on the efficiency of the illumination system, this could set an upper limit to magnification, because of loss of brightness of the image. However, in cases where the ultimate performance is being sought, special precautions will in any case have to be taken to remove this more arbitrary limit on the choice of magnification (by the use of pointed filaments, for instance) because a high magnification is an essential for high resolution operation.

5.9.1 MAGNIFICATION CALIBRATION

The final magnification of the electron optical image is the product of the magnifications of the imaging lenses. Thus

$$M_T = M_0 \times M_{p1} \times M_{p2} \times M_{p3}$$

where M_0, M_{p1}, M_{p2}, M_{p3} are the magnifications of the objective lens and the three projector lenses respectively. Each lens has a strength determined by the magnetic field in the lens gap, and hence, to a first approximation, by the current flowing through the lens windings which excites the field. Since it is not easy to obtain a direct reading of the lens field strength, the currents in each lens are measured, and the magnification read out is in fact set with reference to a known set of currents in the different lenses. Owing to hysteresis effects in the iron circuits, the actual strength of the field will depend on whether the current arrives at a given value from a lower or higher value. Thus, there is a considerable uncertainty surrounding the actual magnetic field strength even though the currents are at a predetermined value. If one takes no particular care about the procedures, the actual image magnifications may have an uncertainty of between 5 and 10% of the nominal value. Some instruments are equipped with an automatic system for normalizing the magnification by cycling the lenses in a standard way and the reproducibility of magnification is then better than 2%. If this is not supplied, then to obtain this 2% reproducibility figure, the magnification setting to be used must always be approached from the same direction of current each time it is used.

Nearly all instruments now have direct-reading magnifications which are probably only accurate to within 5% of the nominal value. This, although sufficient for many purposes, is too large an error for accurate work. For such work it is necessary to have an independent calibration of magnification, and variation in the position of the specimen plane is the most likely cause of error in magnification settings. Certain precautions need to be taken.

1. The specimen holder used for the calibration must be the same as that used for the subsequent accurate work, and it must be an excellent fit into the stage. For a microscope with a focal length of 2 mm, a 2% accuracy in magnification calls for a height setting accuracy of the specimen grid of

± 0.04 mm. Different specimen holders could easily vary by this amount, especially after use and repeated cleaning.

2. The grid on which the calibration specimen is mounted must be perfectly flat since a damaged grid might easily exceed the height tolerance.

3. During the calibration procedure, the currents flowing in each of the lenses should be measured for each magnification setting, using a meter of adequate accuracy to make the readings meaningful. The objective lens current can serve as a monitor of the specimen height. Any divergence from the normal value may indicate a badly fitting specimen cartridge, or a bent specimen grid.

4. The lens currents should be cycled in a standard fashion either by the automatic control, or manually before each record of magnification is made.

When the measurements are complete, a graph of the projector currents

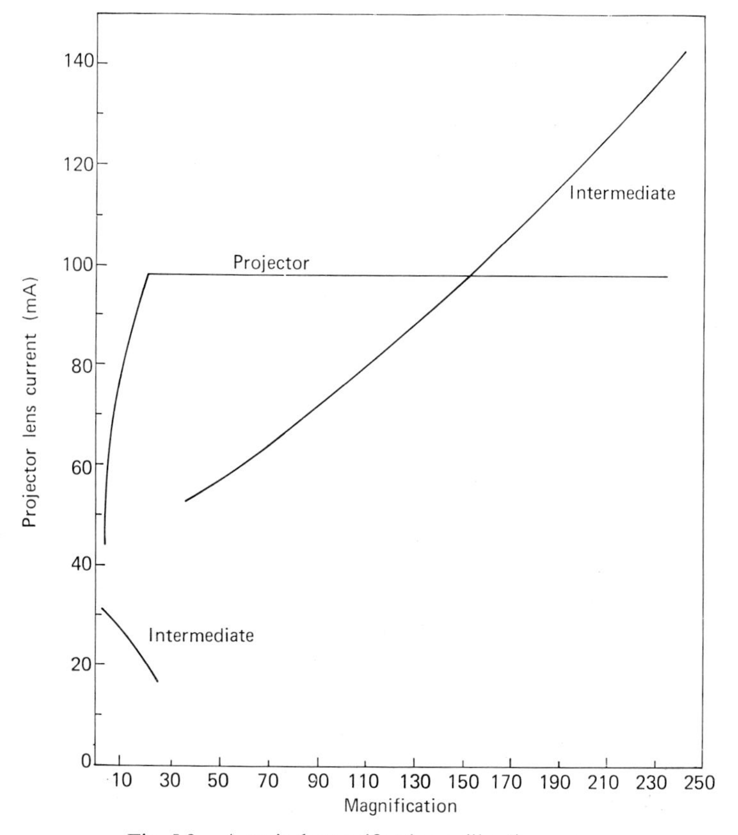

Fig. 5.8. A typical magnification calibration curve.

against measured magnification should be drawn. A typical calibration curve is shown in Fig. 5.8. It will be noted that most of the magnification variation at low magnifications is, in this case, obtained by varying the strength of the final lens. The current variation in the intermediate lens in this part of the range is no longer a sensitive indicator of the magnification.

A calibration is strictly only accurate at the kilovoltage at which it was carried out. On instruments operating at different kilovoltages, the lens currents are compensated for the different voltage values to retain an accuracy of about 5%.

Although good enough for most purposes, it is again inadequate for the most accurate work. A useful check on the instrumental calibration with change of kilovoltage is to observe the focus of the diffraction pattern, because this is a very critical setting. Once adjusted to focus at either maximum or minimum voltage, it should remain in focus over the accelerating potential range. The reference voltage can normally be adjusted to achieve this condition and one can then feel confident that the magnification is within a few per cent over the kilovoltage range.

The magnification is calibrated by having a specimen with known dimen-

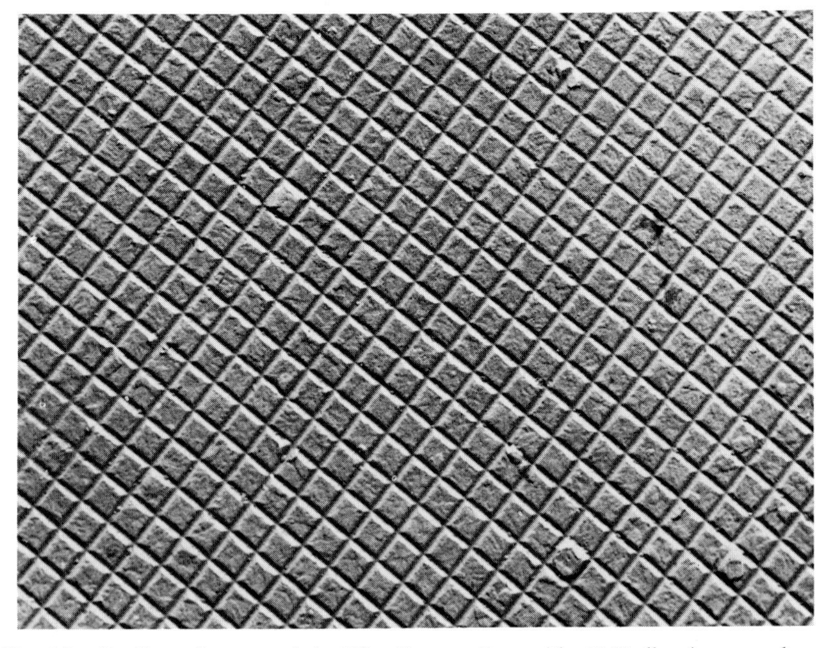

Fig. 5.9. Replica of cross-ruled diffraction grating with 2160 lines/mm used as a calibration specimen.

sions. The earliest effective calibration was in terms of a preparation of polystyrene latex spheres of very uniform size. This type of calibration can be very convenient, because these spheres may be added to any other specimen and hence provide an internal standard of magnification. However, it has been found that these spheres can shrink under certain preparative conditions, or in the microscope itself, and they should not normally be relied on to give an absolute calibration of high accuracy (not much better than 10%), although they can of course be used to check the consistency of the magnification calibration over the range of operation of the instrument.

A more favoured method now is to prepare a replica from a diffraction grating of known ruling spacing, and a very useful one is a fine grating of 2160 lines/mm prepared by Bausch and Lomb (Fig. 5.9). The replicas of this grating are made by the operator or obtained by purchase (see Appendix). These gratings cannot give a direct magnification calibration at magnifications above about 50,000 ×. For these higher magnifications a specimen with

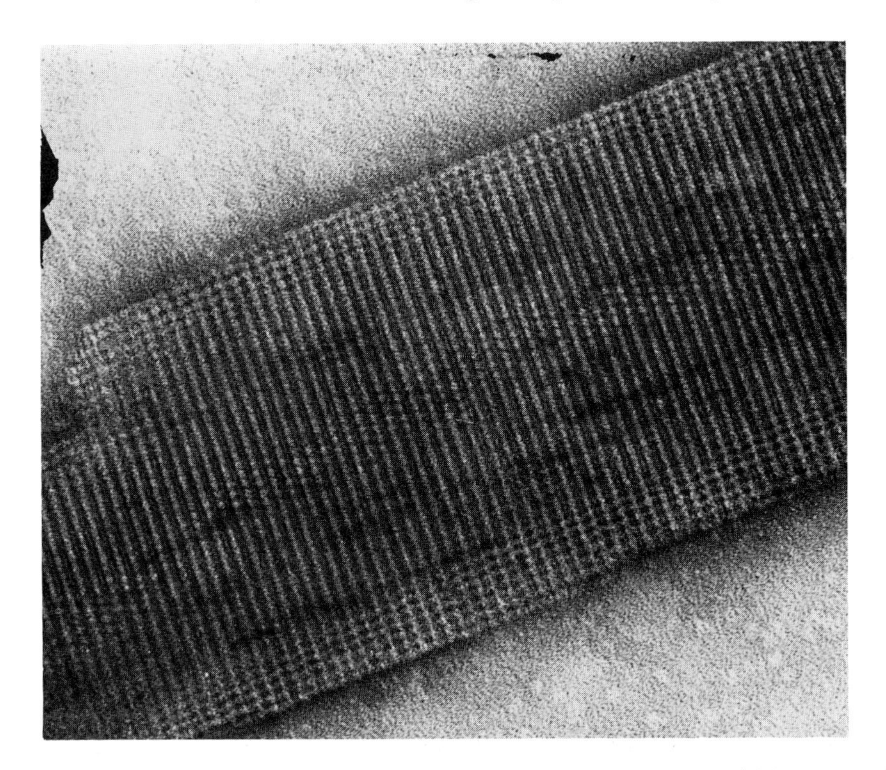

Fig. 5.10. Catalase crystal negatively stained with ammonium molybdate. Magnification 200,000 ×.

a known crystal lattice spacing, easily prepared, is preferable, e.g. beef liver catalase crystals (Wrigley 1968). These crystals have cross-lattice spacings of 8.75 nm and 6.85 nm and are prepared by floating a grid face downwards on the catalase, for a few seconds, and then transferring to ammonium molybdate to negatively-stain the specimen (Fig. 5.10).

5.9.2 GEOMETRICAL DISTORTION

A magnification calibration assumes a uniform magnification over the photographic plate, and as discussed in §1.10.3, there is always some image distortion (barrel or pincushion, plus some spiral distortion). Usually these

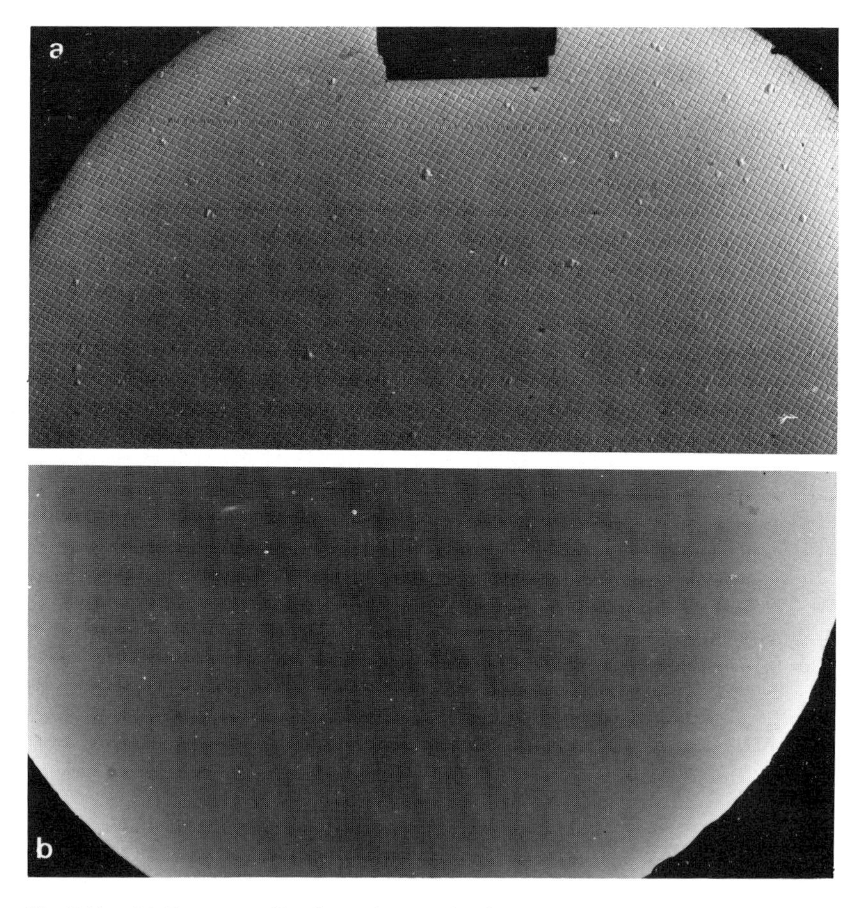

Fig. 5.11. (a) Low magnification micrograph of a cross-grating replica. Some slight barrel distortion can be detected. (b) Same field without a specimen in position.

defects are small enough to pass unnoticed, but they must not be forgotten if measurements are to be made near the edge of a plate, or if a series of plates are to be matched together to form a montage. In such a case, small differences of magnification across a plate can make an accurate match of the detail impossible. A magnification must be chosen at which the whole plate (or very near it) can be utilised. A match can never be made with the edge of one plate and the centre of another.

A further effect of barrel or pincushion distortion is that, since the magnification changes across the plate, and since the specimen is normally uniformly illuminated with electrons, there is an increased intensity in the regions of lower magnification. The effect is often much more visible as an intensity change than as a geometrical distortion since the intensity is proportional to the inverse *square* of the magnification. Fig. 5.11a shows a low magnification micrograph of a cross-grating and Fig. 5.11b is at the same magnification but with no specimen. The variation in intensity can be detected.

REFERENCES

Alderson, R. H. (1974), The design of the electron microscope laboratory, in: Practical methods in electron microscopy, A. M. Glauert ed. (North-Holland, Amsterdam). To be published.

Beeston, B. E. P., R. W. Horne and R. Markham (1972), Electron diffraction and optical diffraction techniques, in: Practical methods in electron microscopy. Vol. 1, A. M. Glauert, ed. (North-Holland, Amsterdam).

Dowell, W. C T., J. L. Farrant and R. C. Williams (1966), The attainment of high resolution. Proc. 6th Int. Congr. Electron Microscopy, Kyoto p. 635.

Heidenreich, R. D., W. M. Hess and L. L. Ban (1968), A test object and criteria for high resolution electron microscopy. J. appl. Crystallography *1*, 1.

Haine, M. E. and T. Mulvey (1954), The regular attainment of very high resolving power in the electron microscope. Proc. 3rd Int. Congr. Electron Microscopy, London, p. 698.

Menter, J. W. (1956), The resolution of crystal lattices, Proc. 1st Eur. Conf. Electron Microscopy, Stockholm, p. 88.

Wrigley, M. G. (1968), The lattice spacing of crystalline catalase as an internal standard of length in electron microscopy. J. Ultrastruct. Res. *24*, 454.

Chapter 6

Operation of the electron microscope

This Chapter describes the choice of the operating parameters for the electron microscope to achieve the best working conditions for different circumstances. Before the instrument is operated, however, it must be realised that it is potentially dangerous, in that it can generate penetrating beams of harmful X-rays. All commercial instruments have to be designed to be safe when operated normally. However, it is possible for the operator to override these safe conditions by withdrawing the aperture in the second condenser lens. This allows a high electron beam current to strike parts of the column and to generate excessive X-rays. *The condenser aperture must not be withdrawn from the beam during operation*, except with the precautions outlined in § 6.6.

6.1 Choice of accelerating voltage

In many microscopes there is a choice of accelerating voltage, usually from 30 or 50 kV up to 100 kV.

The following are the factors governing the choice of voltage.

1. The higher the accelerating voltage, the better the penetration of a specimen of a given thickness.

2. The higher the accelerating voltage, the lower the amplitude contrast from a given specimen.

3. The brightness of the electron gun (β – A/cm^2/steradian) increases with increased accelerating voltage.

4. The efficiency of the phosphor on the viewing screen increases with voltage up to 80–100 kV.

5. The efficiency of most photographic plates is a maximum at about 80 kV.

6. Inelastic scattering decreases with increasing incident beam energies, and hence the energy loss in the specimen decreases. This has two effects: (a) the loss in resolution due to the specimen itself causing chromatic aberration decreases with increased voltage; (b) the radiation damage in the specimen decreases with increased accelerating voltage.

7. The higher the accelerating voltage, the more sensitive the electron gun will be to any particles or unwanted vapour in the chamber, and the less the high voltage stability will be. Higher voltage operation therefore calls for more care in the maintenance of the gun of the microscope.

In general, therefore, it is desirable to work at the higher end of the voltage range of a conventional microscope and to use the objective aperture or staining to secure increased contrast for specimens of low intrinsic contrast. When examining metal foils or extraction replicas, 100 kV operation is more or less essential, because selected area diffraction is used continuously to provide supplementary information, and most precipitate particles require maximum voltage to permit sufficient electrons to penetrate them to yield adequate diffraction information.

6.2 Choice of condenser aperture

As shown in §1.7.1 the condenser aperture only defines the *maximum* illumination aperture which can be used; it is always possible to use a smaller effective aperture by defocusing condenser 2. Factors affecting actual choice of the aperture are:

1. A large aperture may be chosen if the specimen to be examined is very thick (e.g. a metal foil). This will permit a very large electron flux to be used on the specimen to gain a visible image on the screen at the desired magnification. Note that a 'large' condenser aperture for electron microscopy is only about one-tenth of the optimum aperture for the objective lens. This is quite different from light microscopy where the illumination and objective apertures should be matched. The difference arises from the different mode of image formation in the electron microscope; the scattering of the electrons by the object will ensure that the objective aperture is filled.

2. There is a good deal of evidence that the best micrographs require extremely small illuminating apertures. These small aperture angles are obtained in various ways; by use of a pointed filament; by operation of condenser 1 at maximum strength for minimum spot size; and by using very small condenser apertures. It is undisputable that a very small illuminating aperture is necessary to obtain the maximum phase contrast from a specimen.

Such illumination is required for imaging Fresnel fringes, crystal lattice planes and regular structures as discussed in § 3.8.

3. A very small condenser aperture (physical size say 50–100 μm) may be required when it is desired to restrict the maximum illumination that may fall on a specimen which is sensitive to irradiation by the electron beam.

A good choice for general purposes is the smallest aperture size consistent with comfortable focusing at the highest magnification which will be used in the investigation. Since the aperture carrier usually accommodates three or four aperture discs, it may be loaded with a range of aperture sizes so that an appropriate aperture can be chosen to meet different specimen requirements. A typical aperture diameter for routine work is 200 μm.

6.3 Choice of objective aperture

This is probably one of the least constructively used parameters in the electron microscope. The main factors influencing the choice of aperture size are well understood.

1. A small objective aperture will increase the contrast of a given specimen by increasing the amplitude contrast.

2. A small objective aperture contaminates more quickly than a larger one, and therefore many operators choose one of convenient size for fairly long life in the beam (say 50 μm).

This size (50 μm) is often near the 'optimum aperture' for the microscope. The 'optimum aperture' is defined in § 1.3.1. as being the aperture size for which the limitation of the performance of the microscope due to diffraction is balanced by the spherical aberration limitation. However, this aperture is only optimum for the very thin specimens necessary for maximum resolution.

The limiting factor for resolution with most specimens is the specimen itself (§ 3.2), because of the large chromatic energy spread due to energy losses in the specimen. Since the chromatic loss of resolution is also proportional to the objective aperture ($d_c = c_c\alpha(\Delta V/V)$), it follows that it is often advantageous to reduce the objective aperture angle well below the so-called optimum aperture size. Thus, for many specimens, an objective aperture of 25 μm or less is the best that can be chosen. Apertures as small as 10 μm diameter can be used but they require rather frequent cleaning and the effective cleaning of platinum may be difficult. It is advisable always to use an anti-contaminator since this prevents contamination of the objective aperture, as well as of the specimen.

More recently, very thin apertures have come into use in a number of laboratories. A very thin plastic film perforated with a hole of suitable size is coated with a layer of silver by vacuum evaporation. This composite film is mounted on a thick annulus of copper which is inserted in place of the normal aperture disc. An alternative design uses a thin film of gold. Since the disc material is so thin, the scattered electrons striking it cause it to heat up and it achieves a temperature high enough to prevent the formation of a contaminant layer. Such apertures can therefore be used for long periods without replacement, provided they do not suffer mechanical damage. The thin metal layer must of course be thick enough to prevent the scattered electrons from penetrating it. Suitable apertures can be purchased commercially (see Appendix).

When a very thin specimen is being examined for very fine structure, a different consideration arises in using an objective aperture. Most of the contrast arises from phase effects, and the objective aperture, which enhances contrast due to elastically scattered electrons, is relatively ineffective. Therefore, if a physical aperture is inserted in the objective it must be large enough to pass the information about the fine structure in the specimen. If the fine structure has a spacing d, the required angular aperture θ is given by

$$\theta = \lambda/d,$$

where λ is the electron wavelength. (This is the Bragg condition; see Beeston et al. 1972.) Therefore, for an objective focal length f_0, the aperture diameter D must be

$$D \geqslant 2f_0 \cdot \frac{\lambda}{d}$$

As an example, to visualise the carbon lattice of 0.34 nm, in a microscope of $f_0 = 2.5$ mm, working at 100 kV ($\lambda = 0.004$ nm), the objective aperture diameter must be greater than 60 μm. For such thin, phase objects, it will be found that the objective aperture makes very little difference to the contrast observed and it can well be withdrawn completely. In practice, however, this could cause fracture of the specimen film (§ 6.9), and the use of a large objective aperture is preferable.

6.4 Mounting the specimen

This is a mundane task but nevertheless important. Firstly, care should be taken never to handle the specimen holder with the bare hands, since they secrete grease which will give rise to severe contamination in the microscope in the most sensitive region. If the specimen holders are never touched by hand, they remain clean for long periods. Some holders (e.g. AEI EM 801) are designed specifically for specimen loading and transfer into the microscope without handling. Fig. 6.1 shows a multiple specimen holder with the insertion and handling tool clipped on to it. For holders not so designed, a plastic clothes-peg has been found to be a useful handling tool, and nylon or disposable gloves should be worn.

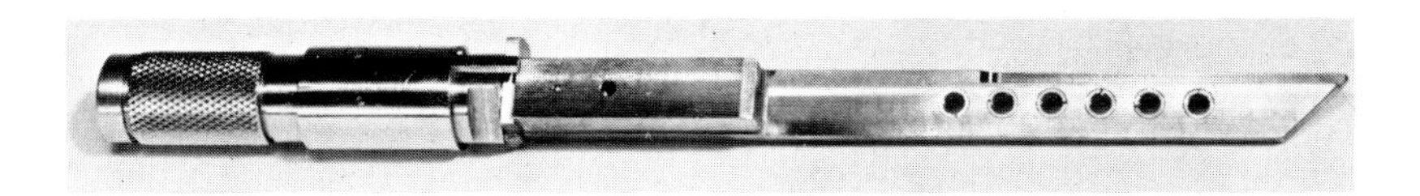

Fig. 6.1. Multiple specimen holder for AEI EM 801 with extractor tool, which enables holder to be extracted and loaded without handling.

Secondly, care must be taken to mount the specimen flat in the holder so that it is properly gripped all round the edge. This ensures good thermal contact; otherwise, it is very probable that thermal drift will ruin any pictures taken. Also, a specimen grid which is buckled may lose contact with the support film which will then heat or charge up and split. There will also be significant changes of focus over the width of the grid, not only giving rise to inconvenience but also causing changes in magnification.

6.5 Inserting a specimen and switching on the microscope beam

The procedure for starting up the instrument in the morning is best taken from the Instruction Manual. The operational procedures here assume an instrument ready to operate, with all the lenses and the H.T switched on and the condenser aperture in position.

1. The specimen, loaded in its cartridge, is introduced into the stage via the specimen airlock. Check *first* that the stage controls are near the centre of their traverse.

2. Operate the pump-out sequence for the specimen airlock.

3. Check that the high vacuum in the column has been restored, following the opening of the airlock.

4. The anti-contaminator reservoir should be checked to be full of liquid nitrogen, to ensure that the specimen will be protected against contamination.

5. Check that the second condenser is well over-focused (to prevent accidental damage to the specimen when the beam is switched on), and that the instrument is set at its lowest magnification (to assist in finding a suitable area). Switch on the filament heater, with the heater control at zero (the H.T will have been left on; this is necessary to establish a steady running temperature in the H.T tank, and hence to achieve improved stability).

6. Check the alignment of the electron gun (§ 4.1), then slowly turn up the filament heater until the saturation point is reached (as described in § 1.6). Early in the day, it may be noticed that the beam current meter is 'kicking'. This will be due to small high voltage discharges in the electron gun, and it will normally achieve stability fairly soon. However, it may be due to a foreign particle in the electron gun and the problem may sometimes be cleared by increasing the kilovoltage on the gun for a few moments. A heavier discharge may then occur, which clears the particle giving rise to the instability. If this treatment fails to work, and the vacuum has had time to achieve its normal level without the gun becoming stable, the gun will have to be opened and cleaned.

Once the emission is steady, then examination of the specimen can begin.

6.6 Finding the beam

It occasionally happens that when the microscope is switched on, no illumination can be seen on the final screen. Before moving a lot of controls which can make it more difficult later to locate the beam, the following points should be checked, since the problem is usually due to something very simple.

(a) Is the voltage supply on and the filament intact as indicated by the beam current reading on the meter?

(b) Is the magnification low? – a very high magnification can result in so low a screen intensity as not to be noticed.

(c) Is the second condenser lens too far overfocused causing very low intensity of illumination? – swing it through its fine control range.

(d) Are the apertures obstructing the beam? Withdraw the objective and intermediate apertures.

(e) Has the specimen been properly inserted into the stage, or is the airlock mechanism obstructing the beam?

(f) Is a grid bar obstructing the beam? At a low enough magnification, a very small stage motion should quickly reveal some part of a hole in the grid.

(g) Is the stage at the end of one set of traverse controls?

(h) Are the beam traverse electrical controls (below the condenser lenses) near the centre of their indicated range? If not, note the readings and then try properly centring them according to the dial reading.

If all the above measures fail to locate the beam, switch off the imaging lenses to see if the illumination is visible at a very low magnification. Then remove the object. Then reduce the emission current by turning down the filament control, remove the condenser aperture, and switch off the condenser lenses. There should always be a visible electron beam through the column, with all lenses off and all apertures removed. It should be noted however that the condenser aperture *should only be withdrawn when other measures have failed.* This aperture acts to remove a good deal of the excess electron beam which would not be used for imaging, and if it is withdrawn a heavy beam current may strike some lower part of the microscope and give rise to X-radiation above the safe levels for normal operation. While a very brief exposure may be tolerable it is best to avoid operating in this way longer than is necessary to locate the electron beam. The gun should not be operated at maximum efficiency (i.e. high beam current and full filament heating) while the aperture is removed.

If all these measures fail, it is just possible that some foreign body has fallen down the column and is obstructing the beam, but this is a very infrequent occurrence and opening the column should be a last resort. It should be noted through all the above procedure (except for (h)) that the alignment controls are not adjusted. It is highly unlikely that these have moved spontaneously and any movements made when the beam is not visible will only make it much more difficult to find later.

6.7 Preliminary examination of the specimen

It is always desirable to minimise the time the specimen is in the electron beam. It is therefore worth minimising the search time. A very low scanning magnification of about 100–200 × is valuable in showing up most of the specimen at once (the objective aperture must be withdrawn, as it limits the field of view). This enables regions of interest in the specimen to be located quickly. The region of interest is moved to the centre of the screen and the instrument switched to one of the lower imaging magnifications.

The beam intensity should be kept as low as is consistent with comfortable viewing conditions, and the magnification should be kept as low as is consistent with seeing the structure of interest.

6.8 Choice of magnification

The very wide range of magnification available in the electron microscope allows a very flexible approach in its use. A number of criteria determine the choice of magnification.

(a) Mapping an extensive area of a specimen to form a montage. This will often be within the light microscope range of magnification, but at much improved resolution. The lowest usable magnification is best, so as to minimise the number of joins in the montage. However, in some instruments, the distortion of the picture increases markedly at the lower magnifications and the maximum usable area may not coincide with the minimum magnification. Likewise, some low magnifications are obtained by limiting the field of view, and it is not useful to use them for this purpose.

(b) Recording a specific region of the specimen – the magnification will normally be the maximum which will include the whole of the region of interest within the photographic format.

(c) Recording a number of fields of view for statistical study. Here the most productive approach is to use the minimum magnification consistent with easy measurement of the smallest feature of interest.

(d) High resolution studies. A very high electron optical magnification is required to avoid confusion from electron noise (discussed in Chapter 7). As a rough guide, a magnification of 300,000 × is desirable for micrographs required to show 0.5 nm resolution.

A general rule on selecting magnification should be that it should be as low as is consistent with the required resolution. Every doubling of magnification results in a four-fold increase in required illumination intensity at the specimen, with the resultant increase in radiation damage (§ 6.13). It also requires four times as many photographs to record a given area of the specimen.

6.9 Focusing

The most important function the operator must exercise, apart from the selection of regions of the specimen to record, is the focusing of the image prior to photography. Focusing the image requires different techniques at low magnifications and at high magnifications.

For magnifications of 10,000 × and below, the simplest method is to use the focusing aid (wobbler) (Le Poole and Stam 1954) now fitted to many microscopes (Fig. 6.2). This produces a cyclical electrical deflection of the

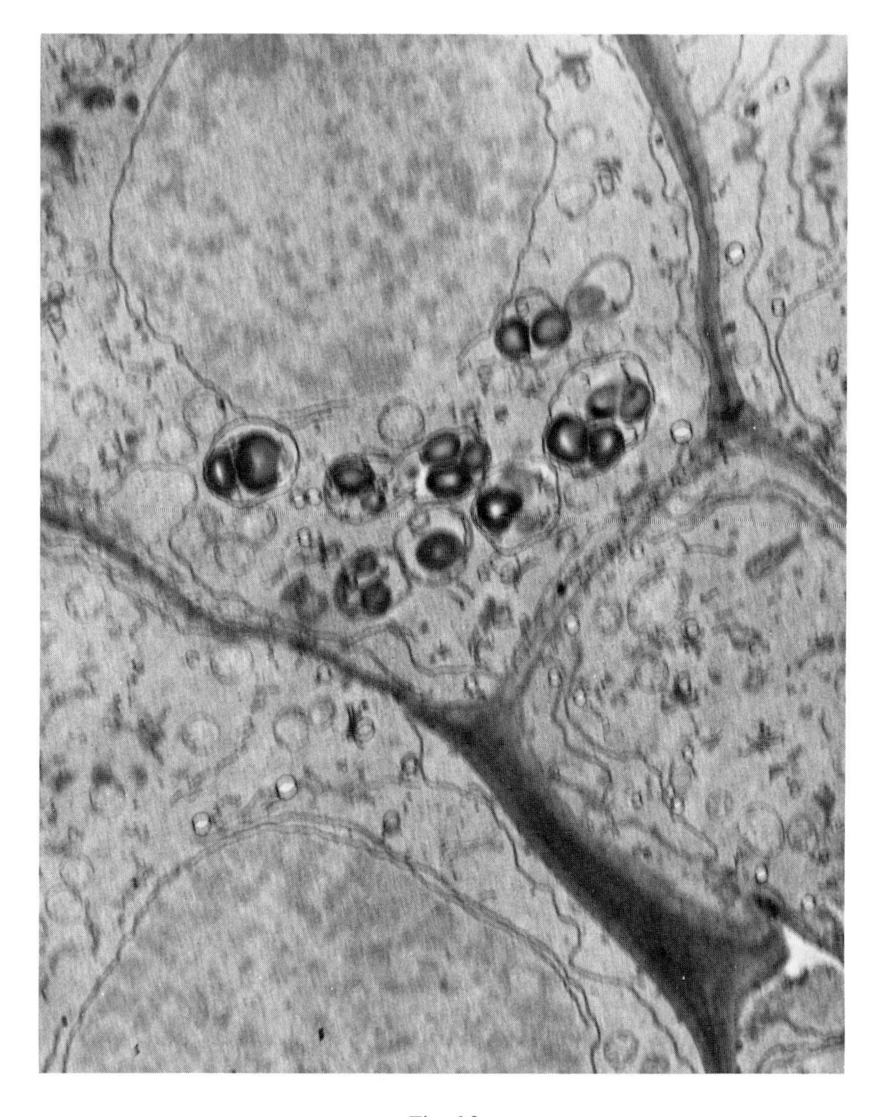

Fig. 6.2a

Fig. 6.2. Micrographs of a section of maize root tip. Magnification 7500 ×. (a) With focus wobbler in operation, objective lens defocused. (b) with wobbler on but objective lens correctly focused.

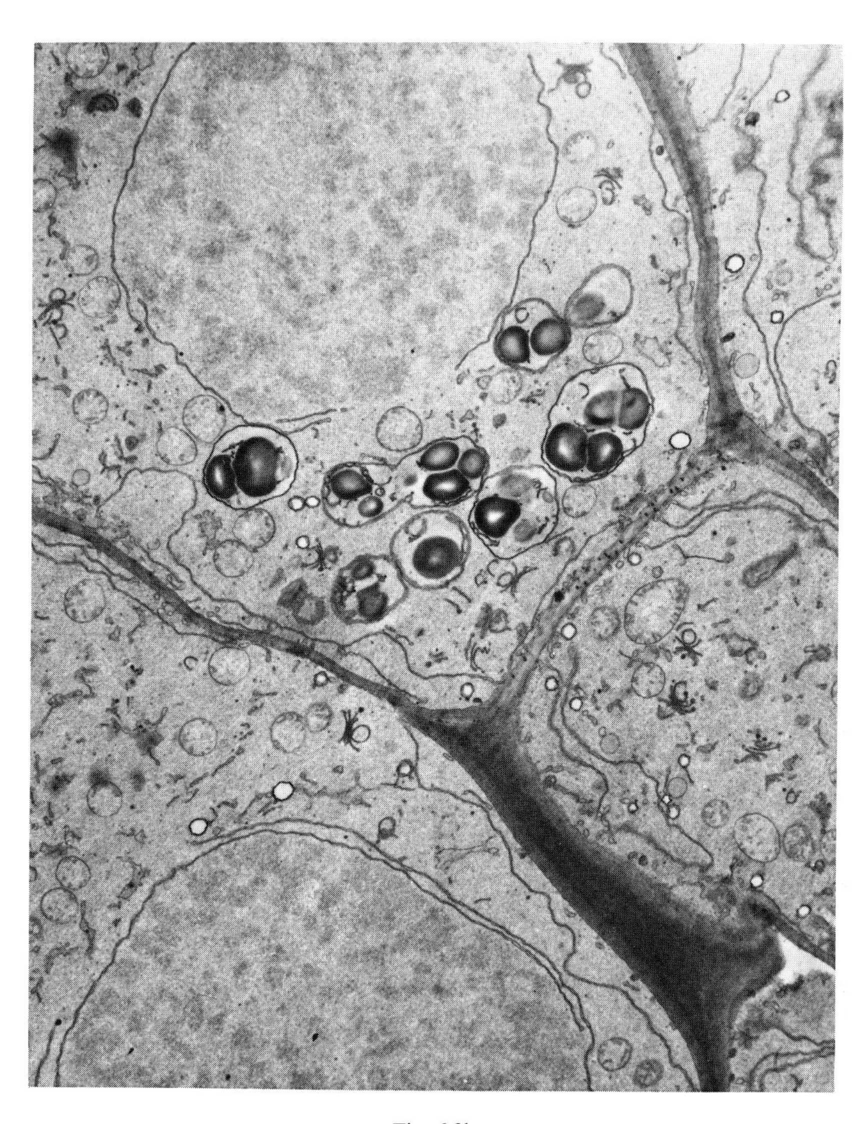

Fig. 6.2b

incident beam first to one side of the axis and then to the other. When the objective lens is focused on a plane above or below the specimen, there is an apparent image movement giving rise to a directional blur. When the objective lens is focused on the object plane, the image remains sharp. This type of focusing aid is simple to operate and results in a sufficiently accurate focus for all normal purposes. It is not normally sensitive enough to be used for higher magnifications (say above 15,000 ×).

For instruments not equipped with a focus aid, the method of minimum contrast focusing can be employed. The objective aperture is withdrawn, and the objective lens current varied up and down through focus moderately quickly. A sharp dip in contrast is seen at focus and a thin specimen may lose almost all contrast. The aperture is re-inserted and centred before photography. This method is again only valid at low magnifications (say below 10,000 ×). Minimum contrast is most easily seen with direct viewing of the whole screen; use of the viewing telescope makes the estimation of the minimum contrast position impossible. If a minimum contrast position is not seen the specimen plus the support film is too thick.

It is sometimes observed that when the objective aperture is withdrawn from the beam, the specimen film suddenly breaks. This is due to a sudden change in the electrostatic forces on the film. When the objective aperture is in the beam, there are many back-scattered electrons from the aperture reaching the specimen film, setting up a quite strong field. When the aperture is removed, this field is suddenly changed, and a very thin film, or one which does not adhere well to the specimen grid may fracture. Consequently, the beam should be de-focused before removing the aperture.

At high magnifications, the viewing telescope must be used to observe both sharp features in the object, and background structure in the support film. The appearance of the background structure depends mainly on phase contrast effects and it has least contrast at focus (§ 3.2). The further out of focus, the coarser the fine structure appears. One must therefore look for the minimum structure and contrast in the background. If the edge of a hole appears in the field of view it is relatively easy to see the Fresnel fringe at the edge and adjust the focus to secure the disappearance of the fringe.

From many points of view, the most difficult magnifications lie between 15,000 × and 30,000 ×, when the low magnification criteria cease to be accurate enough and the phase-contrast effects cannot be seen clearly. One has to use the criterion of sharpness of some well defined piece of structure. However, it is almost always true that the point of maximum contrast in the image does not coincide with exact focus, and for most purposes a degree of underfocus will be found advantageous. The criteria by which the amount of underfocus should be decided were discussed in § 3.8.

6.10 Taking the photograph

When an interesting area of the specimen has been located, the following steps are taken to record the photograph.

1. Check that the whole of the region of interest will fall on the photographic plate adjusting the magnification appropriately. (The area corresponding to the plate format is usually marked on the viewing screen.)
2. Focus the image, as already described. Check that the illumination is uniform.
3. Check that the image is stationary by viewing it through the viewing telescope. Sometimes there may be temporary stage drift after the specimen has been moved, or the grid square being examined may have a broken support film which could cause specimen movement. Wait until the image is stationary.
4. Adjust the exposure time, as indicated by the exposure meter. (On some instruments the time is adjusted automatically.) Note that for the highest quality work the condenser lens should not be adjusted after focusing (see § 7.2.3).
5. Check that all room lights except safelights are extinguished.
6. Advance the plate from the storage cassette under the viewing screen. Raise the screen and operate the shutter. Keep all other parts of the body away from contact with the microscope column.
7. Lower the screen. Return the plate into the cassette box of exposed plates. Record details of the kilovoltage, magnification and relevant specimen details in the record book.

When the plates have all been exposed, the camera chamber is brought up to atmospheric pressure (preferably with dry nitrogen). The new plates must be transferred from the vacuum desiccator as quickly as possible and the vacuum in the camera immediately re-established. This minimises the moisture introduced into the microscope vacuum system.

6.11 Tilting the specimen

6.11.1 TILTING FOR STEREOSCOPY

Owing to the large depth of field of an electron microscope (due to the very small objective aperture), it can successfully portray sharp images in depth.

Where these overlay one another, the image is confused. However the fact of the very great depth of focus makes stereoscopic viewing very suitable as an aid to interpretation. Since the illumination source and the camera are fixed, the specimen has to be tilted between successive exposures in order to obtain a stereo picture of a given field of view. The operational procedure is as follows:

1. Select a suitable field of view, focus, and record a micrograph.
2. Tilt the specimen through an appropriate angle; for choice of angle, see below.
3. Re-focus, and record another micrograph (on a separate plate).

When the two plates are correctly oriented, a simple binocular stereo viewer will yield a striking three-dimensional image.

The angle of tilt chosen must take account of the approximate vertical separation of the structures to be examined (take the probable thickness of the specimen here), and the final magnification at which the stereo prints will be viewed. Hudson and Makin (1970) plotted two sets of curves as a guide to the required tilt angle, reproduced here as Fig. 6.3 and Fig. 6.4. The eye can form a stereoscopic impression from the images received by each eye, provided the parallax Δy is less than about 5 mm when the prints are viewed at a distance of 25 cm.

$$\Delta y = 2\Delta h M \sin \tfrac{1}{2}\theta$$

where Δh is the vertical separation of the images and M the magnification of

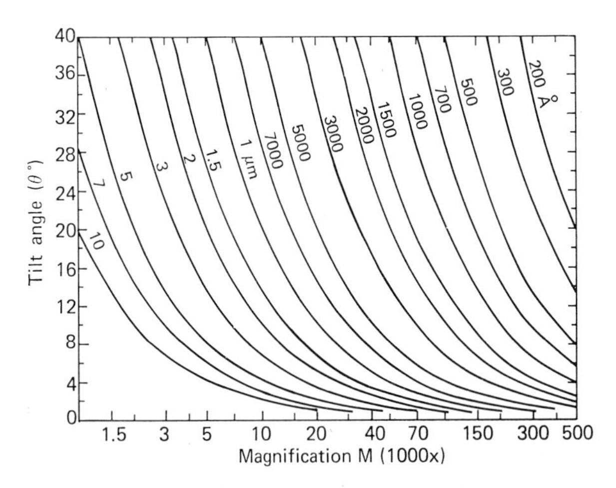

Fig. 6.3. Optimum tilt angle against magnification for various foil thicknesses from 200 Å to 10 μm (Hudson and Makin 1970).

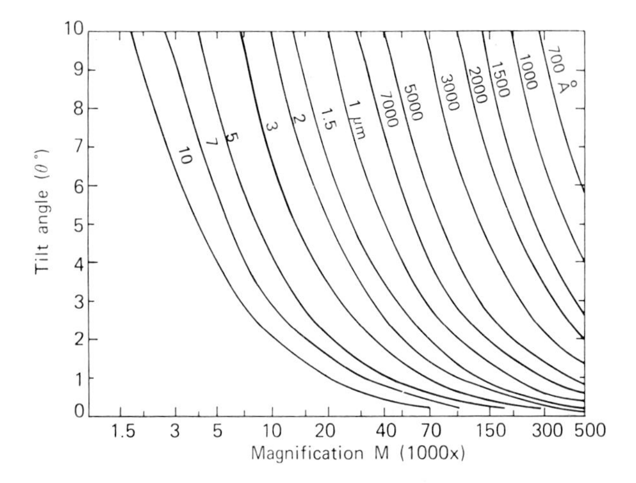

Fig. 6.4. Optimum tilt angle against magnification for various foil thicknesses from 700 Å to 10 μm (Hudson and Makin 1970).

the final print and θ is the stereo tilt angle. Since the specimen thickness may not be accurately known, Hudson and Makin recommend an electron optical magnification of $M/5$ and a print enlargement of $\times$ 2 to secure some stereo effect. This enables a reasonable estimate of Δh to be made, and an optimum enlargement can then be chosen to maximise the stereoscopic effect.

It will be seen from Figs. 6.3 and 6.4 that for specimens of moderate thickness, the tilt angle will normally be less than 10°. Only for very thin objects (presumably for high resolution) does a high magnification and a high tilt angle become necessary.

In spite of careful and detailed early work (Helmcke 1954, 1955) which showed that the technique could be used quantitatively, only limited use has been made of stereo techniques until recently. The advent of high voltage microscopes (§ 9.2), which enable much thicker specimens to be examined, has rendered stereo viewing an important and necessary technique for sorting out three-dimensional structures.

6.11.2 TILTING FOR CRYSTALLOGRAPHY

When examining a crystalline specimen, it is essential to have comprehensive tilt facilities available. The detailed procedures are given by Beeston et al. (1972). It must be possible to orient a known crystal plane precisely with respect to the electron beam, and this will usually be done by observation of the electron diffraction pattern. For this type of work large tilt angles may be required (up to 45°), and a eucentric stage (Fig. 2.17) is a great advantage since then the crystal grain under observation will not be lost during the tilting and observation of the electron diffraction pattern, if the stage is carefully set up.

6.11.3 TILTING FOR INTERPRETATION

This is a function relatively recently developed to aid in the interpretation of biological structures. There are frequently complex membrane structures running at an angle to the plane of a section, and this can lead to difficulties in interpretation. By tilting the specimen into different orientations, one can often obtain an end-on view of membranes or tubes, and elucidate their connections with other structures. It is clearly desirable in this case also to be able to tilt the specimen through large angles, since the direction of the structure may be quite random within the section. If the tilt stage has a $\pm 60°$ movement, it is possible to view some structures from two directions at right angles.

6.12 Photographing a diffraction pattern

The procedure is as follows:

1. Select a region of the object for which diffraction information is required and place it in the centre of the field of view.
2. Insert an intermediate aperture of suitable size to define the region of interest.
3. Withdraw the objective aperture from the beam.
4. Focus the intermediate aperture sharply by adjustment of the intermediate lens (or diffraction lens, in a three projector lens system).
5. Focus the image with the objective lens. The image of the specimen is now in the same plane as the intermediate selector aperture.
6. Adjust the intermediate (or diffraction lens) to focus the back focal plane of the objective lens and to obtain the sharpest image of the diffraction pattern. The second condenser lens must be strongly overfocused to produce a very fine beam (coherent illumination).
7. Adjust the camera length to achieve a suitable size for the pattern obtained.
8. Insert the diffraction stop over the central spot of the diffraction pattern, to prevent undue halation.
9. Photograph the diffraction pattern. A long exposure will be required – generally of the order 30–45 sec.

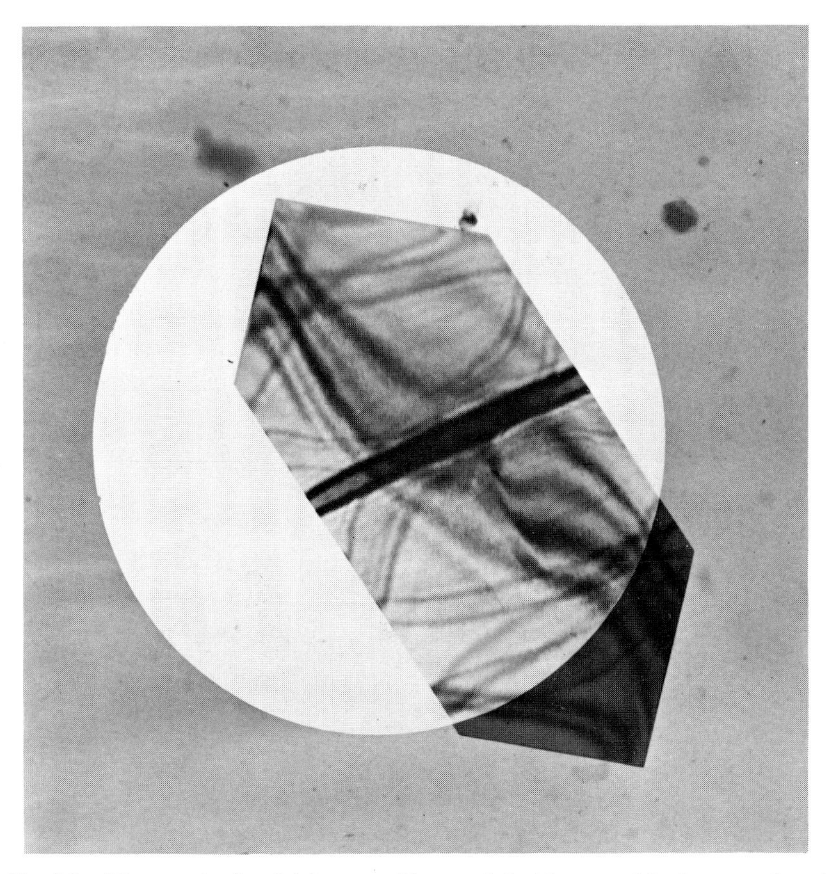

Fig. 6.5. Micrograph of molybdenum oxide crystal double-exposed in the area selected for selected area diffraction.

10. Return to the imaging condition. Photograph the image with the selected area aperture in position for half the normal exposure time. Remove the selected area aperture and complete exposure time. The selected region will then be defined on the plate as a double exposed region in the context of its surroundings (Agar 1958) (Fig. 6.5).

6.13 Radiation damage and ways to minimise it

It is now realised that most specimens are affected in some way by irradiation in the electron beam and it is important that users of electron microscopes realise what may be happening to their specimens. The effects and the possible mechanisms have been widely discussed (e.g. Cosslett 1969; Stenn and Bahr 1970; Thach and Thach 1970, 1971; Glaeser 1971).

When electrons in the beam knock off orbital electrons from atoms in the specimen, they cause ionization and the resulting forces may cause changes in the specimen by inducing bond linkages or chemical chain scission. They may also encounter atomic nuclei, and, if the beam energy is high enough can displace such nuclei, causing 'displacement damage'.

The effect of radiation damage can be seen very strikingly when crystalline polymers are observed, since they lose their crystallinity almost immediately in an electron beam of the intensity commonly used in the electron microscope (Agar et al. 1959). Changes are observable in the plastic materials used for embedding biological material, when they are irradiated. In metal specimens dislocations in the crystal lattice can be seen to form or move under electron bombardment. Almost all specimens are therefore potentially subject to radiation damage, and it is prudent to minimise exposure to the electron beam as far as possible.

There are obvious ways of reducing the electron beam current density onto the specimen, such as reducing the emission current, and fitting a small condenser aperture. There is a limit to how far this can be pursued, because some specimens (e.g. viruses) require very high magnifications to resolve their structure, and therefore require a fairly high beam intensity. Metal foil specimens will appear rather opaque with a low beam current because of the large number of elastically scattered electrons.

6.13.1 MINIMAL EXPOSURE OPERATION

The first technique for reducing electron exposure of sensitive specimens involved focusing on one region of the specimen but photographing another (Agar et al. 1959). By this means, the exposure to the electron beam only starts when the photograph is being taken. The method used involved focusing the image on one grid square and then moving to an adjacent one for the photograph. It has the disadvantage that it may only be used at low magnification, since the stage movement cannot be expected to remain planar to the required accuracy (say 1 μm) at higher magnifications.

The method was improved by Williams and Fisher (1970), who moved the effective field of view by misalignment of the projector lens of a Siemens Elmiskop I, and who therefore had no problem with stage stability. They now use an even simpler system – a very small illuminating spot is focused on one side of the viewing screen, where focusing is carried out. Then, with the camera plate already advanced, the beam is spread out to cover the whole area of the screen and the photograph immediately recorded. An example of this operating technique is shown in Fig. 6.6. The first micrograph (a) was taken by the minimum radiation method just described, and the banded structure of the tobacco mosaic virus rods can clearly be seen. The same field of view was irradiated for 30 sec at the normal intensity for focusing and photography at a magnification of 40,000 ×, and the second micrograph (Fig. 6.6b) was recorded. It can be seen that the fine banded structure is now hardly discernible. It will also be noted that there has been a dimensional change in the virus rod diameter.

6.13.2 MAGNITUDE OF RADIATION EFFECTS

Lest it be thought surprising that damage can occur so quickly with an electron beam of only a few micro-amps – and maybe a defocused one at that, it might be worth quoting from the paper of Grubb and Keller presented at the EM Conference in Manchester (1972). They showed that a rough conversion from electron beam flux to radiation dose could be given by $1\ C/m^2 = 40\ Mrad = 4 \times 10^7$ rad.

As shown in §1.10.1 a comfortable viewing intensity on the viewing screen is $3 \times 10^{-11}\ A/cm^2$. If the magnification is only 10,000 ×, this corresponds to a condition where the condenser lens is well defocused. For a magnificaton of 10,000 ×, the flux at the specimen is then $3 \times 10^{-3}\ A/cm^2$, or $30\ C/m^2/sec$ (= 1200 Mrad/sec).

Keller quotes the radiation dose for destruction of the crystallinity of polyethylene single crystals as 4000 Mrad at room temperature. This dose is delivered in just over 3 sec under the viewing conditions postulated above, and is hence in good agreement with practical experience.

Keller then quoted: 4000 Mrad dose of radiation can be obtained by irradiating for:

0.01 sec in an electron microscope at high beam current;
20 min in an electron microscope at very low beam current;
or 5 weeks inside a nuclear pile;
or 5 years near a 1 Curie Cobalt 60 X-ray source;
or by exploding a 10 MT H-bomb about 30 yards away.

No further comment is necessary!

In view of these depressing numbers, we have to consider what else might be done to reduce the irradiation of the specimen.

6.13.3 USE OF AN IMAGE INTENSIFIER

One possibility would be to use an exceedingly low current density and to

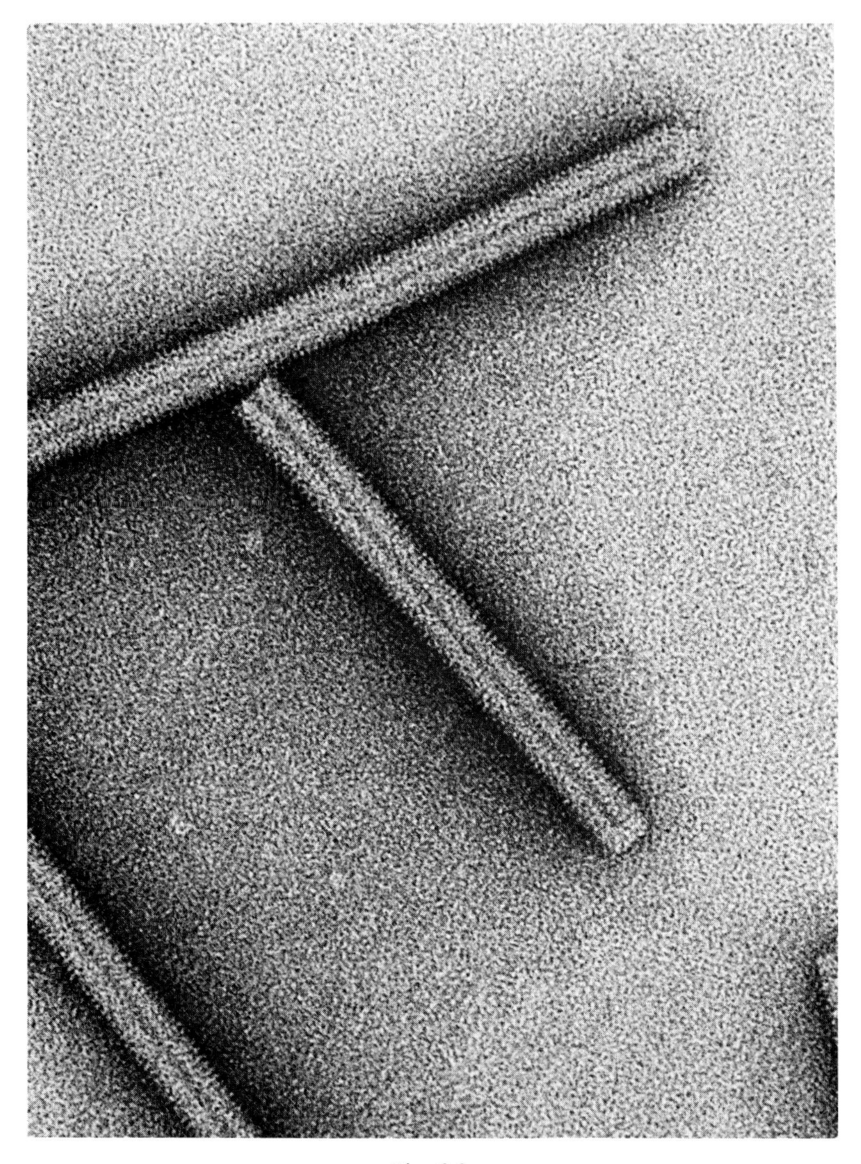

Fig. 6.6a

Fig. 6.6. (a) Tobacco mosaic virus rods, negatively-stained with PTA. Photographed with minimal beam exposure. (b) Same field of view as (a) but after 30 sec of exposure to beam intensity needed for focusing. Electron optical magnification 40,000 ×, print magnification 440,000 ×. (Courtesy Professor R. C. Williams.)

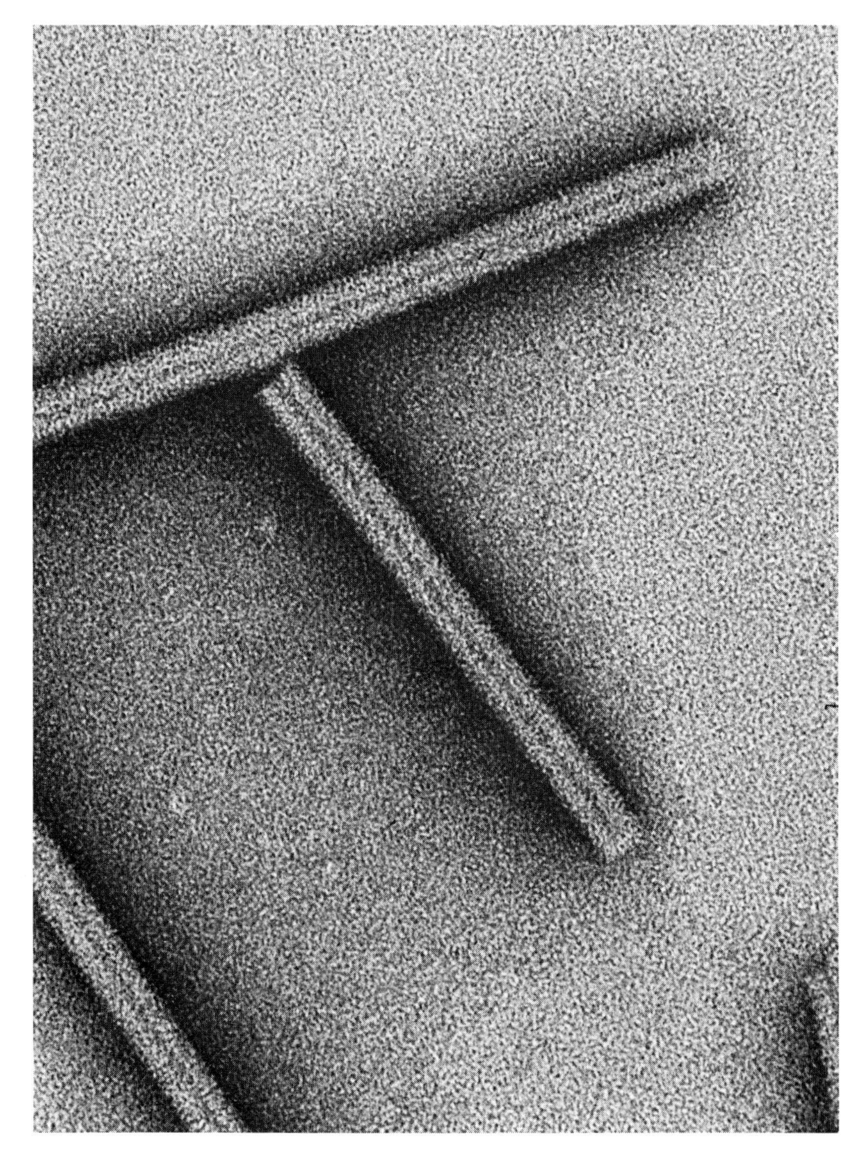

Fig. 6.6b

view and photograph the image by the aid of an image intensifier. However, Glaeser et al. (1970), using as a test the disappearance of the diffraction pattern of crystalline l-valine, showed that a log-log plot of lifetime versus beam current density was a straight line of slope -1: in other words, the operative parameter is the total dose, not the rate of irradiation. Since the number of electrons required to yield a picture of given signal/noise ratio with a given resolution is determined by statistics, no saving can be made by the use of an image intensifier.

6.13.4 LOW TEMPERATURE OPERATION

Another idea is that radiation damage might be minimised by cooling the specimen to a very low temperature, since one might hope that the larger molecular fragments produced by radiation damage would not have sufficient thermal energy to move from their original locations (Glaeser et al. 1971). Grubb and Groves (1971) found a significant improvement by working at 77 °K and Siegel (1970) reported an increase in lifetime of tetracine and paraffin by a factor of about five when examined at 10 °K as compared with 30 °C. As expected, the damage had indeed occurred at the low temperature, and when the specimen was restored to room temperature, it became visible because the molecular fragments were again able to move. A considerable range of organic materials has now been tested and a lifetime increase of between 2 and 5 times seems well established for operation at liquid helium temperature.

6.13.5 ULTRA HIGH VACUUM

It seems possible that a reduction in the number of gas molecules in the neighbourhood of the specimen might reduce the radiation damage. This was suggested by early experiments by Pashley and Presland (1961) who showed that defects appearing in a metal foil specimen ('black death') must have originated during examination in the atmosphere of the microscope column. Qualitative results from a study by Hartman and Hartman (1971) do indeed suggest striking increases in lifetime once the specimen chamber vacuum is better than 10^{-8} Torr. More evidence of these possibilities is clearly needed.

6.13.6 VERY HIGH VOLTAGE OPERATION

The decrease in inelastic scattering as the accelerating voltage is increased suggests that high voltage microscopes may be important in reducing radiation damage. Unfortunately, as shown in § 9.2, the improvement, at least up to 1000 kV is only by about a factor of three.

The onset of displacement damage due to the primary electron beam is important mainly in the context of high voltage electron microscopy (§ 9.2).

6.14 Reduction of contamination

Since the vacuum in the electron microscope is relatively poor, there are a considerable number of organic molecules in the system. These are continuously striking the solid surfaces within the microscope and then leaving again as enough energy is imparted to them. When an electron from the beam encounters such a resting molecule, however, it may decompose it to a form of carbon, so that it cannot move again. Hence a layer of carbon begins to build up on any surface exposed to the electron beam. It is most troublesome of course on the specimen itself, but is also found on apertures, the specimen holder, and parts of the electron gun.

The effect was first extensively studied by Ennos (1953, 1954) who showed that a substrate temperature of about 200 °C was sufficient to suppress the effect, presumably because the molecules were thereby given enough energy to leave the surface almost instantaneously when they arrived. The only way found to prevent the appearance of contamination on a cool surface was to protect it from the rest of the vacuum system by a cold baffle, at a lower temperature, which would trap the organic material. In a normal electron microscope system in which no particular precautions are taken, the rate at which contamination builds up on a specimen may be from 0.1–0.5 nm/sec, so that a relatively thick layer of carbon (say 10 nm) can form in as little as 20 sec.

Apart from improvements to the vacuum system of the electron microscope to reduce the pressure from all gases and vapours near the specimen, the main solution has been to place a surface cooled by liquid nitrogen around the specimen (an anti-contaminator), as first demonstrated successfully by Heide (1958, 1962, 1963, 1964, 1965).

The design by Heide aimed to surround the specimen as completely as possible by a cooled surface, so that a very minimum of solid angle was subtended by surfaces at room temperature. This concept was followed for side entry stages by sandwiching the specimen holder between two thin blades, each with a small hole to allow passage of the electron beam. This design proves inconvenient in practice because, if the holes are small (as they should be for maximum effectiveness), there is a danger of the blades interrupting the electron beam as they cool down, because the contraction of the blades can be greater than the radius of a small hole. Furthermore, the blades occupy valuable space in the objective lens gap, and may restrict the angle through which the specimen may be tilted. Fortunately, it has been found that an alternative system can prove quite effective. This consists of a large area of cooled metal near the specimen, which acts as a cryo-pump for any condensable vapours and reduces the partial pressure of organic vapour to a low enough level to minimise the contamination rate. Results from Forth and Loebe (1969) show that an even colder surface (20 °K) surrounding the object is more effective in reducing contamination since all the light hydrocarbons and permanent gases, except for He and H_2 are condensed. An alternative method (Bryner et al. 1969), which has not been so widely adopted, is to oxidise away the contamination with a gas flow. This system has been used in electron microprobe analysers, where mainly metal specimens are examined. However, in the transmission electron microscope, the presence of oxygen or water vapour can cause, not merely removal of the contamination, but etching away of the specimen itself (§ 8.5).

Although precise figures are often quoted for contamination rates, they are largely meaningless unless accompanied by many experimental details. Because the rate of contamination decreases dramatically with increased temperature of the specimen, it follows that a high beam current (which may however cause other damage to the specimen) will reduce the contamination rate.

The rate will also be decreased if single condenser illumination is used, or if the first condenser lens is weak, giving rise to a relatively large focused spot at the specimen. Although the actual illumination intensity at the centre of the beam is unchanged compared with that in a small spot the larger spot causes the surrounding area to be hotter and hence the heat transfer from the region examined is much slower. This results in a higher temperature at the centre of the illuminated region.

For similar reasons, the contamination rate varies if the specimen is a poor conductor, or if it is thick or thin, or if the region examined is adjacent to a grid bar (good conductor) or not. The parameters affecting contamination rate have been studied in some detail by Hart et al. (1970).

6.15 Switching off the instrument

1. Check that a clear image is showing on the fluorescent screen and that no apertures are obscuring the beam path.
2. Turn the magnification to its lowest setting.

3. Overfocus condenser 2 until there is a low image intensity on the final screen.

4. Reduce the filament heater control to its zero position, and switch off the filament.

5. If the instrument is being shut down, close the valve between diffusion pump and microscope column, and switch off the diffusion pump heater.

6. Switch off the power unit.

7. Allow about 20 min for the diffusion pump to cool before turning off the cooling water.

8. Switch off the rotary backing pump, turn off the water and disconnect the mains supply to the instrument.

The above precautions will ensure that when the instrument is switched on again, there should be a beam through the instrument uninterrupted by any obstacles, and that the initial electron flux onto the specimen will be low and so will not damage a sensitive specimen.

REFERENCES

Agar, A. W. (1958), Selected area microdiffraction in the electron microscope, Brit. J. appl. Phys. *9*, 419.

Agar, A. W., F. C. Frank and A. Keller (1959), Crystallinity effects in the electron microscopy of polyethylene, Phil. Mag. *4*, 32.

Beeston, B. E. P., R. W. Horne and R. Markham (1972), Electron diffraction and optical diffraction, techniques, in: Practical methods in electron microscopy, Vol. 1, A. M. Glauert, ed. (North-Holland, Amsterdam).

Bryner, J. S., J. J. Kelsch and A. G. Holtz (1969), A simple gas anti-contaminator technique for electron transmission microscopy, Rev. scient. Instrum. *40*, 1648.

Cosslett, V. E. (1969), High voltage electron microscopy, Q. Rev. Biophysics *2*, 95.

Ennos, A. E. (1953), The origin of specimen contamination in the electron microscope, Br. J. appl. Phys. *4*, 101.

Ennos, A. E. (1954), The sources of electron induced contamination in kinetic vacuum systems, Brit. J. Appl. Phys. *5*, 27.

Forth, H. J. and W. Loebe (1969), A liquid helium cooled device for producing ultra high vacuum in the specimen chamber of the Elmiskop 101, Siemens Review *36*, 11.

Glaeser, R. M. (1971), Limitation to significant information in biological electron microscopy as a result of radiation damage, J. Ultrastruct. Res. *36*, 466.

Glaeser R. M., R. F. Budinger, P. M. Aebersold and G. Thomas (1970), Radiation damage in biological specimens, Proc. 7th Int. Congr. Electron Microscopy, Grenoble *1*, 463.

Glaeser, R. M., V. E. Cosslett and U. Valdrè (1971), Low temperature electron microscopy: radiation damage in crystalline biological materials, J. Microscopie *12*, 133.

Grubb, D. T. and G. W. Groves (1971), Rate of damage of polymer crystals in the electron microscope; dependence on temperature and beam voltage, Phil. Mag. *24*, 815.

Grubb, D. T. and A. Keller (1972), Beam induced radiation damage in polymers and its effect on the image formed in the electron microscope, Proc. 5th Eur. Conf. Electron Microscopy, Manchester, p. 554.

Hart, R. K., T. F. Kassner and J. K. Maurin (1970), The contamination of surfaces during high energy electron irradiation, Phil. Mag. *21*, 453.

Hartman, R. E. and R. S. Hartman (1971), Residual gas reactions in the electron microscope IV. A factor in radiation damage, Proc. 29th Ann. Meeting EMSA, Boston Mass., p. 74.

Heide, H. G. (1958), Die Objektverschmutzung und ihr Verhütung, Proc. 4th Int. Congr. Electron Microscopy, Berlin *1*, 87.

Heide, H. G. (1962), The prevention of contamination without beam damage to the specimen, Proc. 5th Int. Congr. Electron Microscopy, Philadelphia 1, A4.

Heide, H. G. (1963), Contamination of the object in the electron microscope and the problem of irradiation damage by carbon removal, Z. angew. Phys. *15*, 116.

Heide, H. G. (1964), Composition of the residual gas in an electron microscope, Z. angew. Phys. *17*, 70.

Heide, H. G. (1965), Contamination and irradiation effects and their dependence on composition of residual gases in the electron microscope, Lab. Invest. *14*, 1134.

Helmcke, J. G. (1954), Theorie und Praxis der elektronenmikroskopischen Stereoaufnahmen, Optik *11*, 201.

Helmcke, J. G. (1955), Theorie und Praxis der elektronenmikroskopischen Stereoaufnahmen. III. Methoden zur Tiefenausmessung elektronenmikroskopischer Objekte und Forderungen an ein photogrammetrisches Auswertgerat, Optik *12*, 253.

Hudson, B. and M. J. Makin (1970), The optimum tilt angle for electron stereo microscopy, J. Phys. E. (Sci. Instrum.) *3*, 311.

Le Poole, J. B. and P. Stam (1954), An objective method of focusing, Proc. 3rd Int. Congr. Electron Microscopy, London, p. 666.

Pashley, D. W. and A. E. B. Presland (1961), Ion damage to metal films inside an electron microscope, Phil. Mag. *6*, 1003.

Stenn, K. and G. F. Bahr (1970), Specimen damage caused by the beam of the transmission electron microscope: A correlative reconsideration, J. Ultrastruct. Res. *31*, 526.

Siegel, G. (1970), The influence of low temperature on the radiation damage of organic compounds and biological objects by electron irradiation. Proc. 7th Int. Congr. Electron Microscopy, Grenoble *2*, 221.

Thach, R. E. and S. S. Thach (1970), Damage to biological samples caused by the electron beam, Proc. 7th Int. Congr. Electron Microscopy, Grenoble *1*, 465.

Thach, R. E. and S. S. Thach (1971), Damage to biological samples caused by the electron beam during electron microscopy, Biophys. J. *11*, 204.

Williams, R. C. and H. W. Fisher (1970), Electron microscopy of tobacco mosaic virus under conditions of minimum electron beam exposure, J. molec. Biol. *52*, 121.

Chapter 7

Image recording and display

7.1 Introduction

The output of the electron microscope is a visual image, often with a wealth of complex detail, and in the majority of investigations it is necessary to have a permanent record of the image in the form of an electron micrograph, for eventual detailed assessment, measurement, duplication, publication, or exhibition. From the resolution point of view, much more information can be obtained from a photographic record; the information seen on the fluorescent screen is limited by the resolution of the phosphor and the relatively poor contrast of a reflection screen, and, at low screen intensities, by the acuity of the eye. In addition, prolonged direct viewing of the image on the screen may be limited by electron-induced contamination and radiation damage.

In pointing out the advantages of a permanent record and the limitations of direct viewing, it must be remembered that a great deal of information, other than resolution of fine detail, can be obtained by direct observation of the image. With many specimens, the significance of a given field of view can only be understood when it is related to the whole of the specimen. With crystalline specimens, many of the features of the image are dynamic, i.e. they move or change in contrast, or even disappear, as small adjustments to the specimen orientation are made, and again, a full appreciation of the situation can only be obtained when the various movements have been directly observed on the screen. The examination of a batch of micrographs is no substitute for an actual operating session at the microscope. It will be clear that direct observation and image recording are complementary.

The traditional method of recording in the transmission electron microscope is to expose a photographic plate or film directly to the electron beam in the vacuum of the microscope. The electron microscope designed and constructed by Marton (1935) was the first to incorporate the direct method of recording together with the great convenience of a camera airlock, although according to Gabor (1953) internal photography was already a long-established practice in cathode-ray oscillography, and a camera airlock was described in 1929 by Hochhäusler.

The choice of this method for electron microscopy was fortunate; it is only comparatively recently, except for a (then) comprehensive study by Von Borries in 1942, that the photographic response to electrons has been fully investigated from the point of view of the electron microscopist. Although photographic methods have been replaced by electrical detection and recording in many scientific instruments, no completely satisfactory alternative has yet been devised for the conventional transmission electron microscope. Indeed, as a near-perfect recorder (see § 7.2.3) the photographic emulsion cannot be bettered, at least as far as information is concerned.

Recording the image in a permanent form is the last step in the operation of the electron microscope prior to the extraction of information from the image. It is a most important step, because the value of everything that has gone before it – consideration of the scientific problem, acquisition of the instrument, establishment of the laboratory, specimen preparation and the acquisition of expertise – is considerably reduced if the recorded image is not a faithful reproduction of the final electron image.

Image recording by photographic means is simple and straightforward. It is perhaps unfortunate that this simplicity often results in abuse of the process, and it must be stressed that a little care and attention to detail will make all the difference between a first-class micrograph containing maximum information and with visual appeal, and a result of indifferent quality for which the microscopist has constantly to apologise.

In this chapter, the practical aspects of the various factors involved in the photographic recording of electron images are discussed. Emphasis is placed on those factors which can be controlled or influenced by the electron microscopist. A number of non-photographic methods of recording and viewing the image have been developed in recent years, and the apparatus has become commercially available. These methods are described. Observation of the image on the fluorescent screen is a necessary preliminary step to image recording, and the optimum conditions for image viewing are discussed.

7.2 Basic aspects of photographic recording

7.2.1 PRINCIPLES OF PHOTOGRAPHIC RECORDING

The transmission electron microscope image is made visible by allowing the electron beam to fall on a fluorescent screen, and an obvious way of recording the image would be to photograph the fluorescent screen through the viewing window using a camera. Although this method has certain applications (see § 7.7) it does have limitations of contrast and resolution set by the fluorescent screen, and in general transmission electron microscope images are recorded by lifting the fluorescent screen, and allowing the electron image to fall directly onto a photographic plate or film within the vacuum of the electron microscope. The exposed photographic material is removed from the electron microscope, usually via an air-lock mechanism, and chemically processed to yield a permanent silver image called a *negative*. The negative can then be *contact-printed* or more usually *enlarged* to produce a positive print on a white paper base – the *electron micrograph*.

The photographic material consists of a thin layer of radiation-sensitive silver halide crystals dispersed in gelatin, called the *emulsion*, supported on a glass plate or flexible film (Fig. 7.1). When the grains of silver halide are hit by quanta of radiation, an invisible *latent image* is formed by the exposed silver halide crystals. A chemical process called *development* will convert the exposed silver halides to metallic silver, i.e. the invisible latent image is transformed into a visible image of black metallic silver grains.

The unexposed silver halide crystals would darken on exposure to light during subsequent handling of the negative, and therefore the unexposed

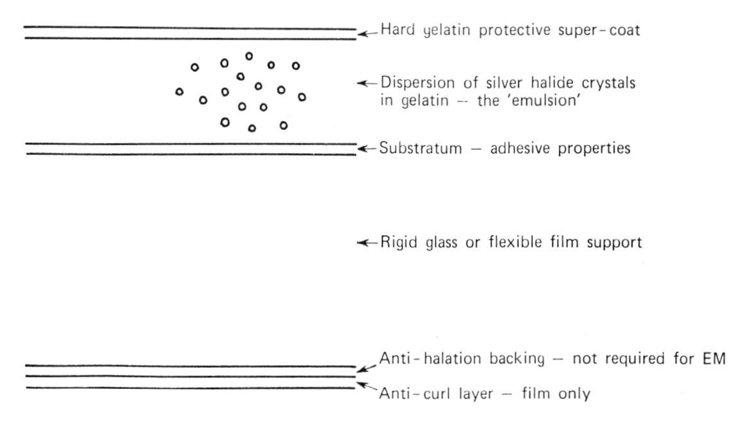

Fig. 7.1. Structure of the photographic material.

silver halide must be removed from the developed image by a further chemical process called *fixing*. After fixing, the chemicals used in the processing procedures, and their by-products, are removed from the negative by thorough washing in water. The negative is dried, and is then ready for direct examination, enlargement into a positive print, or storage.

To minimise contamination of the fixer by developer carried over in the gelatin of the emulsion, and also to ensure a more precise termination of the development process, a *stop-bath* of acid, or plain water, is used between development and fixing.

The photographic *print* material consists of a paper base, similarly coated with a layer of light-sensitive silver halide crystals. The photographic paper is exposed to a light image produced by the negative held in contact with the paper – for *contact-printing* – or more usually, to a magnified light image produced by the negative held in an *enlarger*. The exposed paper is developed, rinsed in a stop-bath, fixed, and washed, in a similar way to the negative, and then dried, usually on a *glazer*, in order to produce the final permanent print or micrograph.

Until photographic material has been fixed, the silver halide crystals remain sensitive to radiation. For this reason, films and plates are supplied in light-tight foil, paper, and cardboard wrappings, and are loaded into light-tight dark-slides or cassettes for electron microscopy.

During chemical processing a *safe-light* has to be used, i.e. illumination with a spectral distribution outside the sensitive range of the photographic material. Thus, photographic methods require the use of a so-called *darkroom* suitably light-tight and fitted with appropriate safe-lights and chemical processing and washing facilities. After fixing, photographic material can be handled and examined in normal artificial light or in daylight outside the darkroom.

A large number of factors affect the choice and behaviour of photographic materials, the methods of processing, the subsequent manipulation of the materials, and the choice and use of ancillary apparatus. The practical aspects of these factors are considered in more detail in the following sections.

The design of photographic darkrooms for electron microscopy is discussed in detail by Alderson (1974) in another book of this series.

7.2.2 DENSITY AND EXPOSURE

The optical density D of a photographic negative is defined as

$$D = \log_{10} \frac{I_0}{I}$$

where I_0 is the intensity of the incident and I the intensity of the transmitted radiation. Thus, a part of the negative transmitting 10% of the incident light has a density of 1.0. In densitometry a distinction is made between specular density and diffuse density (Fig. 7.2) but this is not significant in the present discussion.

A top density of 1.0 is useful to aim at for normal electron microscopy, but for most purposes actual measurements of density are quite unnecessary. A useful guide can be obtained by laying the negative, emulsion side down,

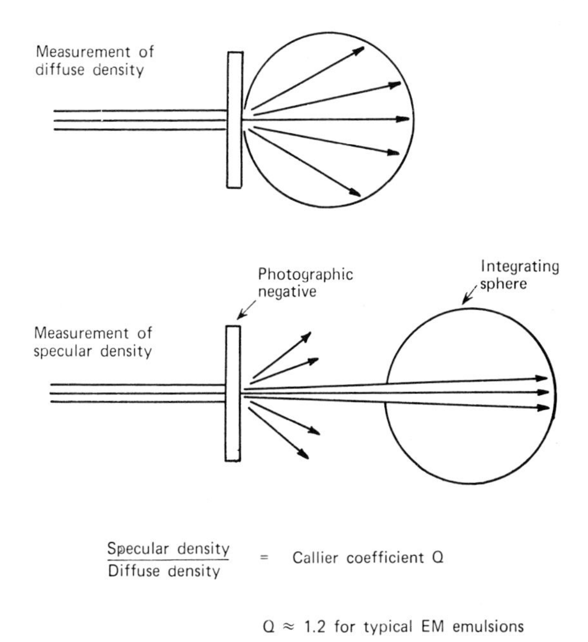

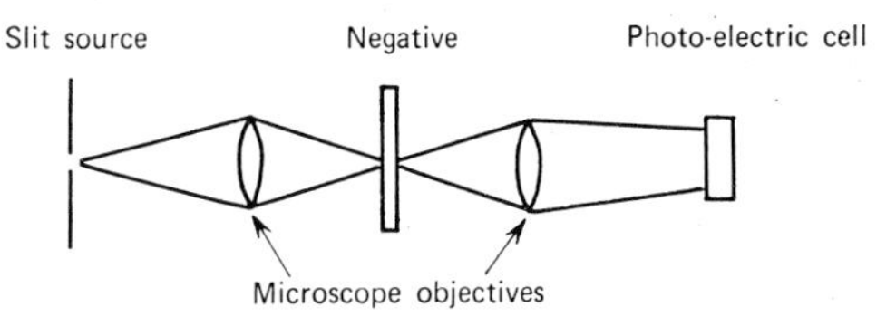

Fig. 7.2. Diffuse and specular density.

on a sheet of clean newsprint. A density of 1.0 allows the newsprint to be just about readable through the negative in good daylight.

The exposure (E) to electrons is the quantity of electricity per unit area and is measured in C/cm^2

$$E = \frac{Q}{a} = \frac{it}{a}$$

where the quantity Q (coulomb) flows into area a (cm^2) in time t (second). i is the current (ampere). Actual measurements of exposure are not necessary for routine electron microscopy, but where a measurement is essential it can be obtained by using a Faraday cage and electrometer amplifier, or by counting electron tracks in nuclear emulsions (Valentine and Wrigley 1964). Some electron microscopes have apparatus for the direct measurement of exposure built into the instrument; others have devices for indicating exposure which can sometimes be modified or calibrated to give a measurement, as distinct from an indication. For the purpose of determining the exposure to be given to the negative to reach the desired density, many electron microscopes make use of indirect methods, such as photometry or integration of the current incident on the fluorescent screen. Various devices for indicating exposure, some manually operated and others fully automated, are provided on different commercial instruments, and are fully described in the manufacturer's operating manuals.

It has been known from the early days of electron diffraction, where the emphasis has always been on intensity measurement, that the density of a photographic emulsion exposed to electrons is directly proportional to the exposure up to a value of D of a least 1.0. This is a consequence of the single hit law of electron exposure (Frieser and Klein 1958), i.e. that each incident electron causes at least one silver halide grain to become developable. Thus the photographic response to electrons is a very efficient process, unlike the relatively inefficient response to light where several tens of photons may be required to make a grain developable.

The response to electrons can be represented by

$$D = D_s(1 - e^{-naE})$$

where D_s is the saturation density,
n is the number of grains hit by an electron,
a is the area of the developed grain,
and E is the exposure.

D_s, n, and a are constant for a particular emulsion.

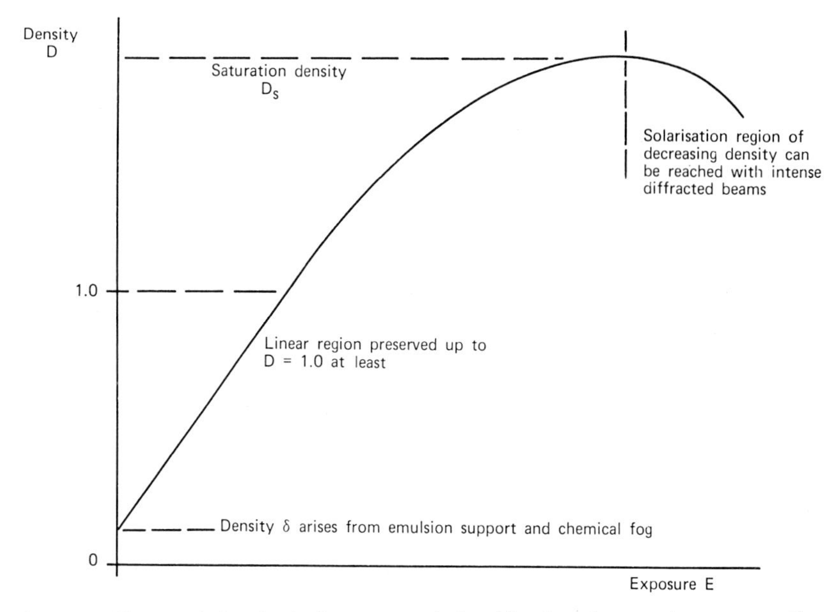

Fig. 7.3. Characteristic density/exposure relationship for electron image recording.

A small density δ (Fig. 7.3) is present for all emulsions and arises from the density of the glass or film support, plus a so-called fog, usually of chemical origin.

For small exposures the relationship becomes

$$D = D_s\ n\ a\ E = kE.$$

The linear relationship holds approximately for most combinations of emulsion and developer used in electron microscopy, often up to a density of 1.5, but the relationship breaks down at higher densities approaching the saturation density of the emulsion, or with under-development, or low electron energies, or when the emulsion has a thick super-coat.

The linear relationship is of importance for quantitative work where intensities have to be determined, e.g. for structure analysis or mass determinations. The linear relationship also leads to an interesting consequence when considering contrast in the negative (see § 7.2.6).

7.2.3 SPEED AND GRANULARITY

The speed or sensitivity of an emulsion refers to the exposure required to produce a given density. Thus, an emulsion requiring more exposure to reach a given density will be slower in speed than an emulsion requiring less exposure for the same degree of blackening.

Speeds for emulsions designed primarily for use with visible light are expressed as numbers on arbitrary but defined standard scales, notably the ASA (American Standards Association) and BS (British Standard) scales which coincide, and the DIN (Deutsche Industrie Norme) scale. For example, the internationally popular Kodachrome II colour positive film is rated at 25 ASA/BS or 15 DIN. These scales, however, are not relevant for electron exposures (see § 7.2.4) and they give no guide to the speed of the emulsion when exposed to electrons.

Valentine (1966) has pointed out that a convenient expression of the sensitivity to electron exposure is given by the number of electrons per square micrometre required to reach a density of 1.0. One electron per square micrometre is equivalent to a charge density of 1.6×10^{-11} C/cm^2. (The charge on the electron is 1.6×10^{-19} C.) Since in practice, a current density at the negative of about 2×10^{-11} A/cm^2 is typical for an observer adapted to dim light (see § 7.9) viewing fine detail on the fluorescent screen at 100 kV, the sensitivity figure gives an approximate indication of the required exposure time in seconds. For example, a sensitivity of 4 electrons/μm^2 is equivalent to $4 \times 1.6 \times 10^{-11}$ C/cm^2 which is

$$\underbrace{4\ \text{sec}}_{\text{the approximate exposure time}} \times \underbrace{1.6 \times 10^{-11}\ \text{A/cm}^2}_{\text{the approximate current density for normal viewing}}$$

The speed S can then be defined as the reciprocal of the number of electrons/μm^2 required to reach a density of 1.0, i.e. S is given by the slope D/E of the linear region of the density/exposure plot (Fig. 7.3).

A number of plots for currently available emulsions are shown in Fig. 7.4. The slope of the linear region (and to a lesser extent the overall shape of the curve) is markedly affected by the concentration and formulation of the developer used (Fig. 7.5). The development conditions are frequently used deliberately as a method of controlling the emulsion speed. Therefore, in any discussion of density or speed, the emulsion and developer must be regarded as a *combination*, and not as separate entities.

For a given image intensity it is desirable to choose an emulsion/developer combination which will require an exposure time not longer than say, 5 sec, to give the desired density. Longer exposure times may allow specimen or stage drift, contamination, and radiation damage, to affect adversely the image quality, although of course, there are circumstances where long

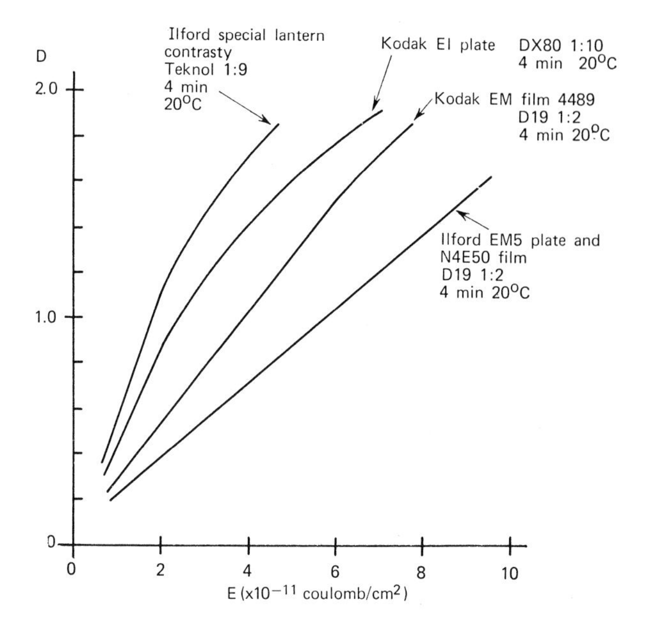

Fig. 7.4. The variation of density with exposure for four emulsion/developer combinations.

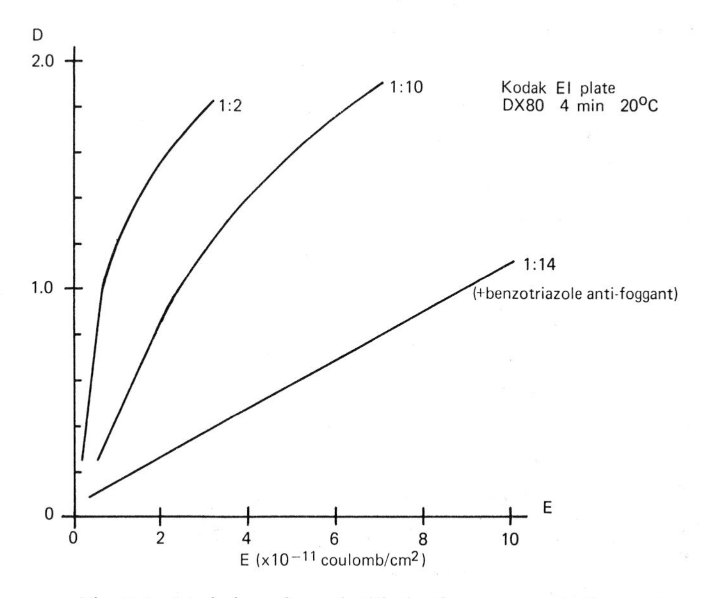

Fig. 7.5. Variation of speed with developer concentration.

exposures may be necessary and unavoidable such as in dark-field microscopy, or diffraction experiments with weak beams.

It is natural to think that the fastest possible emulsion/developer combination should be used in order to minimise the exposure time, but two difficulties can arise here. The first is the mechanical problem of the actual timing of the exposure; electron microscope timing devices and mechanisms are not normally consistent or reliable at exposure times shorter than 0.5 sec. The second problem is more fundamental and arises from *granularity* in the recorded image.

The recorded image consists of a number of patches or adjacent areas of different density. On a very fine scale, the image can be thought of as an array of small areas or elements, each with a uniform density across the element and with a sharp boundary to its neighbours. The assembly of elements of different densities gives an impression of a continuous tone image when viewed at a large enough distance.

For any element of the image there will be a statistical fluctuation in the rate of arrival of electrons, arising from the particulate nature of the beam. Thus, adjacent elements, although nominally exposed to the same electron intensity, may receive a different number of electrons in a given time, and if the fluctuation is sufficiently large, the resulting variation in density, element to element, becomes noticeable. These density variations arising from the electron fluctuations – the *electron noise* – interfere with the variations in density arising from the structure in the specimen, and the image is said to be *noisy* or *grainy*. The physical measurement corresponding to visual graininess is called *granularity* (G).

If the wanted density – the signal – results from N electrons in an image element, the unwanted density fluctuations – the noise – will be $\sqrt{N}$ (from an elementary law of statistics and assuming a Poisson distribution), and therefore the signal-to-noise ratio S/N will be $\sqrt{N}$. A ratio of 5 is often stated in the literature to be the minimum acceptable for visual work. However, most quality conscious microscopists would regard an image with a signal-to-noise ratio of 5 as only just acceptable and in practice, much higher ratios can often be obtained, in which case the granularity is small, and the graininess imperceptible.

Granularity can also arise in the emulsion itself. The density in an image element is directly proportional to the number of grains n hit by each electron, and the area a to which each exposed grain grows during development. Both n and a are subject to statistical fluctuations, as with E, and give rise to noise in the image, in addition to the electron noise.

The *detective quantum efficiency* (Jones 1955) gives a measure of the relative contribution of electron noise and 'emulsion' noise, and is defined as

$$DQE = \frac{(\text{Signal/Noise})^2 \text{ in recorded image}}{(\text{Signal/Noise})^2 \text{ in electron beam}}.$$

A perfect recorder with no noise contribution has a detective quantum efficiency of unity. Manufacturers now quote *DQE* figures for emulsions designed for electron microscopy. Burge and Garrard (1968) give the *DQE* for a number of different commercial emulsion types, Hamilton and Marchant (1967) show *DQE* and granularity figures for a variety of unnamed emulsions, and Valentine (1965) gives figures (actually $\sqrt{DQE}$) for a range of emulsions then in current use.

Except for underdevelopment, or exposure to low voltage electrons, where *DQE* is expected to be low, all the published data show the *DQE* to be at least 0.5, and with many emulsion/developer combinations, *DQE* approaches unity. This gives rise to the statement that the photographic emulsion is a near-perfect recorder for electron images at the energies used in the conventional transmission electron microscope.

Thus, visual graininess in the recorded image is almost entirely due to electron noise in the electron image, and can only be reduced by increasing the exposure. Modifications to the photographic processing will have a very small effect on emulsion noise but may well reduce emulsion speed and allow the exposure to be increased, thus reducing the electron noise.

This situation can be compared with light photography where the *DQE* may be as low as 0.001 (Mees and James 1966) and extensive use is made of fine grain development techniques. Fine grain developers are occasionally used in electron microscopy but this is primarily for the purpose of reducing emulsion speed, rather than for producing a fine grained image. Emulsions used in electron microscopy are inherently fine grained, with a typical grain size of 1 μm or less. The graininess seen arises from the irregular arrangement of the grains, and not from a resolution of the individual grains. This can be demonstrated with the aid of Fig. 7.6 which shows two squares, each containing the same number of black spots, but with an even distribution in the left-hand square and an irregular distribution in the right-hand square. If

Fig. 7.6. Demonstration of graininess arising from an irregular pattern of grains.

the viewing distance is gradually increased until the spots in the left hand square can no longer be separately resolved by the eye the graininess in the right-hand square will still be apparent.

It can be shown (e.g. Hamilton and Marchant 1967) that

$$G^2 = \frac{1}{DQE} \cdot D \cdot S$$

(for D proportional to E and neglecting possible grain yield per electron variations),
where G is the granularity,
D is the density,
S is the speed.

The fact that a given speed determines the granularity can be seen, for emulsions of high DQE, from a consideration of electron noise alone, for if the speed is fixed, the exposure E or the number of electrons N is fixed for a given D; hence the electron noise or granularity is fixed. Farnell and Flint (1973) have shown that the granularity/speed relationship is closely approximated for a range of practical emulsions exposed to 75 keV electrons.

The granularity, perceived as graininess in the recorded image, has contrast c where $c = \sqrt{N}/N$ and may be thought of as producing a granularity 'particle' size d.

Now, if

$$N = nd^2$$

where N is the total number of electrons falling on the image element (or granularity 'particle') of size d, and n is the number of electrons per unit area, it follows that the particle size is given by

$$d = \frac{1}{c} \cdot \sqrt{\frac{1}{n}}$$

Alternatively, if n is measured in electrons/μm^2

$$d = \frac{1}{c} \sqrt{S}\ \mu\text{m} \qquad \text{for } D = 1.0 .$$

As an illustration of the use of this relationship consider the granularity on Kodak EM Film 4489 developed in D19 according to the manufacturer's recommendations. The film speed S is then 0.4. The minimum detectable contrast in an image is usually considered to be 2–3% for the normal eye. If the granularity is at the minimum detectable contrast level, for a good quality image, c is 0.02. Then from the relationship above, the granularity or noise particle size d_{noise} is 30 μm.

Now the resolution of the eye (d_{eye}) is usually considered to be 100 μm for normal viewing conditions. Therefore the film can be enlarged photographically by d_{eye}/d_{noise}, i.e. by about 3.5 times without the noise becoming visible. If greater enlargements are necessary, perhaps in order to reach a particular print size, the noise will become visible. (The degree of enlargement to the point where the structure is at the resolution limit of the eye is called the *blending magnification.*)

Whether or not the noise interferes with the imaged specimen structure depends on the recorded size of the structure. Clearly, the size of the imaged specimen structure must exceed the noise particle size by at least a factor of 2, and preferably a factor of 5, if the noise is not to appreciably modulate the true structure on the one hand, or be mistaken for true structure, on the other hand. Thus if the structure of interest is say, 3 nm (30 Å) in size, the necessary electron microscope magnification

$$M = 5 \times \frac{30\ \mu\text{m}}{3\ \text{nm}} = 50{,}000$$

to ensure that the structure is clearly visible above the noise. Similarly if structure near the resolution limit of the electron microscope is being investigated, perhaps for the purpose of testing resolution, microscope magnifications of 500,000 may be required. An adequate microscope magnification as described above, is a necessary criterion for resolving fine structure, but whether or not the structure will be visible depends on the contrast of the structure, which is determined by electron scattering at the specimen, and is considered in § 3.3. The visibility will also depend on the type of structure; a regular or extended structure is much more easily perceived than a random pattern.

Hamilton and Marchant (1967) have concluded that for a given exposure time and constant intensity *at the specimen* the final quality of a micrograph is the same, whether it is made by using a fast photographic material, a high microscope magnification, and moderate photographic enlargement, or by using a slower photographic material, a lower microscope magnification and a greater photographic enlargement, and that (within limits) the microscope magnification can be adjusted to achieve optimal recording for every specimen. In practice, however, the microscope is not used in this way. The magnification is chosen with reference to the field of view, the structure size and the aesthetic appearance, and the intensity at the viewing screen is set to allow comfortable viewing and focusing. Ideally the *viewing screen intensity* should be constant, independent of the magnification. On some commercial instruments this is achieved by coupling the condenser lens current with the projector lens currents so that a change of magnification automatically keeps the screen intensity constant, at least approximately (see § 2.14.3).

The emulsion/developer combination can then be chosen so that at this image intensity, and with an exposure time in the range 1–5 seconds, say, (for a normal bright field image) the desired photographic density is achieved.

It is sometimes the practice to adjust the exposure by reducing the intensity below the level set for comfortable visual observation and focusing, but this is *not recommended* for work of the highest quality. The specimen may drift after focusing if the electron current loading is reduced, and electron noise may become a limiting factor if the intensity is reduced appreciably.

Speed of recording has to be a balance between the avoidance of drift and other instabilities at long exposures, and the preservation of an adequate number of electrons per image element at short exposures.

7.2.4 SPECTRAL RESPONSE

The sensitivity or speed of an emulsion varies with the energy (or wavelength) of the incident radiation. The electron microscopist is concerned with the photographic response to the direct image-forming electron beam in the microscope, the response to indirect ionising radiations such as X-rays, and the response to visible light which may fall on the emulsion during manipulation of the photographic material both inside and outside the electron microscope, and during subsequent processing.

All types of emulsion are sensitive to ionising radiations; photographic materials are particularly vulnerable to X-radiation because of the high penetration, and particular attention must be given to this point if X-ray apparatus has to be used in the vicinity of the electron microscope or the photographic darkrooms and stores.

The early emulsions were sensitive to light in a range of wavelengths extending from the ultra-violet to the blue region of the visible spectrum; these were called *ordinary* emulsions. The inclusion of sensitising dyes led to the *ortho-chromatic* type of emulsion, sensitive into the green region of the spectrum, and later, *pan-chromatic* emulsions were produced with a sensitivity extending into the red region to cover the whole of the visible spectrum. However, the sensitising dyes used in emulsions designed for light recording play no part in determining the response to electrons. The emulsions chosen or designed for electron microscopy have the characteristics, when exposed to light, of the ordinary blue-sensitive emulsions. The speed to visible light is very low – only a few degrees at the most on the ASA scale – and this is particularly fortunate for both the designer and user of the electron microscope. The low sensitivity to light simplifies the construction of light baffles in the camera and viewing chamber, and allows the use of relatively bright safe lighting in the manipulation and processing operations. Unwanted exposure to light radiation is not usually a problem in electron microscopy but some precautions are worth taking. Room lighting (preferably from a safelight – see Alderson 1974) should be arranged so that the source does not shine directly into the viewing chamber. When working with very weak electron beams, and hence long exposure times, extraneous light should be kept out of the viewing chamber by fitting light-tight caps over the chamber windows. These are usually provided by the manufacturer.

The vast majority of electron microscope exposures are made in the energy range 60 keV to 100 keV. In recent years as high voltage microscopy has developed and instruments have become more readily available, an increasing amount of work is performed in the range 200 keV to 1 MeV. Since the early days of electron microscopy there has also been a small but continuing interest in low voltage microscopy in the range 5–25 keV.

Response data for a wide range of commercial emulsions has been given by Digby et al. (1953) for the energy range 25–78 keV, by Valentine (1966) for the range 42–80 keV, and by Burge et al. (1968) for the range 7–60 keV. Frieser and Klein (1958) investigated 29 unnamed experimental emulsions in the range 40–100 keV. Much of the published data refers to materials which unfortunately are no longer readily available.

Ilford EM5 plates, Ilford N4E50 film, Kodak EI plates, Kodak 4489 film, and Kodak Fine Grain Positive film, are currently available and recommended (see § 7.3.1).

In the operating range 60–100 keV, changes of accelerating potential are normally infrequent; the biologist will choose an accelerating potential of 60 or 80 kV for example, and the materials scientist 100 kV or perhaps 120 kV if this is available (see § 6.1). For this range it is not difficult to remember a simple multiplying factor for use either in reading or setting the exposure meter.

At energies less than 40 keV the sensitivities and saturation densities of the ‘conventional’ emulsions fall appreciably, and at 5 keV these emulsions are quite unsuitable for electron image recording. Van Dorsten and Premsela (1960) found that a lithographic emulsion – Gevaert Litholine 082 (now Agfa-Gevaert) – had a good sensitivity for low voltage microscopy. Burge et al. (1968) recommend the use of a nuclear emulsion – Ilford G5 – at energies greater than 7 keV. Schumann-type emulsions are also shown to be satisfactory from the recording point of view but are somewhat difficult to process uniformly.

As the electron energy decreases, the detective quantum efficiency decreases. The decrease is small from 100 keV to 60 keV and is unlikely to have an appreciable effect on resolution. Below 60 keV, *DQE* falls rapidly, and Burge and Garrard (1968) show that for a conventional emulsion (typified by Ilford Special Lantern Contrasty) *DQE* falls by a factor of about 30 between 60 keV and 7 keV. For nuclear emulsions, however, the factor is only 2 over the same energy range, which is a further reason for the choice of nuclear emulsions for low voltage microscopy.

At the present time high voltage microscopy is much more experimental in nature, and the whole range of energies from 100 keV to 1 MeV, or even up to 3 MeV, is being explored in many different applications. The photographic response varies appreciably over the energy range 100 keV to 1 MeV often by a factor of 15, and an appropriate calibration curve for the exposure meter should be to hand, if consistent densities are to be achieved. Fig. 7.7 shows the variation in speed for Kodak 4489 film over the energy range 100 keV–1 MeV. Response data for the Fuji FG plate, ST plate and FG film, widely used in the Japanese electron microscopes, has been given by Iwanaga,

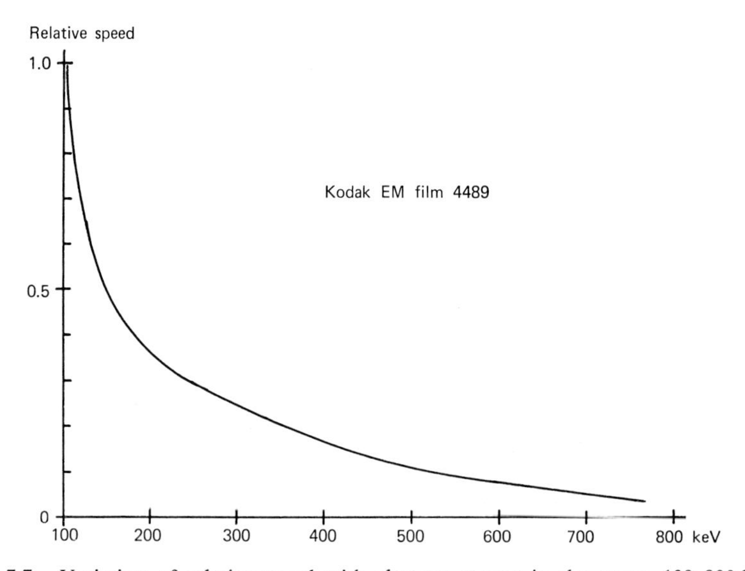

Fig. 7.7. Variation of relative speed with electron energy in the range 100–800 keV.

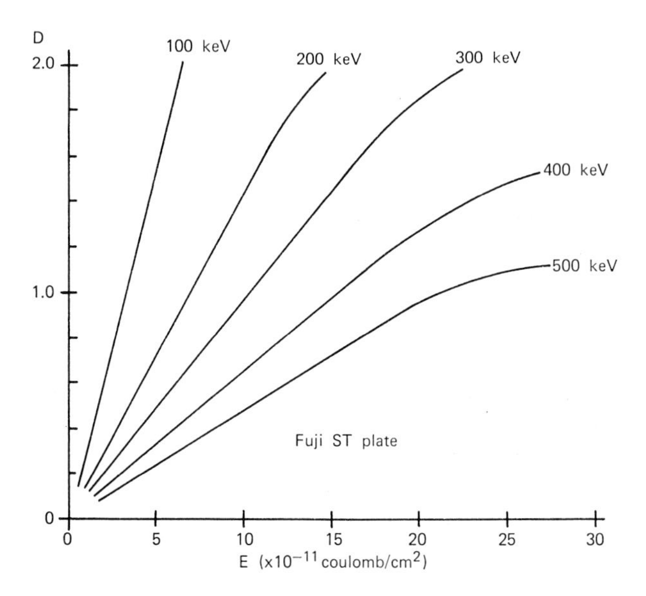

Fig. 7.8. Linear relationship of density and exposure at high energies (Iwanaga et al. 1968).

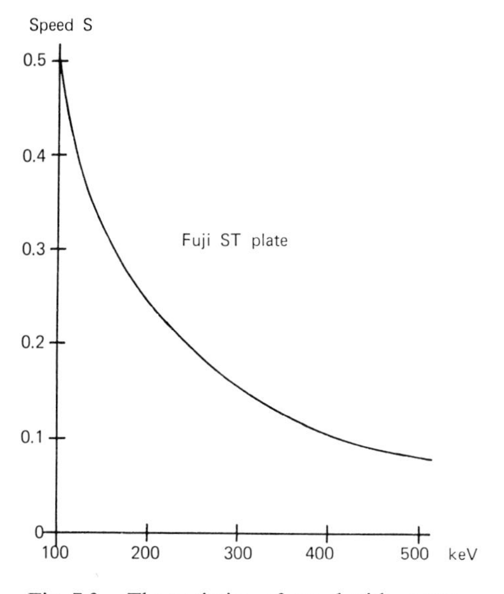

Fig. 7.9. The variation of speed with energy.

Ueyanagi, Hosoi, Iwasa, Oba and Shiratsuchi (1968) for the range 100–600 keV. Fig. 7.8 (reproduced from Iwanaga et al. 1968) shows that the linear relationship between density and exposure is well preserved up to 500 kV. Fig. 7.9 (constructed from the data of Iwanaga et al.) shows the speed S as a function of electron energy.

In principle, improved sensitivities to the higher energy electrons could be achieved by using thick emulsions with a high density of silver halide grains, i.e. nuclear emulsions. Iwanaga et al. (1968) have shown that the Konishiroku NR-E1 nuclear plate has a sensitivity at least 10 times greater than the conventional emulsions over the range 100–500 keV. Unfortunately, the resolution is appreciably impaired with thick emulsions, as a result of lateral electron diffusion (see § 7.2.5).

7.2.5 RESOLUTION

The effect of granularity on resolution has been discussed in § 7.2.3. A further limitation to the resolution of the photographic emulsion arises from the diffusion or scattering of the electrons as they pass through the emulsion itself.

The effect of lateral diffusion can be seen schematically in Fig. 7.10 which shows an opaque edge illuminated with electrons and in contact with an emulsion. In the vicinity of the edge, electrons are scattered out of the illuminated region, as they pass through the emulsion, into the shadow region. When developed, the emulsion will have the density profile shown.

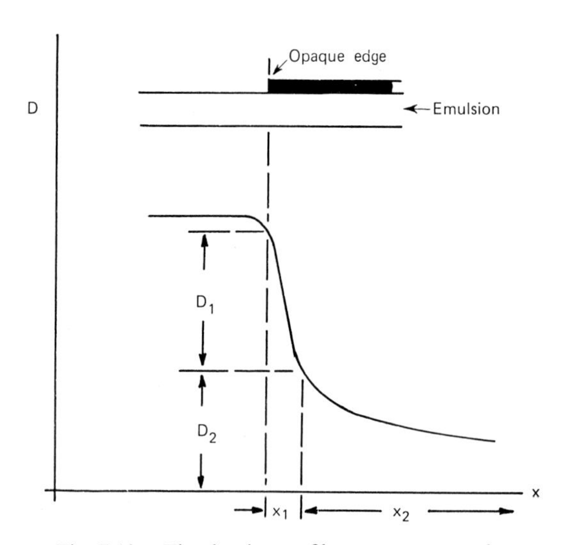

Fig. 7.10. The density profile at an opaque edge.

The lateral extent of the scattering can be described in terms of a *line-spread function* or an *edge-spread function* which relates the radiation intensity at a point in the shadow to its lateral position. Various functions based on different mathematical models for the profile are defined in the literature (e.g. Ingelstam et al. 1956; Frieser and Klein 1958; Jones 1958; Kelly 1960). Frieser and Klein (1958) and Burge and Garrard (1968) show that an exponential model corresponds most closely to the experimental data.

For exposure to electrons in the linear D/E region, the various spread function intensities correspond exactly to the density profile. Thus,

$$D = e^{-ax}$$

where D is the density at point x and a is the exponential diffusion constant. It is convenient to divide the observed density profile into two regions D_1 and D_2 with corresponding distances x_1 and x_2 (Fig. 7.10). For exposure to electrons at energy levels up to 100 keV the region D_2 is very small, comparable with the fog and base density (~ 0.01). Up to 100 keV, therefore the resolution is affected only by diffusion in the region x_1. Objects appearing dense on the micrograph are seen to be surrounded by a *halo* of lesser density, whose width is independent of the object size. The width of the halo may be defined as the distance from the shadow edge to a point where the density has decreased by a factor of 10 (Fig. 7.11).

Since

$$D = e^{-ax}$$

$$\log_e D = -ax$$

and for a density range of 10:1 (i.e. $\log_e D = 2.3$) the width of the halo is $2.3/a$ (Fig. 7.11).

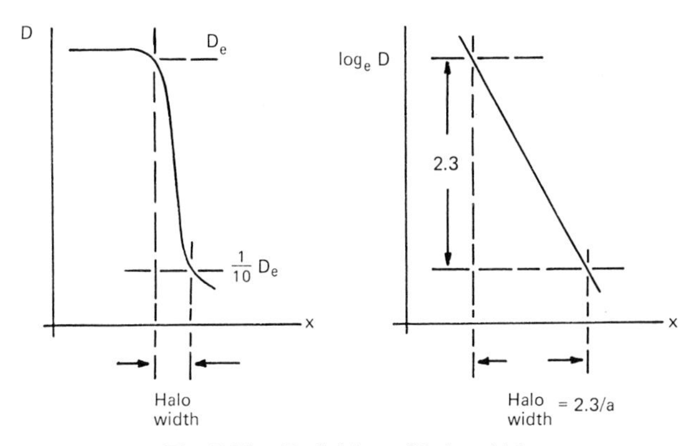

Fig. 7.11. Definition of halo width.

Thus, from the data of Frieser and Klein (1958) who give a parameter $k(=4.6/a)$ and from Burge and Garrard (1968) who give $1/a$ it is found that the diffusion widths do not exceed 30 μm up to 80 keV and can be as low as 7 μm for conventional emulsions, and only 1.2 μm for the 'low voltage' emulsions.

These widths do not exceed the noise particle size considered in the granularity example on page 202, and it seems reasonable to suggest that if the minimum magnifications are worked out on the basis of granularity alone, no additional deterioration will arise from diffusion effects, up to 100 keV.

For energies exceeding 100 keV, the situation is different. The diffusion widths increase with electron energy, and at 500 keV, D_2 in Fig. 7.10 can be as high as 0.5. At this density, the contribution from the x_2 region becomes significant. The x_2 region includes the tail of the exponential diffusion curve together with a contribution from electrostatic charging and fluorescent effects. From the results of Iwanaga et al. (1968) the diffusion width appears to be about 700 μm at 500 keV for a conventional emulsion, although this figure seems surprisingly large.

At high energies, therefore, diffusion effects can exceed considerably the granularity, and become the limiting factor on resolution.

Many specimens give rise to images consisting more of undulating variations in intensity, rather than of sharp, high-contrast edges. For such images, the limitations to resolution can perhaps be better considered in terms of the *contrast* or *modulation transfer function* (CTF or MTF). The contrast transfer function is based on the concept that a given structure can be resolved into a spectrum of spatial frequencies. The concept has physical reality, as can be readily demonstrated with an optical diffractometer. An outline of the theoretical background is given by Zeitler and Hayes (1965) and an extensive account of contrast transfer and image formation is presented by Hawkes (1972).

When all the component frequencies are transferred to the emulsion (i.e. recorded photographically) with their original contrast, the recorded image will be a faithful reproduction of the incident image. If the transferred contrast of certain spatial frequencies is reduced, or completely lost, the recorded image will lack fidelity and the resolution will be impaired. The contrast transfer function is a plot of the contrast transfer, expressed as a ratio of the incident contrast to the recorded contrast, versus the spatial frequency measured in lines/mm. (The CTF is considered here in relation to the photographic emulsion; clearly, it can be applied to every stage of information transmission in the electron microscope § 3.8.)

In order to resolve a structure of size λ, it is necessary that at least the fundamental spatial frequency $1/\lambda$ is present in the image. However, the presence in the image of only one spatial frequency, or *sub-structure*, may give a very poor reproduction of the object structure, and if this should be the situation in a practical example, it would be more appropriate to speak of *detection* of the object rather than of resolution.

The electron image consists of an array of intensity maxima and minima, and diffusion in the emulsion will always result in some electrons from a peak of intensity falling into an adjacent trough, thereby reducing the contrast. Thus, unless diffusion is negligible, the contrast in the recorded image will always be less than the contrast in the incident electron image. If troughs are completely filled in this way the contrast in the recorded image will be eliminated. If diffusion is increased further, new peaks may be formed in place of the original troughs and *pseudo-contrast* will result.

It can be shown that

$$CTF = \frac{1}{1 + \left(\dfrac{6.21}{\lambda a}\right)^2}$$

where a is the exponential diffusion constant used previously, and λ is the periodicity of a sub-structure of frequency f lines/mm $(f = 1/\lambda)$. The CTF is plotted in Fig. 7.12. The maximum value of the function is, of course, unity.

As an illustration of the use of the function, consider the Kodak EM Film 4489 which has a diffusion constant a (estimated by Farnell 1973 from the known structure of the emulsion) of 0.14 μm^{-1} at 100 keV. The lower axis in Fig. 7.12 is plotted for this value of a. Suppose that a specimen structure of 3 nm (30 Å) is to be resolved in the recorded image. To *detect* a 3 nm structure, an image sub-structure of 3 nm must be recorded with good contrast, and if the structure is to be resolved rather than detected, some additional sub-structures, say, down to 0.3 nm, must be recorded in order to achieve reasonable fidelity. Suppose that the contrast in the electron image is such that a further degradation by a factor of 0.4 is just about acceptable. Reference to Fig. 7.12 then shows that for a contrast transfer of 0.4, the sub-structures must be at least 35 μm in size at the emulsion. The required minimum magnification is therefore,

$$M = \frac{35 \times 10^{-6}\ \text{m}}{0.3 \times 10^{-9}\ \text{m}} \fallingdotseq 120{,}000$$

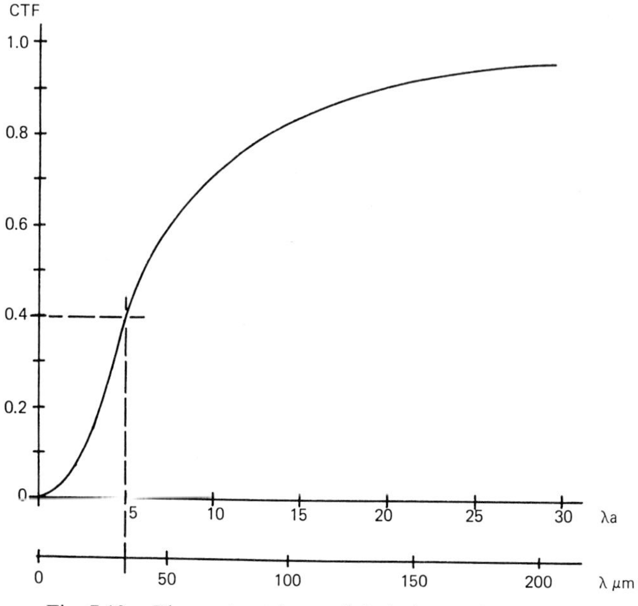

Fig. 7.12. The contrast (or modulation) transfer function.

If more fidelity is required, smaller sub-structures must be transferred with good contrast and a higher magnification will be necessary.

In making use of the CTF, it is important to remember that in many practical circumstances the contrast in the electron image may already be low, perhaps even approaching the minimum detectable limit, and that further degradation by the emulsion may be unacceptable. If this is so, the first few sub-structures must be transferred with a contrast ratio approaching unity, and very high magnifications may be necessary. (Mathematical analyses in the literature often assume an incident contrast of 100%, which is far from the case in practical electron microscopy.)

For structures approaching the limit of resolution of the electron microscope the best that can be hoped for is detection of the structures. Measurements or deductions of shape and size, other than average spacings, cannot be made with any certainty.

When working at the ultimate performance level, every effort must be made to keep the intensity and contrast in the electron image as high as possible. It will be clear from the discussions of granularity and diffusion, that no modifications to the photographic procedures will compensate for inadequate exposure or magnification of the image.

7.2.6 CONTRAST AND TONAL RANGE

Over much of the visible spectrum the eye responds to changes in brightness in a logarithmic manner, and when visual observations are made on a negative or positive print, it is more convenient to consider density related to the logarithm of the exposure. Hurter and Driffield (1890) showed the significance of a $D/\log E$ plot (Fig. 7.13) and this is the usual form of density/exposure relationship shown in the photographic literature. In theory, and in

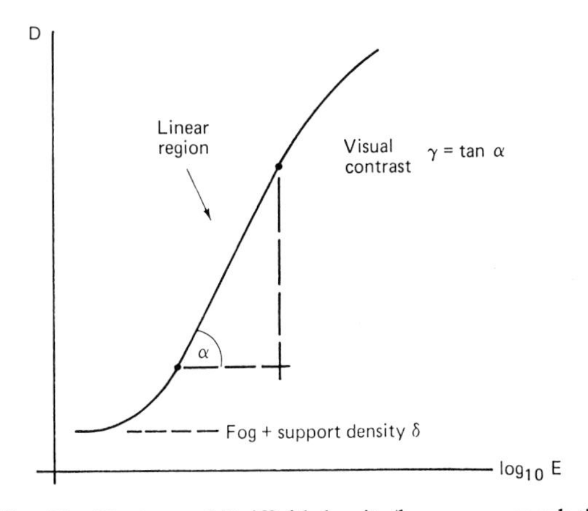

Fig. 7.13. The Hurter and Driffield density/log exposure relationship.

practice with a rather restricted range of materials, the plot has a straightline portion with a slope, gamma ($\gamma = \tan\alpha$, Fig. 7.13) which is a measure of the visual contrast in the photographic material. Contrast here refers to separation of tones, i.e. of different density levels in the negative.

Thus, by definition,

$$\gamma = \frac{dD}{d\log_{10} E}.$$

Valentine (1966) has pointed out that since $D = kE$ for electrons (a consequence of the single-hit law)

$$\frac{dD}{dE} = k \qquad \text{a constant}$$

and therefore

$$\gamma = 2.3\,D.$$

(A similar situation exists for exposure to X-rays (Van Horn 1951).) The relationship $\gamma \propto D$ holds up to a value of

$$D = \frac{D_s}{4}$$

approximately, where D_s is the saturation density of the emulsion (Fig. 7.3) and $\gamma_{max} = 0.8\ D_s$, at $D = 0.65\ D_s$ approximately.

Thus it can be concluded that all emulsions exposed to the same density will have the same visual contrast. This of course, is not true when emulsions are exposed to light.

This 'gamma' method of expressing contrast is not now widely used, since it does not take into account the long toe region of a practical $D/\log E$ plot, nor is it satisfactory when the plot does not exhibit a straight-line portion, which is frequently the case. Consequently, 'gamma' has largely been replaced by the 'contrast index', defined as the slope of a line joining two points on the $D/\log E$ curve representing the maximum and minimum densities used in practice (usually $D = 2.2$, and $D = 0.2$ above fog plus support density).

Nevertheless, it is useful to remember that for electron exposures, visual contrast will increase with the density of the negative, and that at the same density all emulsions have the same contrast. For high resolution microscopy, and also when the specimen contrast is inherently low, a high contrast negative is required, and the negative should be as dense as possible. The maximum useful density will be limited only by the difficulty of printing a very dense negative (see § 7.4.2).

In the photographic printing operation, the dense parts of the negative, corresponding to brightness in the electron image, are converted into bright regions in the print, and similarly, less dense parts of the negative are converted into dark regions on the print. Ideally, the photographic print should have its darkest parts at the full blackness obtainable on the print, and its lightest parts just perceptibly darker than the white base of the paper, with a number of intermediate tones between black and white. Van Dorsten (1950) has suggested that 5 or 6 tones in the print will give maximum visual impact and aesthetic appeal.

A photographic print is viewed by reflected light, and for a paper with reflectivity R, illuminated with light of intensity I_0, the reflection density D_R is given by

$$D_R = \log_{10} \frac{I_0}{R}$$

The maximum reflection density that can be obtained is usually not more than 2.0 for a glossy surface and only about 1.2 for a matt surface (Fig. 7.14).

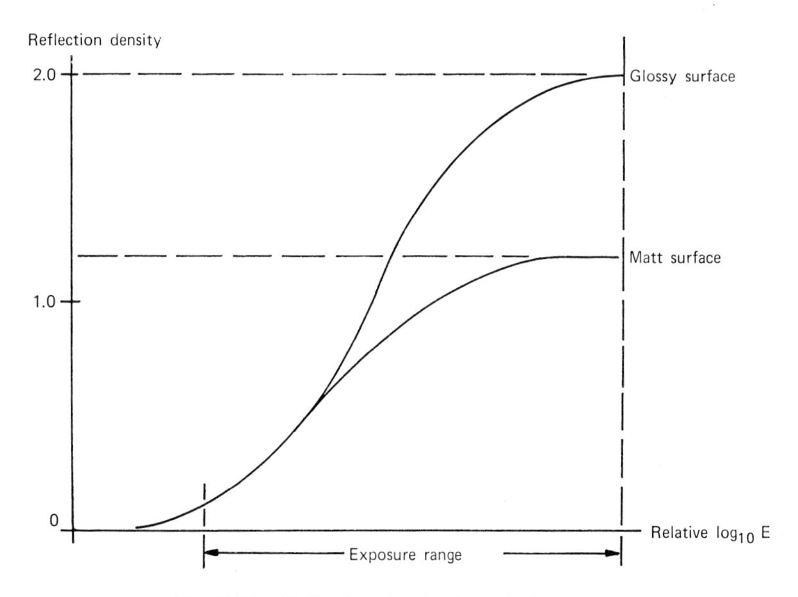

Fig. 7.14. Reflection density for printing paper.

Glossy surfaces are usually used for electron microscopy where observation of fine detail is important. The problem, then, is to make the range of densities on the negative fit as closely as possible to the exposure range of the print, so that the full density range of the print is used. If only a few closely separated tones are present on the negative, i.e. the negative is of low contrast, the full exposure range of the paper will not be used and the print will be of poor contrast showing only a few tones of grey, more or less dark depending on which part of the range is used.

Conversely, if a very wide tonal range is recorded, i.e. the negative is of high contrast, the exposure range of the print may be inadequate to cope with the density range of the negative, and either tonal separation or detail in the dark parts of the negative ('highlights' in the print), or detail in the light parts of the negative ('shadows' in the print) will have to be sacrificed. Thus, although the print will have adequate overall contrast from black to white, tonal separation will be lost in parts of the image.

To overcome these difficulties photographic printing papers are made in a range of contrast grades i.e. with different exposure ranges (Fig. 7.15).

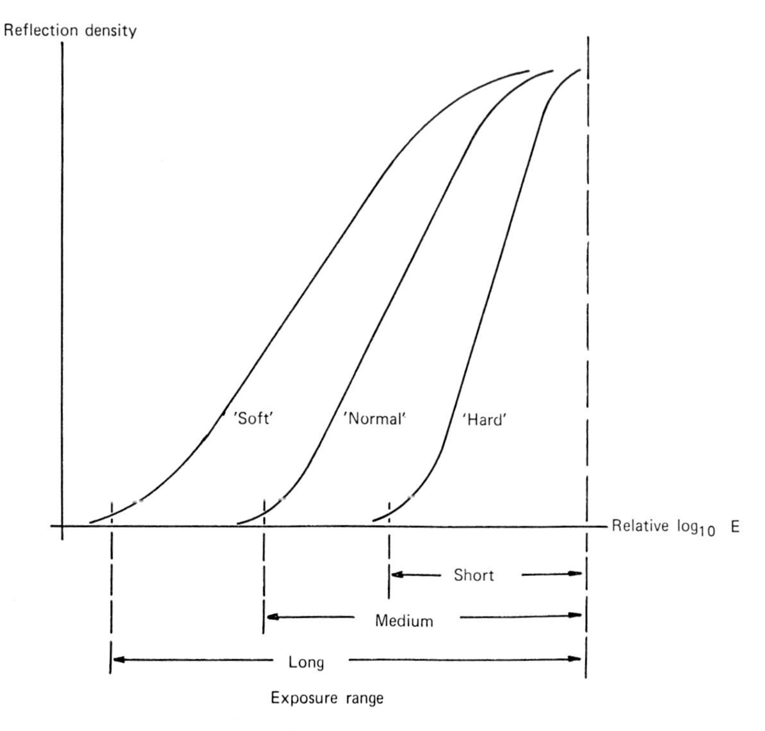

Fig. 7.15. Exposure range and printing paper contrast grades.

So-called 'soft' papers have a long exposure range, and 'hard' papers have a short range (§ 7.3.7).

In practice, therefore, if a negative has a long range of densities, it will be necessary to use a soft grade of paper with a long exposure range in order to produce a print which shows clearly all the tonal separations between black and white. Similarly, a negative with only a short range of densities will require a hard paper for a satisfactory print. Note that in these considerations, it is the range of densities that is important; the absolute level of density (within the linear D/E range) affects only the exposure time in printing.

It is usual to choose the negative exposure and processing in such a way that a 'normal' grade of paper can be used for routine work. This will then permit some latitude in choosing the contrast grade for more unusual conditions of exposure and processing.

The type of light source used in the enlarger also affects the contrast in the print (see § 7.4.2).

The processing conditions for the chloride papers normally used for making contact prints, and for the bromide and chloro-bromide papers used for enlargements do not seem to affect the contrast grade to any noticeable extent. However, a small but sometimes useful increase in contrast grade, i.e. reduction in exposure range, can be achieved with a bromide paper by extending development.

In much electron microscopy the number of brightness levels in the electron image is severely restricted and in certain images, for example Figs. 3.7 and 3.12b, the print shows only one or two tones between black and white. Information is more important than aesthetic appearance in a micrograph, but the two are not incompatible, and whenever possible, a quality print should always be sought from a given negative, by correct choice of the printing paper, the exposure in enlargement, and the subsequent processing conditions.

With some images, a very long range of densities may be recorded on the negative, which even the softest grade of paper will be unable to cope with without loss of tonal separation. An obvious remedy would be to reduce the negative exposure in order to restrict the density range, but this could result in a loss of visibility in the low-intensity detail of the image. Farnell and Flint (1969) have described a useful low contrast development technique for dealing with this situation.

Where contrast is used to compare or interpret electron scattering from different regions of the specimen, it is important that the regions are recorded over the same density range on the negatives, if the contrast comparisons are to be valid. Careful control of the exposure and subsequent processing is required. It is sometimes possible to compensate for small errors in exposure by varying the contrast in printing, i.e. by using different grades of paper (§ 3.3). However, only a limited correction is possible, and this practice is not recommended unless facilities are available for making density measurements. If exposure errors are made, it is safer to record the image again at the correct density level.

The reflection density, and hence the contrast, of selected portions of a print can be controlled by a number of special printing techniques (see § 7.5.13).

7.3 Photographic materials

7.3.1 EMULSIONS FOR ELECTRON IMAGE RECORDING

The choice of a particular emulsion for electron image recording depends on scientific and technical factors, discussed in § 7.2.2–7.2.5 and also on a number of economic and convenience factors.

The cost of photographic material is of great importance in most electron microscope laboratories, the majority of which seem to be financed adequately in terms of capital (equipment) but inadequately in terms of running costs. Although in a scientific environment quality and performance will have first priority, cost is an important consideration, nevertheless.

The convenience aspects include consistency in characteristics, uniformity of coating and freedom from blemishes, adequacy and convenience of packing, and ready availability. In the past these factors have been very variable and have often led to the choice of emulsion being a matter of opinion rather than of fact. The selection was usually made from emulsions primarily designed for lantern slide making, nuclear particle detection, X-ray or photomechanical work, and the necessary data on which to base the choice was not always readily available.

In recent years, however, the photographic manufacturers have formulated some of their products specifically for electron microscopy and most of the requirements of the electron microscopist have been met very adequately. It must be remembered that notwithstanding an estimated 50 million electron microscope exposures a year, by normal production standards the electron microscopist is so small a consumer as to be a nuisance, and it is fortunate indeed that the manufacturers have provided so well for the numerically trivial requirements of Science and Industry.

The technical aspects affecting a choice of emulsion have been discussed in § 7.2.2–7.2.5 but to summarise briefly the choice is made on the following basis. The structure in the specimen and the nature of the scientific problem will determine the magnification range over which images are required. Control of the illuminating system of the electron microscope will allow a comfortable viewing intensity to be set over the required magnification range. Corresponding to the viewing intensity on the fluorescent screen will be a certain electron current density at the photographic film or plate. The exposure time is set in the range 1–5 sec (for normal bright-field imaging). The exposure is now fixed, as the product of the current density and the time. The granularity is also fixed, since for photographic emulsions of high *DQE*, the granularity depends only on the number of electrons incident on the emulsion, i.e. the exposure. The minimum magnifications for resolution of structures of particular sizes can now be calculated.

If it is felt that more exposure is required, perhaps to reduce the granularity at a given magnification, some adjustment to the illumination intensity will be required, probably by increasing the gun brightness. Limits to the maximum exposure will be set by the maximum attainable brightness, by thermal, mechanical and electrical instabilities, and by radiation damage to the specimen.

If less exposure is required, perhaps to reduce radiation damage, the gun brightness can be reduced, with some gain in filament life, or other appropriate adjustments to the illuminating system can be made, provided that the granularity requirements are satisfied, and that the image on the fluorescent screen is bright enough for focusing and observation. (Special techniques may be required for *very* sensitive specimens (see § 6.13) for weak-beam diffraction experiments and dark-field microscopy (see § 3.10.3) and for diffraction patterns (see § 6.12).)

The emulsion/developer combination can now be selected so that with the chosen exposure, a top density of about 1.0 can be achieved. A speed in the range 0.3 to 0.5 is satisfactory for most purposes. For a given emulsion the speed can be modified by altering the development conditions. If greater contrast is required, more exposure must be given by increasing the electron intensity or the exposure time, and vice versa.

At one time a very wide range of emulsions was used in electron microscopy; the literature on the photographic response to electrons will give some idea of the extent of the range. The number of emulsions available has dwindled appreciably in the last few years, as economic factors have become more important, and not least, as the photographic manufacturers have set out to cater for the needs of the electron microscopist. Some of the currently available materials which have proved to be very satisfactory for electron image recording, in a wide range of applications in different laboratories, are listed in Table 7.1. For many workers, an emulsion/developer combination chosen from the table will prove to be a permanent choice. For others, who may wish to modify their choice for special purposes, one of these combinations will be an excellent starting point.

It is natural and desirable that each emulsion manufacturer should recommend a particular developer (usually of their own manufacture) for use with each emulsion. There are good reasons for following the manufacturers' recommendations, not the least being that technical data for the combination is published and if any difficulties are encountered the user can be assured of help from the manufacturer if the recommendations have been followed.

Other developers can be used, however, with excellent results. The author has used Kodak D19 with most of the emulsions currently available, and also Ilford PQ Universal and May and Baker Teknol (see § 7.6) which have

TABLE 7.1

Some curently available electron image recording materials

Negative material	*Manufacturer's recommended developer*	*Approx. speed (100 keV)*	*Comments*
Ilford EM4 plate EM5 EM6	PQ Universal	0.6 0.3 0.1	Metric and inch sizes available
Ilford N4E50 roll film N4E50 sheet film	PQ Universal	0.3	Polyester base 70 mm rolls metric and inch sizes in sheets
Kodak Electron Image plate	D19 HRP	0.2–2.0	Wide range of speed with different development conditions
Kodak EM 4489 sheet film	D19	0.4	Polyester base
Kodalith LR 2572 roll film	D19	0.4	35 and 70 mm mm rolls special order for EM sizes polyester base
Kodak Fine Grain Positive roll film	DG-10	0.05	35 mm rolls triacetate base
Eastman (Kodak) Fine Grain Release Positive roll film	D19 Dektol	0.06 0.1	35 mm and 70 mm rolls triacetate base special order for EM sizes

the merit of being equally useful, when suitably diluted, as print, slide, and general negative developers.

As discussed in § 7.2.3 the principal effect of a change in developer is an alteration in the emulsion speed. Other effects may occur, e.g. a change in fog-level, or *DQE*, but these effects do not normally introduce appreciable visible effects on the final print except perhaps in extreme enlargements.

The practical factors, and also many of the theoretical factors, in the formulation of emulsions for electron microscopy are now well understood by the manufacturers, and there would seem to be little point in further experimentation or deviation from present day recommendations, unless of course, some particular effect or property is required for special purposes. The improvements still sought by electron microscopists are more concerned with matters of cost and availability than with the technical performance.

In an emergency, almost any emulsion can be used; at the least a usable image should result, and at the best a good quality image may be obtained. However, the spectral sensitivity is likely to seriously restrict the type of emulsion usable in an emergency to the photo-mechanical and similar categories. Panchromatic films used for ordinary photographic purposes are too sensitive to visible light to allow of their use in most electron microscopes.

7.3.2 PLATES AND FILMS

The traditional base material for supporting photographic emulsions is the glass plate. The rigidity and dimensional stability of photographic plates are particular virtues for scientific work, and the majority of electron microscope exposures have been made on plates. However, plates are comparatively bulky, heavy, and fragile, and recent increases in the cost of plates together with the difficulties of storing large numbers, have led to an increasing interest in the use of flexible film.

In the past, the use of film in the electron microscope has presented some difficulties. A long length of film is difficult to desiccate effectively, and if desiccation is carried too far, the dry film accumulates static charges which give rise to discharge patterns on the emulsion. Acetate-based films liberate the relatively volatile plasticiser and contaminate the vacuum system, and the dimensional stability can be rather poor. With the introduction of polyester film as a base material, these disadvantages have largely been overcome. For most purposes in electron microscopy, the polyester-based film is entirely satisfactory, and in some respects, notably storage and handling, is superior to the glass plate.

Film can be obtained in the form of rolls several metres in length – *roll-film* – or as individual sheets with the dimensions of the dark-slide or plate – *sheet-film*. Sheet-film can be fitted directly or with the aid of adaptors to most of the cassettes or dark-slides used in commercial electron microscopes. Many workers have devised simple techniques for cutting roll-film into sheets to suit particular instruments; film cuts easily with scissors or a sharp knife, but care must be taken to avoid touching the area to be exposed (see § 7.3.4).

The use of sheet-film offers the greatest advantage and presents the minimum of change to the microscopist accustomed to glass plates.

The use of roll film offers special advantages when long series of exposures are necessary. Roll film cameras are available for most of the commercial electron microscopes, to accept 35 mm or 70 mm film in various lengths. In some cameras up to 250 exposures can be accommodated. If extensive use is to be made of roll film, particularly in the 70 mm size, the use of professional mechanised processing equipment is strongly recommended.

For exacting work where the operator needs to make sure that the information is properly recorded before proceeding further, the comparatively long length of roll film is perhaps a disadvantage. A camera with a built-in guillotine will alleviate the difficulty.

Many of the plates primarily designed for light microscopy, slide making, or lithographic work, but which are used for electron microscopy, have an anti-halation backing on the glass. The backing is usually a red dye, which reduces reflection or halation of light from the back surface of the glass by modifying the refractive index at the interface. This backing is, of course, ineffective for electron exposures and whilst having no deleterious effect on the quality of the processed plate, the deposition of the dye in the processing equipment can be a nuisance. Plates for electron microscopy, if not primarily designed for electron microscopy, should be specified as 'unbacked'.

Variations in thickness or refractive index of the glass support can introduce spurious contrast in optical diffraction experiments (Horne and Markham in Beeston et al. 1972).

7.3.3 PLATE AND FILM DIMENSIONS

In the U.K. the $3\frac{1}{4}$ in × $3\frac{1}{4}$ in and $3\frac{1}{4}$ in × $4\frac{1}{4}$ in plate sizes are in common use, the chosen sizes being determined by the electron microscope manufacturers' cassette sizes. In the rest of Europe the 6.5 cm × 9 cm size predominates, and in the USA the 4 in × $3\frac{1}{4}$ in 'lantern' size is widely used; 10 in × 2 in 'spectrographic' size plates were used in early RCA instruments.

There is, unfortunately, no standardisation amongst instrument manufacturers in regard to plate sizes for electron microscopy, and at the present time, every size of plate available seems to be in use somewhere. Ultimately, standardisation will be enforced by economic factors; correspondence with the manufacturers (Agfa-Gevaert 1970; Ilford 1970; Kodak 1973) has indicated that general use of the 6.5 cm × 9 cm size would be to the advantage of the microscopist.

The situation with roll film is more fortunate; only two sizes, 35 mm and 70 mm, are in widespread use in electron microscopy, and both sizes are international standards. The 35 mm size is used mainly in cameras which are located in the upper part of the viewing chamber, close to the final projector lens, whereas the 70 mm size is used in cameras which are fitted in place of the conventional plate cameras below the normal viewing screen. Both systems have advantages (see § 2.12). The 70 mm film is usually transported by friction rollers and the film is unperforated, i.e. there are no sprocket holes. Sprocket holes are a potential cause of uneven development, for work of the highest quality.

The plate and film sizes quoted are nominal; photographic manufacturers will supply dimensional tolerances to microscopists or designers wishing to modify or make cameras or handling devices.

7.3.4 DEFECTS OF PLATES AND FILMS

For a given electron exposure, the density of a processed photographic material is dependent to some extent on the thickness of the emulsion, hence any non-uniformity in thickness will give rise to a modulation of the actual image densities. Such defective coatings are normally rejected by the manufacturer. However, it appears that certain plates, cut from the edge of the large coated glass sheet, have a band of thickened emulsion along one, or if cut from the corner, two edges of the plate. These plates show a dark band after development, up to 1 cm in width and superimposed on the electron image. This defect can be avoided by specifying 'no coated edges'. Recently, those manufacturers who supply plates specifically designed for electron microscopy (as distinct from plates which are used for electron microscopy but primarily designed for some other purpose) have announced that the emulsions are uniform in thickness on all plates and the 'no coated edge' specification should no longer be necessary.

A band of abnormal density at the edge of a plate can also result from a lifting or separation of the emulsion from the plate.

Difficulty has occasionally been experienced with the appearance of a shower of pinholes on a batch of plates. This defect arises during manufacture and if many plates are affected the suppliers should be contacted.

Pinholes can also result from the processing procedures and chemicals if inadequate attention is paid to agitation and the cleanliness of solutions (see § 7.5.8).

Occasionally, processed emulsions may be found to have a mottled appearance and the emulsion may be blistered or partially stripped from the glass or film backing. These defects can usually be attributed to inadequate storage (see § 7.3.9), over-desiccation (see § 7.3.5), or excessive fixing and washing (see § 7.5.8).

Glass plates sometimes have a small hook or burr on a corner which may prevent an easy fit in the cassette. The burr will usually break off with gentle pressure. The edges of glass plates are supplied 'as cleaved' and unpolished, consequently tiny fragments of glass dust sometimes lodge in the cassettes or stick to the plates. Before loading plates it is worth tapping an edge of the plate lightly on the side (not the top) of the bench to dislodge such particles. Cassettes and dark-slides should also be cleaned in a similar way. Blowing out the dust often results in the dust being transferred from one unwanted place to another, or sometimes into the eyes, and is not recommended.

Thin slivers or whiskers of film are sometimes found when using cut-film or roll-film. These whiskers can charge electrically in the electron beam, and can interfere with camera mechanisms; the whiskers must be removed.

When exposed to the electron beam, the plate or film surface is also exposed directly to any particulate matter which may be present in the lower part of the instrument. Camera housings collect a surprising amount of debris.

In addition to being radiation sensitive, the silver halide emulsion is also pressure sensitive, and if the emulsion is mechanically disturbed by abrasion with dust or the fingers, or by damaged cassettes and dark-slides, dark marks will appear when the emulsion is developed. To reduce the sensitivity to abrasion, plates and films are usually *super-coated* with a thin layer of tough gelatin on top of the halide-carrying layer. Care must be taken not to touch the emulsion surface, nevertheless. Always handle plates and films by gripping two opposite edges, and not by gripping the front and back surfaces.

This sensitivity to pressure can be put to advantage, in that emulsions can be numbered or otherwise marked before processing, by using a pointed scriber. A rounded stylus is often recommended but a medium grade (HB) pencil has the advantage that the writing can be seen under the safe-light of the darkroom before processing. After development, of course, the mark will appear as a permanent density in the emulsion.

7.3.5 DESICCATION OF PLATES AND FILMS

The photographic emulsion contains a proportion of water in normal ambient conditions, and when exposed to the vacuum of the electron microscope will liberate water vapour in sufficient quantity to impair the vacuum. Etching of the specimen material may also be caused, and the filament life may be reduced by the reaction of water vapour with the hot tungsten emitter. Flexible film usually contains a proportion of plasticiser which also has a high vapour pressure. It is necessary, therefore, to partially dry photographic material before it is loaded into the camera of the electron microscope, by a period of storage in a vacuum desiccator.

Some electron microscopes have built-in desiccator chambers; others require the use of a separate free-standing desiccator. It is an advantage to have the desiccator adjacent to the electron microscope so that the instrument can be kept almost continuously loaded with desiccated plates or film with a minimum period of exposure to the atmosphere and with minimum effort on the part of the operator.

The effectiveness of a vacuum desiccator is improved considerably if a chemical desiccant is also loaded into the vacuum chamber. The traditional desiccant is phosphorus pentoxide powder. Care should be taken to prevent dry powder blowing into the pump or onto the cassettes. The phosphoric acid formed is corrosive. A recommendation is often made to breathe moisture onto the phosphorus pentoxide powder so as to form a protective crust which will prevent the powder blowing about. Unfortunately this procedure reduces the effectiveness of the desiccant.

Vacuum desiccators can be run without an auxiliary chemical desiccant, and gas-ballasted rotary pumps are usually fitted to desiccators to allow the pumping of a large amount of water vapour without severe emulsification of the pump oil. It is not necessary to run a gas-ballasted pump continuously in the gas-ballasted mode, but when it is in this mode, some oil vapour will almost certainly be expelled. A suitable exhaust pipe should be fitted, of about 20 mm bore, to lead the fumes out of the laboratory. This is a desirable precaution for any of the rotary pumps used in electron microscope laboratories, as the hydro-carbon vapours are injurious to health.

Plates and sheet-film can usually be adequately dried in about 30 min, but the period will depend appreciably on the degree to which the emulsion surface is effectively exposed to the vacuum, and on the basic pumping speed (for water vapour) of the desiccator unit. Cassettes or dark-slides should be partially opened if this is possible. Film on a tightly wound spool is difficult to desiccate effectively. Short lengths of a few metres can be wound on the stainless steel processing spiral and desiccated in the processing tank provided that this is also made of metal. Plastic tanks are not as satisfactory for this particular purpose. It is, of course, useless to attempt to desiccate photographic material in the manufacturer's wrappings.

Although some microscopists successfully use the vacuum desiccator as a 'store' for unexposed photographic material, it should be noted that prolonged desiccation can sometimes give rise to mottling or fogging of the material, particularly if the temperature in the desiccator is allowed to exceed 20 °C. Blistering and stripping of the emulsion can also occur with prolonged desiccation of film, and for Kodak polyester-based film, for example, the manufacturers recommend a maximum desiccation period of 4 hr.

Desiccated photographic material should be exposed to the atmosphere for as short a time as possible, before loading into the electron microscope, otherwise moisture from the air will be rapidly re-absorbed.

7.3.6 DIMENSIONAL STABILITY AND DISTORTION OF PLATES AND FILMS

The dimensional stability of photographic films and plates is important in electron diffraction and microscopy as measurements often have to be made directly from the negative material.

Changes in humidity and temperature may cause temporary dimensional changes. Glass plates are unaffected by humidity changes, but the polyester based films have humidity coefficients of about 0.002%/% relative humidity change. The older types of cellulose acetate and nitrate bases have coefficients up to 4 times this value. Thus for a maximum change in relative humidity of say 50%, the dimensional change will be zero with glass plates, 0.1% with polyester film, and up to 0.4% for acetate film.

The temperature coefficient of expansion for glass plates is about 0.001%/°C and is about 3 and 6 times this value for polyester and acetate films respectively. Thus for an ambient temperature change of say, 10 °C, the expansion will not exceed 0.06% even with acetate film.

Magnification, dimensional and positional measurements, at least for all normal electron microscopy purposes, are seldom made to better than 0.2%. Thus the dimensional stability of glass and polyester-based material is more than adequate for practical purposes as far as ambient temperature and humidity effects are concerned.

Of greater importance are the permanent changes which take place during processing and subsequent natural ageing. The extreme conditions of wetness and dryness during processing operations may lead to complex changes and the resulting expansion or contraction is not predictable and difficult to analyse. For this reason, for the most exacting work, calibration standards for either microscopy or diffraction should be exposed onto the plate or film carrying the image to be measured, either adjacent to the image (which is facilitated by the 'half-mask' or 'half-frame' in some electron microscopes) or superimposed on the image.

7.3.7 PRINTING PAPERS

Photographic printing papers are of generally similar construction to plates and films, except that the support for the emulsion is a tough paper. Two thicknesses of paper are commonly available – single-weight and double-weight. Single-weight paper is adequate for prints which will be mounted, or sandwiched in reports, but for unmounted prints which may have frequent handling, the double-weight paper is recommended.

The longest range of print densities will be obtained with a white base paper; the tinted papers (ivory) are not much used in electron microscopy.

A range of surface textures is available from most manufacturers. For almost all electron micrographs a smooth glossy surface is chosen, since the object is to display fine image detail with maximum sharpness and brightness range. The brightness range can be extended by glazing a smooth glossy surface (see § 7.4.2). For very large exhibition prints, however, where reflections from a glossy surface could be troublesome, the possibilities of some of the slightly more textured papers should not be overlooked.

Three principal types of emulsion are available for printing papers; chloride papers are for contact printing (see § 7.5.11), and bromide and chloro-bromide papers are for enlarging (see § 7.5.10). Bromide papers usually give a neutral black image; chloro-bromide papers typically produce a warmer image tone. The warmth of tone can be influenced by the exposure and processing conditions (see § 7.6).

The speed of both bromide and chloro-bromide papers is such that exposure times in the range 5–50 sec are usual for EM negatives enlarged and processed with the apparatus and developers described in § 7.4 and 7.6.

Each type of paper is available in a range of contrast grades to match the density range of the negative, as shown in Table 7.2. The contrast grades shown are for Kodak Bromide paper, and are typical for most bromide papers, although the exposure ranges (contrast scales) of different manufacturers' products may not coincide exactly, grade for grade. Special variable-contrast papers are also available, with which the contrast is varied by using one of a range of colour filters during the enlarger exposure (e.g. Ilford 'Multigrade' and Kodak 'Polycontrast'). An advantage is that only one 'grade' of paper needs to be stocked.

For use with rapid stabilisation processors (see § 7.4.3) special printing papers containing a development agent are required (e.g. Ilford 'Ilfoprint'

TABLE 7.2

Contrast grades of printing paper

The negative		match with	Printing paper		
Contrast	*Density range*	⟶	*Exposure range* (Log E units)	*Grade*	*Number*
Very high	1.4		1.7	Extra soft	0
High	1.2–1.4		1.5	Soft	1
Medium	1.0–1.2		1.3	Normal	2
Low	0.8–1.0		1.1	Hard	3
Very low	0.6–0.8		0.9	Extra hard	4

and Kodak 'Ektamatic' projection paper). Variable-contrast stabilisation papers are also available (e.g. Kodak 'Ektamatic' selective contrast paper). Stabilisation papers, although primarily designed for processing with the stabiliser/activator apparatus, can also be processed with conventional developer and fixer to give prints of bromide paper quality.

Printing papers are available in a range of sizes, with both inch and metric dimensions (see the manufacturers' catalogues). The 10 in × 8 in size, and the DIN standard A4 and A5 sizes (297 × 210 and 210 × 148 mm) are popular for electron micrographs.

7.3.8 DIMENSIONAL STABILITY OF PRINTS

Precise measurements are best made from the plate or film negatives, but in some circumstances, especially where features in the image have to be annotated with reference marks or other information, it may be more convenient to make measurements from photographic prints. Unfortunately, printing paper suffers anisotropic dimensional changes during processing and especially when the print is dried on a hot glazing drum. The author has measured dimensional changes up to 5% between extracting unexposed paper from the manufacturers box and dry-mounting the print. Paper is very hygroscopic and further dimensional changes can take place during storage. Thus, distortion is an important factor, as well as straightforward expansion and contraction.

If it is essential to make precise measurements from prints, the necessary dimensional stability can be achieved by using a special paper of sandwich construction with an aluminium foil filling, e.g. Kodak Bromide Foil Card. These materials are exposed and processed exactly as for normal printing paper, but the presence of the stabilising layer ensures that dimensional changes during processing and storage are less than 0.05%.

In considering distortion in printed images, a possible contribution from a defective enlarger must not be overlooked (see § 7.4.2).

7.3.9 STORAGE OF UNEXPOSED PHOTOGRAPHIC MATERIAL

In general, any environment which is comfortable for an electron microscopist in terms of temperature, humidity, and cleanliness, is also quite adequate for the storage of unexposed photographic material. Ideally, however, and for maximum shelf life of the unexposed material, it should be stored in a cold box or refrigerator kept specially for the purpose at a temperature in the range 4 °C to 10 °C, in a clean atmosphere of relative humidity less than 60%, free from acid or alkali fumes, and in surroundings free from ionising radiations. (Storage next to an X-ray laboratory is not recommended!)

When material is removed from the cold store, a suitable period, sometimes up to 2 hr, must be allowed for the boxes of material to reach ambient temperature, otherwise condensation of moisture may seriously damage the material if the package seals are broken before the warming period has elapsed. In exceptionally humid atmospheres such as in tropical zones, extra precautions should be taken to protect photographic material, both unexposed and exposed (see for example, Kodak Technical Data Sheet no. RS-6).

Many electron microscope laboratories are air-conditioned or have more temperate atmospheres, and in these circumstances it is usually adequate and convenient to store the photographic material on a shelf or in a cupboard. Storage in the darkroom itself is quite satisfactory provided that the darkroom is well ventilated. A darkroom with a humid atmosphere and a strong aroma of fixer or acetic acid stop-bath is a hostile environment both to unexposed photographic material and to the microscopist.

Storage for prolonged periods in moist conditions with temperatures widely fluctuating or in excess of 24 °C may lead to fogging, mottling, loss of speed and other change of characteristics. Polaroid film in particular, as used for external photography, is damaged rapidly in the author's experience, at temperatures greater than 24 °C. If the storage conditions necessitate the use of a desiccant, silica gel is very convenient and can be re-activated from time-to-time by heating at 200 °C for about 1 hr in a laboratory oven. In an emergency dry unused tea leaves can be used as an effective desiccant.

All photographic material should be stored well above floor level to ensure protection against accidental flooding or other water damage.

7.3.10 STORAGE OF EXPOSED PHOTOGRAPHIC MATERIAL

A little consideration will show that the photographic negative is the most valuable item in the electron microscope laboratory. The negative represents the end result of the expenditure of a great deal of money and effort. Indeed, the information on the negative is the *raison d'être* of the entire electron microscope operation.

Prints can be re-made easily; negatives are often irreplaceable. The storage of negatives and other exposed photographic material is of the greatest importance, therefore.

All photographic materials must be completely dry and dust-free before being stored.

Many electron microscope plates (e.g. Ilford and Kodak) are now supplied in boxes with internal walls of corrugated plastic, which hold the plates back-to-back in separated pairs. Such boxes can be re-used to house processed plates. They are ideal for temporary storage but too bulky to use as a permanent store.

For permanent storage, individual plates, cut-films, and short strips of roll-film are usually housed in bags made from cellophane, translucent paper, or polythene. Clear bags have the advantage that negatives can be inspected or identification numbers read without removal from the bag. Transparent polythene bags are available with a matt writing panel. They are waterproof, and if the type with a self-sealing mouth is used, they afford maximum protection to the negative against moisture, fungi, bacteria, insects, and dust, all of which may have a deleterious effect on the emulsion. Cellophane and paper bags are satisfactory but some types have a central seam which occasionally 'leaks' adhesive onto the negative. Negatives should be inserted into this type of bag with the glass or film backing against the seam, rather than the emulsion.

Short strips of roll film can be stored in loose-leaf negative albums which are now available for both the 35 mm and 70 mm film sizes. These albums usually contain index pages and can hold up to 2400 exposures in the 35 mm size. Long lengths of roll-film can be stored in standard aluminium film cans.

Metal card-filing cabinets are ideal for the permanent storage of all photographic materials. They are readily available in a range of sizes from office furniture suppliers, and they are robust, lockable, and fire-resistant. The smaller sizes are stackable and can be mounted on strong shelves. The larger cabinets are designed for floor mounting, and the type with legs should be chosen as a precaution against possible flooding or water damage to the lower drawers.

A metal cabinet filled with photographic plates is very heavy, and after some years of electron microscopy in an active laboratory, a floor loading and accommodation problem can arise. For this reason the longer-term storage of negatives is often relegated to a little-used basement or corridor. This can be satisfactory provided that the environment is favourable. As with unexposed material, negatives must be protected from extremes of temperature and humidity, and kept free from dust and other damaging particulate material. The fine dust generated in building operations is abrasive and particularly penetrating. Negatives, and most other items in the electron microscope laboratory, must be suitable protected if building work is in progress nearby. Combinations of high temperature and humidity are particularly damaging and will favour bacterial and fungal attack of the gelatin of the emulsion. Such damage is usually irreparable.

On occasions negatives of special historical or scientific interest may require to be processed and stored for archival permanence. Details of the special precautions to be taken in these circumstances are given in the 'Kodak Data Book of Applied Photography'.

7.4 Photographic apparatus

The fundamental requirements for any photographic darkroom are that all unwanted light should be excluded, the room should be provided with an adequate distribution and intensity of *safe-lighting*, i.e. light with a spectrum falling outside the spectral response of the photographic materials, and wet processing operations involving chemicals and water should be physically separated from the operations associated with the handling of dry apparatus and materials. This latter requirement usually results in the provision of a separate *wet-bench* and a *dry-bench*.

The design principles of negative and print darkrooms and the layout of the major items of apparatus which are usually fixed in position are discussed in Alderson (1974). In the following sections, the various pieces of photographic apparatus are described in relation to their use in the processing of negatives, prints, and slides.

7.4.1 NEGATIVE PROCESSING APPARATUS

The necessary apparatus can be conveniently considered in relation to the various processing procedures.

The major steps in the processing of negatives are as follows:

(a) unload the cassettes or magazines containing the exposed negative material,

(b) load the plate rack or film spiral,

(c) develop,

(d) rinse,

(e) fix,

(f) inspect briefly, and if necessary, reject,

(g) wash,

(h) dry,

(i) bag,

(j) re-load the cassettes or magazines,

(k) take the loaded cassettes to the desiccator,

(l) remove the processed negatives for evaluation or storage.

Many modern electron microscopes have built-in devices for printing serial numbers and other information on the plate or film. If a numbering device is not provided, an *HB pencil* will be required to identify each plate as it is unloaded from the magazine.

In the early days of photography, negatives were often processed in a dish of developer; some workers still prefer dish development, especially when only a few plates are to be processed at a time. A standard 25 cm × 20 cm (10 in × 8 in) dish will comfortably accommodate up to six EM plates. In this type of processing it is usual to provide *four dishes*, one each for developing, rinsing, fixing, and washing. To facilitate the transfer of negatives from one solution to another, plastic *forceps*, which can grip a negative close to an edge, or some form of plastic *scoop* which can be slid under the negative, will be required. Scoops should be provided with an adequate number of holes to allow the solutions to drain off before transfer.

(It is very bad practice to immerse the fingers in photographic processing solutions. At the very least, it is difficult to remove from the fingers the aroma of the solutions, especially the fixer, and at the worst, contamination of other solutions and pieces of apparatus will occur. Some workers may develop an unpleasant form of dermatitis.)

For the development of large numbers of negatives, dishes have been superseded by developing tanks. For plates, the tanks are rectangular and the plates are loaded into *racks*. For films, the tanks are cylindrical and the film is wound onto a *spiral*. Stainless steel tanks and racks will last almost indefinitely, but they can be expensive. The early plastic tanks were not too satisfactory; the material was brittle and easily damaged, and surface-crazing occurred after prolonged use with processing solutions. The more recent plastics are much more satisfactory in these respects.

For processing plates, at least two tanks are required, one for the developer and the other for the fixer. Rinsing between developing and fixing, and final washing of the plates can be performed in the darkroom sink, but it will be more convenient to provide additional tanks for washing, so that the sink can be used for other purposes. At least three racks are recommended, each accepting a full camera-load of plates, but additional racks should be acquired if possible. These will allow simultaneous processing, drying and general inspection and handling of several batches of plates. No other handling devices are required when racks are used; the racks usually have handles which project above the level of the solutions. Plate tanks are usually provided with lids which are useful for keeping out dust during periods of inactivity, or for the temporary exclusion of light. The lids are not normally used during processing; the safe-lighting will be switched on, and agitation is by means of lifting the plate racks up and down in the tank.

For processing cut film (individual sheets) plate or cut film racks are used, as described above.

For processing lengths of roll-film, the same tank and spiral is normally used throughout the processing operations, with the various solutions poured into and out of the tank as required. Film tanks are always provided with light-tight lids (the pouring apertures are suitably light-trapped) and processing can take place in white light if required, once the loaded spiral is in place with the lid correctly secured. It is useful to have more than one spiral, so that when one film is being washed (with the spiral in a large beaker, for example) another film can be developed.

A suitable length of small diameter hose pipe attached to a cold water outlet, and dipping into the tank, will facilitate washing.

A *white light* will be required for inspecting the negatives after fixing. The main white room-lighting can be used, but it is much better to have a separate white light close to the sink, so that drips from the wet negative do not contaminate the floor and run down the operator's sleeve. The white light can be a simple pearl lamp, or a more sophisticated wall-mounted *negative viewer* of the type specially made for wet materials. Negatives are usually inspected critically with a $10\times$ *hand magnifier*; most microscopists carry such a magnifier in their pockets. A separate magnifier supplied for use in the darkroom only is very likely to disappear!

Negatives which are not rejected at this stage will then be washed and dried. Drying can be by means of natural convection, but it is more usual to provide a photographic *drying cabinet* which will exclude dust, and which has a source of heat and (sometimes) forced circulation. Negatives can be dried very quickly in this way. Plates and cut film remain on the processing racks during drying; roll-film is usually unwound from the spiral and hung vertically in the drying cabinet.

The negatives should be protected from dust and other damage with suitable *negative bags* or *film-strip albums*, as soon as possible after drying is completed. Generally, negatives should be bagged on the drybench before removal from the darkroom. Plates and cut-films are bagged individually; roll-film is usually cut into strips of six frames or more, and a pair of good quality *scissors* will be needed on the dry-bench. On occasions when plates are required for immediate inspection by a group of observers, such as during a demonstration, it will be more convenient to take the rack of plates out of the darkroom for evaluation. Racks which have been out of the darkroom should be rinsed before re-use to remove any dust. Some workers prefer to have a serial number printed on the negative bag as well as on the negative; a *felt-tipped pen* with a fine point gives a clearly visible print on most materials from which bags are constructed.

In modern processing techniques, it is necessary to control the temperature of the solutions and the times of immersion of the photographic material. For EM purposes the permissible temperature range is moderately wide. Close control of the solution temperatures is not required, although the actual temperature must be known. The solution temperatures can be adjusted by the addition of hot (or cold) water during the preparation of the solutions. Tap water is usually rather too cold for direct use in a developing solution, especially if the water source is a bore-hole rather than a surface reservoir, although at the other extreme in a very hot climate, it may be necessary to add ice, as a means of temperature adjustment. A small clearly readable *thermometer* (0–50 °C) will be required. A number of proprietory thermostatic controllers are available for photographic purposes, but these are not normally required in EM darkrooms, unless the room itself is subject to very wide fluctuations in temperature.

The time of development is normally controlled to within a few seconds in a period of a few minutes. A wall-mounted electric *clock* with a centre-sweep second hand should be fixed above the wet-bench or in a position clearly visible from the wet-banch. Some workers like to make use of an additional *photographic timer*, in the form of a clock with a pre-set alarm bell.

Developer is usually bought as a concentrated solution, for subsequent dilution with water in a range of ratios from 1:3 to 1:20. Two *graduated measures* will be required; 100 ml and 1 litre capacities will be convenient. The graduated stainless steel or polythene jugs supplied for domestic use are more stable and easier to manipulate than the conventional cylindrical laboratory measures. These measures can also be used for preparing the fixer, which is now commonly available as a concentrated solution. A large polythene *funnel* will be useful. Filtering of photographic solutions can be performed rapidly and quite satisfactorily with a plug of *cotton wool* (batting). The use of filter paper is unnecessary, and unless a very coarse grade is used, is extremely slow and tedious.

7.4.2 PRINT PREPARATION APPARATUS

As for the negative processing apparatus, the printing apparatus can be conveniently considered in relation to the important steps in the processing sequence, which is as follows:

(a) load the negative into the enlarger negative carrier,

(b) adjust the magnification and focus of the enlarger, and the position of the easel, to give the required image,

(c) number the sheet of printing paper,

(d) position the paper on the easel,

(e) expose the paper,

(f) develop the print,

(g) rinse in the stop-bath,

(h) fix,

(i) inspect briefly, and if necessary, reject,

(j) wash,

(k) dry (and glaze),

(l) remove from darkroom for mounting, trimming, retouching, and final presentation and evaluation.

Most electron micrographs are enlargements of the original negatives, produced with an illuminating and projection system called an *enlarger*. The enlarger is the most important piece of darkroom apparatus, and as it is a link in the chain of events concerned with the extraction of information from the specimen, a high quality enlarger should be acquired, with appropriate excellence of the optical system.

The optical system has two important functions – to project a magnified (or de-magnified) image of the illuminated negative onto the *easel* or *printing frame* where the printing paper is temporarily held, and to illuminate the negative uniformly. Three basic optical systems are available, as shown in Fig. 7.16.

Many enlargers have a cold-cathode light source, (Fig. 7.16a) which is compact and cool in operation, and gives uniform illumination of the negative without the need for adjustment relative to the negative. The cold-cathode source gives a bluish light, not entirely easy to work with at the low intensity levels often experienced at the enlarging easel when working with dense electron microscope negatives. The light source is diffuse and the prints tend to be 'soft', lacking slightly in the contrast and sharpness required for scientific work. Although the cold-cathode source is a very convenient illumination system, admirable for pictorial work, it is not recommended for the majority of electron microscope work.

Crisper, more contrasty results will be obtained with an opal lamp and condenser light source (Fig. 7.16b). The opal lamp is specially made for enlarging purposes, has a uniform white diffusing layer on the lamp envelope, and can be obtained in various power ratings. The 150 watt type is used extensively. (The domestic pearl lamp is not sufficiently uniform for use in enlargers but could be used in an emergency.) For optimum quality of the illuminating system, the condenser is a matched pair of optical glass (not plastic) lenses. usually provided with a heat absorbing filter. Transmitted heat from the lamp is normally of no consequence with glass negatives, but it can cause unsupported film negatives to buckle.

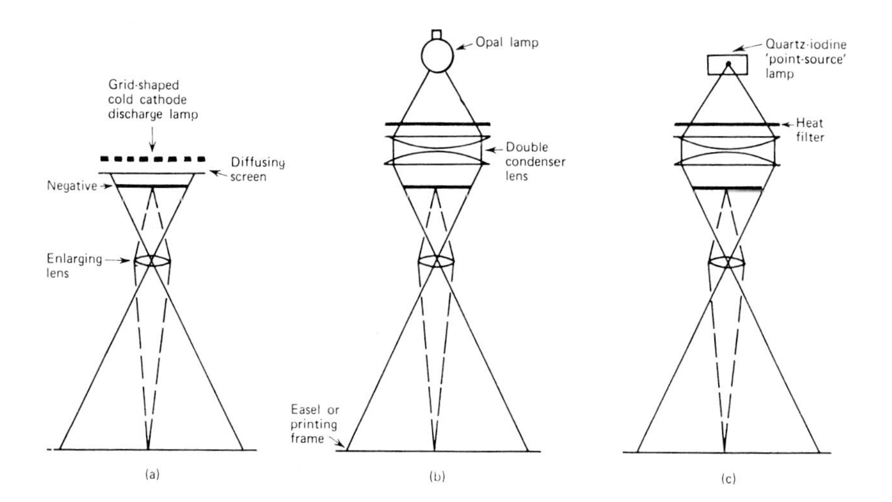

Fig. 7.16. Enlarger optical systems.

To ensure that the negative is uniformly illuminated, the light source must be focused into the plane of the enlarging lens as shown in Fig. 7.16. This is usually achieved by adjustment of the position of the lamp along the optical axis, and in certain instruments by the provision of extra or substitute condenser lenses. The enlarger magnification is altered by moving the entire lamphouse, negative and enlarging lens assembly up and down a column or guide rails, with respect to the fixed easel which is normally at bench height. At each change of magnification the image is re-focused by moving the enlarging lens on a small subsidiary column or screwed concentric tube arrangement, with respect to the negative. For moderate changes in the position of the enlarging lens, it will not be necessary to adjust the position of the light source, as the opal lamp is a relatively large diffuse source. For very wide ranges of magnification, however, it may be essential to re-focus the light source in order to ensure even illumination of the negative. For general purposes a suitable compromise position for the source can usually be found. For the majority of electron microscope work, the opal lamp and condenser illuminating system is entirely satisfactory and probably offers the best compromise between quality and convenience in use.

The use of a point source of light such as a quartz iodine lamp (Fig. 7.16c) will increase the intensity of illumination and the contrast even further, and if much ultimate performance electron microscopy is to be undertaken (in terms of resolution) the use of a point source enlarger is highly desirable if not essential. The low-voltage quartz iodine sources of high luminous efficiency, developed for slide projectors, can usually be fitted to enlargers with a small modification. It is essential with a point source that the source and lenses are correctly aligned; both axial and transverse adjustments of the source are necessary, otherwise difficulties will arise from colour-fringes and non-uniform illumination. Enlargers with point sources are more tedious to use, and compromise is usually not acceptable. However, for maximum sharpness of fine detail in a negative (including, unfortunately, all the defects such as scratches, hairs, and pinholes!) the point source enlarger offers a real advantage.

The influence of the type of light source on the contrast of the micrograph is quite marked. There is approximately one contrast grade difference between a cold-cathode and an opal lamp/condenser source, and almost one grade between the opal lamp and the point source.

The maximum size of negative which can be accommodated in an enlarger is determined optically by the size of the condenser, and mechanically by the negative carrier assembly. Both factors should be checked when purchasing an enlarger; it is especially important that the illuminating beam from the condenser, *in the plane of the negative*, will adequately cover the whole of the negative.

The lens panel together with the actual lens mounting flange, the negative carrier, the enlarging easel or baseboard, and the printing frame (paper-holder) must be exactly parallel to each other and perpendicular to the optical axis if distortion and non-uniform sharpness is to be avoided. Since some 'professional' grade enlargers are deficient in this respect, it is well worth checking with a test negative. Test negatives are commercially available; they can be made for the present purpose by ruling a grid of sharp lines on an exposed and processed plate with a scalpel or needle. This test will probably also reveal that at full aperture the enlarging lens does not have a flat field, i.e. a different focusing setting is required for the edge compared with the centre of the negative. This point must be borne in mind when making enlargements and the maximum usable aperture determined by previous experiment. In general, a very high quality lens will give best performance within one or two stops of the maximum aperture. The lens aberrations are corrected zonally, and further stopping down will remove the contribution of outer zones and may degrade the performance. An inferior lens, on the other hand, will usually improve in performance as it is stopped down.

Generally, the focal length of the enlarging lens should be about equal to the diagonal of the negative. Thus, for 6 cm × 9 cm (or $2\frac{1}{2}$ in × $3\frac{1}{2}$ in) negatives a 10 or 12 cm ($4\frac{1}{2}$ in) focal lens should suffice, and for the larger 9 cm × 12 cm (or $3\frac{1}{4}$ in × 4 in) negatives, a 15 cm lens will be required.

For very high magnifications of a restricted area of a negative, a shorter focal length of say, 7.5 cm, can be used with advantage. Note that if one corner only of a negative is to be enlarged, the focal length must be sufficient for coverage of the whole of the negative and not just the corner, unless, of course, the negative carrier permits the chosen area to be centred about the enlarger axis. Some carriers do not have this adjustment; the negative is located in a shallow well or is held in position with a frame. It is usually possible to have a plain flat carrier, with no restrictive locating device, machined in the local workshop. Some carriers have a plain face on the lower surface, and can be simply inverted in the enlarger. The negative can then be positioned as required on the flat surface of the carrier; it is important that the clear peripheral areas are masked off to avoid flare, as described later.

For enlargements of the highest quality over the full field of view of the negative, first quality lenses are required. Unlike the well-known and reputable camera lenses which often show only marginal differences in performance, 'good' enlarging lenses seem to vary appreciably in performance. The author has observed some surprising and very marked differences, particularly in respect of flatness of field, between enlarging lenses of the highest reputation. A choice of lens should be based on a demonstrated result, preferably in the microscopist's own darkroom.

Automatic focusing enlargers, as exemplified by the Leitz Focomat, are particularly valuable in laboratories where several workers may have intermittent use of the enlarger, but it must be stressed that if the auto-focus facility is not of the highest quality, there is little point in acquiring this type of enlarger. The Leitz instrument, nominally for 6 cm × 9 cm negatives can be modified to accommodate (optically and mechanically) the $3\frac{1}{4}$ in × $3\frac{1}{4}$ in size (Evennett and Oates 1968).

In laboratories where appreciable numbers of very large prints are produced a bench-level focusing control will be very useful. At 10 × magnification for example, a 15 cm enlarging lens would be 165 cm from the paper holder and at this distance the critical focusing required is difficult without a bench-level control.

If it is desired to use the enlarger to make slides in the internationally standard 2 in × 2 in size (35 mm frame size of 24 mm × 36 mm) the enlarger must permit a reduction of 3 or 4 times to be made. This will require the use of extra long bellows or extension tubes with appropriate focusing adjustment. Under these conditions some axial adjustment of the light source will usually be essential to ensure even illumination of the negative.

A rotatable negative carrier is an extremely useful facility.

It is important that the negative carrier adequately masks off any clear or unexposed peripheral regions of the negative otherwise intense light passing through the clear regions will be scattered in the enlarging lens and will give rise to flare with consequent fogging and loss of contrast in the image. Masks can be improvised from black paper or by using opaque adhesive tape.

The photographic exposure given to the printing paper is the product of the light intensity and the exposure time. The intensity is a function of the light source used, and of the aperture of the enlarging lens. The light source intensity is usually fixed in a given enlarger. Although intensity can be reduced by stopping down the enlarging lens, it is best for critical work to use the variable aperture to control the performance of the lens, rather than the transmitted light intensity (modern enlarging lenses have 'click-stops' which give an audible indication of the aperture setting). Exposure, therefore, is usually controlled by varying the time, and an electrical timing device will be found useful in obtaining reproducible exposures. Some form of enlarging exposure meter will help to avoid wasted paper or the need to make frequent test strips. The functions of both devices can be combined in one instrument, e.g. the Melico Enlarging Photo Computer (Crawley 1966).

With a properly exposed negative and an opal lamp or point source there should be no difficulty in focusing the enlarger visually on the easel. However, some workers like to make use of a focusing aid based on the principle shown in Fig. 7.17, which allows a small part of the image to be viewed effectively in transmission, with a magnifier. Focusing aids are available commercially.

Before processing the exposed print, the negative number, the enlarger magnification (or total magnification) and any other desired information can be written on the back of the printing paper with a *soft pencil* (not a hard sharp point). Take care that the emulsion side of the print is not scuffed during the writing; the emulsion is pressure sensitive.

Printing papers are normally developed in *dishes* or *trays* (automatic processors are described later). Three dishes are required to contain the developer, the acid stop-bath, and the fixer. The earlier enamelled steel dishes have largely been superseded by plastic dishes which are lighter and much less susceptible to damage. For most general purposes within an EM department

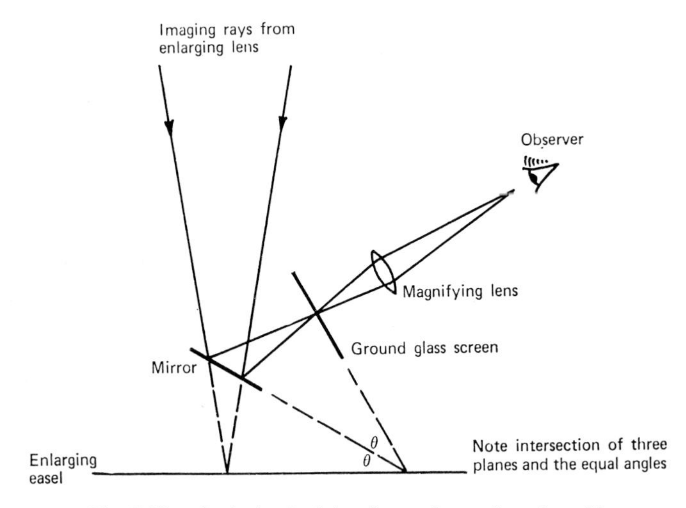

Fig. 7.17. Optical principle of an enlarger focusing aid.

prints are no larger than 25 cm × 20 cm (10 in × 8 in). For ease of manipulation of the prints, a dish should be chosen which allows a few centimetres of clearance all round the print. If larger prints are required frequently, perhaps for exhibition purposes, it will be worth acquiring a set of correspondingly large dishes. If the demand for large prints is only occasional, however, economies can be made by improvising large dishes with a temporary wooden frame resting on the bench (a base is not required) and lined with a sheet of polythene.

Prints are manipulated in the dishes and transferred with the aid of plastic print *forceps*. Prints should be handled at the edges only, to avoid pressure marks on the emulsion.

After fixing, the print can be examined in *white light*. The normal white room lighting in the printing darkroom may be sufficiently strong for critical examination of the print, but it will usually be more convenient to have a separate white light over the sink, shaded so that light falls on the sink and not on the operator's eyes.

Any prints rejected at this stage should be placed in a watertight *plastic bin*, and not in a metal bin which would rapidly corrode.

Prints can be washed quite adequately in a sink, providing that the water is kept moving and is completely changed from time to time. However, in view of the importance of thorough washing, print washing apparatus is often installed close to the sink. Print washing devices are available commercially (although it is not difficult to construct a washer from developing dishes) usually in the form of a horizontal drum or a series of dishes mounted in echelon fashion. An adequate flow of water to ensure constant movement of the prints is an important factor in many types of print washing device.

After washing, excess water can be removed with a plastic *sponge*. Paper with a smooth glossy surface is chosen for electron micrographs to reveal the maximum detail, and for maximum contrast (see § 7.3.7) the prints are glazed. Prints are usually dried and glazed in one operation. Small numbers of prints can be dried and glazed on static glazing apparatus. Large numbers of prints can be handled much more conveniently on a *rotary drum glazer*. Small numbers of prints can be dried without glazing by natural convection after clipping to a line, or by supporting on a layer of clean cloth. The method of drying prints by stacking between sheets of absorbent paper is not recommended; the drying time is very long, and fibres may be embedded in the emulsion surface. The rotary glazer is a very worthwhile investment, even in a small EM department.

For trimming prints and cutting paper in the darkroom a *print trimmer* or *guillotine* will be required on the dry bench. Prints will also be trimmed during the mounting procedure which usually takes place on a separate bench outside the darkroom; the mounting boards may also require cutting and an additional *guillotine* will be required on the dry-mounting bench. The darkroom guillotine, for paper only, can be of light construction, but for the mounting guillotine a robust construction will be required to cope with the thick cardboard mounts.

A variety of rubber-based adhesives can be used for attaching prints to the mounting boards, but the professional method is to use *dry-mounting tissue* as the adhesive. The tissue is tacked onto the print, and then to the mount (after trimming) with a *dry-mounting iron*. The assembly is then inserted in a *dry-mounting press* which melts the adhesive tissue and firmly attaches the print to the mount.

An alternative method is to use a mounting film (e.g. Lomacoll) which is coated on both sides with a contact adhesive. No additional apparatus is required with this method, but the print has to be carefully located on the mount before contact is made; sliding adjustments are impossible with this type of adhesive.

If photographic prints are to be exhibited or submitted for publication, it is usual to remove minor blemishes on the prints by spotting or etching techniques. A fine *camel-hair brush* and a *scalpel* are necessary for these retouching operations. Two or three small bottles of different shades of *retouching dye* are required. A *steel ruler*, *pencils*, and a soft pencil *eraser* for cleaning mounts, will be useful in the final stages of print preparation.

Graduated measures and a *funnel* are required for the preparation of solutions, and a *clock* and *thermometer* for the processing operations, as described for the negative materials in § 7.4.1.

7.4.3 RAPID PRINT PROCESSING APPARATUS

In recent years a number of rapid print processors have become available commercially. These processors make use of a *stabilisation processing* technique, and the two-solution process, in which an activator and a stabiliser replace the conventional developer and fixer, is most suitable for continuous tone electron micrographs. Special printing paper is required for this method of processing. After exposing the paper in the enlarger the paper is fed into the processor. Fifteen seconds later the fully processed print emerges, and after a period of a few minutes at room temperature, the print is completely dry and ready for use.

This method of print processing is ideal when prints are required as soon as possible after the negatives have been processed.

Stabilised prints have limited permanence; as processed by the stabilisation technique, the unexposed silver halide is only temporarily stable. (In conventional processing, the unexposed silver halide is dissolved by the fixer.) Stabilised prints will deteriorate rapidly if exposed to moisture, strong sunlight, or excessive heat, but they can be made permanent if required, by conventional fixing and washing, after their initial purpose has been served.

Stabilisation processors are available from several manufacturers (see Appendix) and they can be accommodated comfortably in a wet-bench space of less than 1 m^2.

An excellent practical description of the use of stabilisation processors in electron microscopy is given in Kodak Publication P-236 (1973).

7.4.4 SLIDE MAKING APPARATUS

The electron microscopist frequently requires slides for projection during lectures and demonstrations.

Slides can be made from EM negatives quite simply by using the enlarger. An image of the negative is projected with the necessary de-magnification onto a standard 50 mm × 50 mm (2 in × 2 in) fine grained plate. The plate is exposed, and then processed in a similar way to an EM negative. After drying, the plate can be bound up with a mask and a cover glass to form a standard slide. 35 mm roll-film has some advantages for this method of slide making. The film can be held in a transporting device, such as the Leitz ELDA printer which is designed especially for this purpose, and each processed frame can be inserted quickly and conveniently into standard plastic slide holders which are now universally available.

Titles and simple diagrams can be prepared for projection by using Kodak Ektagraphic Write-On Slides. These slides consist of a colourless translucent plastic material in a 50 mm × 50 mm card mount. Suitable marking materials include lead pencil, ball-point and fibre-tipped pens, and crayons.

It will be clear that the simple methods outlined above have some severe limitations. Many slides for EM purposes require annotations on the micrographs, and slides showing diagrams, graphs, tables and apparatus are often required. This type of material requires the use of a positive master print, together with a copying camera and associated lighting and ancillary apparatus. A description of the more sophisticated methods of slide making is clearly beyond the scope of this chapter, which is concerned primarily with EM image recording rather than with visual aids. Descriptions of the various slide making and copying techniques are given in photographic manuals (e.g. Horder 1968) and in the technical data sheets of the photographic manufacturers. Hammond (1963a) has described a versatile apparatus for making 50 mm × 50 mm slides, and Farrand (1961), Hammond (1957, 1963b) and Turnbull (1958) have discussed various factors affecting legibility and presentation.

Although the methods and apparatus are not described here, the making of high quality slides is well within the capability of an experienced EM photographic technician.

7.4.5 OTHER PHOTOGRAPHIC APPARATUS

The provision of adequate safe-lighting is an important factor in any darkroom. The safe-light fittings are best regarded as part of the lighting installation, since it is usual to have these fittings permanently wired when the darkroom is constructed. The actual *safelight screens*, however, may be regarded as photographic apparatus, as they may be changed at will by the operator.

Generally one set of screens, usually amber-yellow, will suffice for handling materials in the printing darkroom. Darker screens, usually red, are required in the negative darkroom for handling materials for electron image recording. Safelight screens will deteriorate rapidly if they are overheated and the manufacturers' recommendations for maximum lamp power (usually 25 W) should be followed.

Regular and consistent agitation during development is an important factor in ensuring uniformity of development. Manual agitation is not difficult, but in a negative darkroom with a reasonably continuous throughput, some operator's time can be usefully saved by installing *nitrogen burst* apparatus. With this apparatus, short duration bursts of nitrogen gas are liberated at the base of the developing tank. The rising gas bubbles provide a very efficient method of solution agitation. The timing and duration of the gas bursts are adjustable, and the control units will usually supply a number of tanks.

An inspection of the photographic manufacturers' professional product catalogues will show that there are many hundreds of devices and gadgets, all of which can serve some special purpose during the various stages in the production of electron micrographs. It is well worth studying these catalogues to become familiar with the devices available, but the beginner in electron microscopy is strongly advised to make a start with the basic items of apparatus previously described in this section, which will be entirely satisfactory and sufficiently comprehensive for producing negatives and prints of the highest quality. When some practical experience has been gained, the need, or otherwise, for additional apparatus will be seen much more clearly.

7.5 Photographic methods

There are a number of methods for processing EM negatives and preparing electron micrographs; one or more of these methods will usually be adopted as a standard routine, to suit the amount and type of work in the department. As with all technical activities, some preliminary work is necessary before the actual photographic work can start. In a large EM department the photographic technician is usually responsible for ensuring that the photographic solutions and apparatus are ready for immediate use, and that the darkrooms and benches are clean and tidy. In the small department, it is possible that the microscopists themselves will be required to share the work involved in preparing and maintaining the photographic facilities. Whichever arrangement is adopted, the various responsibilities should be clearly defined and understood, and it is important that adequate records are kept.

The preparations necessary for photographic work, and step-by-step operating procedures for the various practical methods, are given in this section. Frustrating difficulties may be occasionally encountered in photographic work; the more common difficulties and some suggested remedies are included.

7.5.1 INITIAL PREPARATIONS – NEGATIVE DARKROOM

(1) Switch on the roomlighting.

(2) Switch on the safelighting. If the negative darkroom has been temporarily used for non-EM work, check that the correct type of safelight screen is in position in each fitting.

(3) Switch on the negative drying cabinet.

(4) Set out the processing tanks or dishes (trays). Larger tanks and dishes are usually left in position permanently.

(5) Prepare the developer. If a liquid concentrate is used, dilute the concentrate with water in the correct proportions, using the graduated measures. If a dish or tank is in use for the very first time, it will be necessary to establish the total volume of developer required. It is useful to display this type of information on a small notice-board on the darkroom wall. The level of liquid should be about 1 cm above the top of the negative materials, as held in the racks or spirals. Adjust the temperature of the developer during preparation by mixing hot and cold water, as required. The temperature is usually set in the range 18 °C–21 °C. Small deviations (of one or two degrees) from the recommended processing temperature can be allowed for by adjusting the developing time. Large variations may alter the developer characteristics and should be avoided. There will be no problem if the darkroom is properly heated and ventilated (see Alderson 1974). Add the developer to the tank or dish. If roll-film is to be developed on a spiral, using white light illumination during processing, the prepared developer can be stored temporarily in a bottle, ready for pouring into the loaded tank. If safelighting is to be used during roll-film processing, add the developer to the tank.

(6) Prepare the water rinse. The water rinse between developing and fixing can take place in the sink using running tap water. However, it is desirable that the temperature of the emulsion remains reasonably constant during

processing, and for this reason, a separate tank or dish containing water at the temperature of the developer is recommended. In the photographic text books, a stop bath of glacial acetic acid is recommended to follow development. This is entirely satisfactory and desirable for papers, and for many negative materials, but an acid stop bath is *not* recommended for EM negative materials.

(7) Prepare the fixer. Dilute the liquid concentrate with water in the correct proportion, using the graduated measures, and adjust the temperature to be within a few degrees of the developer temperature. (If powders are used for preparing the developer and/or fixer, the mixing is best performed outside the darkroom on a general chemical bench, to avoid the possibility of contamination in the darkroom from particles of the powder.) Add the prepared fixer to the tank or dish, or temporarily store the fixer in a bottle if roll-film processing is to be performed in white lighting.

(8) Mark the time and date of preparation of fresh developer and fixer on the darkroom notice board.

(9) Wipe up any spilt solution, and wash the hands.

(10) Tidy the dry bench, and check that the required number of clean, dry racks, spirals, or sheet film hangers are to hand. A pencil may be required, to identify the negatives.

(11) Check that negative materials are available in the darkroom cupboard for re-loading the cassettes or magazines after processing.

7.5.2 PROCESSING PLATES – DISH (TRAY) METHOD
(Suitable for small numbers of plates only.)

(1) Transfer the exposed plates from the electron microscope to the dry-bench, using the light-tight magazines or boxes supplied by the EM manufacturer.

(2) Switch on the safe-lighting.

(3) If a light trap is not provided at the darkroom entrance, latch the darkroom door. If an 'engaged' signal is fitted outside the door, close the appropriate switch.

(4) Switch off the room-lighting. Pause a few moments, to become adapted to the change in lighting intensity.

(5) Open the plate cassettes in the magazine, and transfer the plates, in order (for convenience), to the developing tray. If the plates have not been automatically numbered in the electron microscope, write an identification number at the edge of each plate, on the emulsion side, using an HB pencil.

(6) Note the time, or start the pre-set timer, and slide each plate gently

into the developer, emulsion side up, and manipulate into place with forceps or a scoop.

(7) Agitate the developer continuously by raising and lowering each side of the tray in turn. The developer should swill smoothly across the plates, with no splashing or violent turbulence.

(8) Develop for the prescribed time, then lift each plate out of the developer with the forceps or scoop, allow a moment for excess solution to drain off, and transfer to the water rinse. Remove each plate in the order in which they were immersed, to ensure consistent development for each plate.

(9) Agitate thoroughly in the water rinse for about $1\frac{1}{2}$ min.

(10) Transfer the plates from the water rinse to the fixer.

(11) Agitate the fixer for about 10 sec each minute. As fixing proceeds, the milky appearance of the negative will gradually disappear. Fix for at least twice the time taken for the negative to clear. A time of 10 min is usually adequate.

(12) Switch on the negative viewer or white light, briefly rinse excess fixer off the plate, and inspect each plate, using a hand magnifier. Reject any obviously unsatisfactory plates and discard into the waste-bin.

(13) Transfer satisfactory plates to a dish or rack and wash for 30 min in running water. The flow should ensure a complete change of water about every 5 min. If there is any doubt about the flow, empty the washing dish or tank every 5 min.

(14) Transfer the washed plates to a drying rack, with the plates in a vertical plane, and place the rack of plates in the drying cabinet. Drying time will depend on the type of cabinet, and the extent of the loading of wet materials, but can be as short as 15 min. The drying temperature should preferably not exceed 40 °C.

(15) Switch off the negative viewer, or white light, or room lighting.

(16) Check that the dark-slides are free from particles of glass or slivers of emulsion, by tapping gently over a waste bin (not over the dry bench or darkroom floor).

(17) Re-load the cassettes, and assemble the magazine.

(18) Switch on the room lighting. The processing sequence is now complete. When the plates are dry, protect with negative bags, and remove from the darkroom for evaluation and storage.

7.5.3 PROCESSING PLATES – TANK METHOD

Steps 1 to 4 – as in § 7.5.2.

(5) Open the cassettes in the magazine, and load the plates one by one

into the slots of the plate rack. If the plates have not been automatically numbered in the electron microscope, write an identification number close to the edge of each plate, using an HB pencil, on the emulsion side.

(6) Note the time, or start the pre-set timer, and lower the plate rack smoothly into the developing tank.

(7) Raise and lower the rack once or twice, with a gentle shaking action, to dislodge any air-bells (i.e. small static bubbles) which may have formed on the surface of the emulsion.

(8) Agitate by raising the plate rack up and down in the solution for 10 sec every minute. A good interchange of solution can be promoted by lifting the rack completely out of the developer during the raising and lowering action, but the rack must be immediately re-immersed in order to avoid oxidation difficulties. Violent turbulence of the solution is not required. If nitrogen-burst apparatus is used for providing solution agitation, switch on the burst valve at the start of development, and leave on until the development is completed. 1 second bursts at intervals of 10 sec will usually suffice. Switch off the burst-valve before removing the rack, to avoid spray out of the tank. Manual agitation is not required when burst agitation is in use.

(9) Develop for the prescribed time, then lift the rack out of the developer, allow excess solution to drain off, and then transfer the rack to the water rinse.

(10) Agitate thoroughly in the water rinse for $1\frac{1}{2}$ min.

(11) Drain off excess water, and transfer the rack to the fixing tank.

(12) Agitate in the fixer for 10 sec every minute, and fix for at least twice the time taken for the milky appearance to clear. With conventional fixers, a time of 10 min will usually be adequate. This time can be reduced to 3 min if rapid fixers are used (see § 7.6).

(13) Proceed to a brief inspection, washing, drying and re-loading, as in § 7.5.2. The plates can remain in the same rack throughout all these operations. When washing, lead the water hose into the bottom of the tank to ensure an adequate change of water; fixer is heavier than water.

7.5.4 PROCESSING SHEET FILM – TANK METHOD

Sheet film is processed in the same way as plates, as described in § 7.5.3. The separate sheets of film, however, are held in sheet-film hangers, the top bar of which rests on the edges of the developing tank.

If nitrogen burst agitation is not used, manual agitation will be made much easier if the separate film hangers are loaded into a standard sheet film processing rack. The assembly is then handled exactly the same as a plate rack.

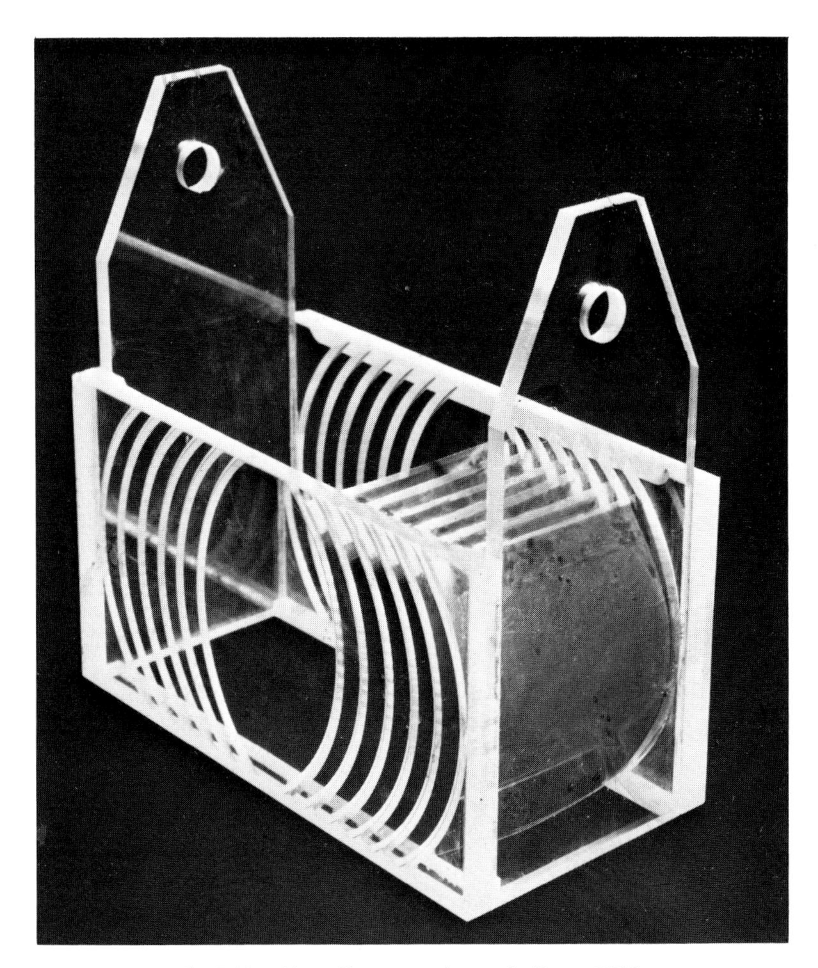

Fig. 7.18. Sheet-film processing rack (Oates 1972).

Alternatively, a sheet film rack of the type devised by Oates (1972) (Fig. 7.18) can be constructed. Separate hangers are not then required.

7.5.5 PROCESSING ROLL-FILM – ROLL-FILM TANK METHOD

Steps 1 to 4 – as in § 7.5.2.

(5) Open the film cassette, to reveal the leading end of the film.

(6) Set the film spiral to the required width. (Most spirals will adjust, to cope with a range of film sizes; some spirals are designed for one width only.)

(7) Feed the leading end of the film into the gate of the spiral, and push

the film gradually into the spiral. Very short lengths will slide into place with one push. To facilitate the loading of longer films, most spirals are provided with a small amount of rotary play in the spiral flanges, so that the film can be wound in by alternate rotations of the flanges. Very long lengths of film (several metres) are best loaded with a proprietary automatic loading device. With all spirals check that the tail end of the film is correctly in place, otherwise the film may tend to unwind during agitation.

Spirals must be *completely dry*; the film will stick if any part of the spiral track is wet.

(8) Place the loaded spiral in the empty developing tank, and secure the lid.

(9) Switch on the room-lighting. Processing may now proceed in white lighting.

(10) Note the time, or start the pre-set timer, and pour in the developer.

(11) Insert the central stirring rod, and agitate for 10 sec every minute, with three spins clockwise, three spins anti-clockwise, and three spins clockwise. (Some roll-film tanks are designed to be agitated by inversion – follow the manufacturer's instructions.)

(12) Develop for the prescribed time, then pour the developer out of the tank. (It is usual to prepare fresh developer for each roll-film, but the used solution may be collected and used again, if required; used solutions require compensating corrections to the development time.)

(13) Pour in the water rinse and agitate thoroughly with clockwise and anti-clockwise spins of the spiral, for about 1½ min.

(14) Pour out the water rinse.

(15) Pour in the fixer. Fix for 10 min (shorter times for rapid fixers) with agitation for 10 sec every minute.

(16) Remove the tank lid.

(17) Pour out the fixer, and briefly rinse with water.

(18) Switch on the negative viewer, and briefly inspect the last few exposed frames. (It is not usual to unwind the whole of the film for inspection at this stage; it is extremely difficult to rewind a wet film. If this has to be attempted for some reason, immerse the film, the spiral and the hands in water, and try to rewind completely under water.) Push the short length of inspected film back into place.

(19) Wash the film in running water for 30 min, with the water hose led down the centre of the spiral to the bottom of the tank. Empty the tank at intervals, to ensure several complete changes of water.

(20) Allow excess water to drain from the spiral, then unwind the film carefully.

(21) Hang the film vertically in the drying cabinet. Attach a clip to the bottom of the film (a heavy clip is necessary) to keep the film taut during drying; an unweighted film will tend to curl on itself as it dries, and the emulsion may be ruined.

(22) Switch off all white lighting.

(23) Re-load the cassette.

(24) Switch on the room lighting.

The processing sequence is completed by cutting the dried film into individual frames, or more usually, into short lengths of 5 or 6 frames, for storage in film-strip albums.

7.5.6 PROCESSING ROLL FILM – OPEN TANK METHOD

When a continuous throughput of roll-film is to be processed, it is more convenient to use a series of open tanks, as described in § 7.5.3 for processing plates. The film spiral is transferred from tank to tank, exactly as for the plate racks. In this method, safelighting must be used, of course.

If it is desired to process several roll-films at the same time, proprietary processing racks are available which will contain a large number of film spirals. The rack of spirals is handled in exactly the same way as a rack of plates; processing tanks of the appropriate size are required.

7.5.7 RAPID DRYING OF PLATES AND FILMS

For most purposes, the speed of drying of negatives in a heated drying cabinet, and of prints in a drying and glazing machine, is quite adequate. Various methods have been devised for accelerating the process; most of the published methods seem to have some disadvantage, and for general EM purposes accelerated drying methods are not recommended. In certain circumstances, however, perhaps in the course of a demonstration, there may be a need for rapid drying of negatives, and the following technique has been used successfully with plates and polyester-based sheet and roll-film.

Prepare a bath of about 70% industrial alcohol in clean filtered water. (De-ionised water has often been recommended for EM processes, but traces of the ion exchange resin are almost inevitably carried over into the effluent from the de-ioniser; the resin will contaminate photographic emulsions, and also EM specimens.) It is unnecessary and uneconomic to use laboratory-grade methanol, but it is a wise precaution to filter industrial alcohol as it is usually supplied from metal drums.

Following the usual washing period for the plates or film, transfer the negatives to a clean water rinse containing a drop of a surface active agent (e.g. Photoflo). Agitate thoroughly, lift out of the rinse, shake off excess water and immerse the plate rack or film spiral in the alcohol bath for 2 min, using moderate agitation. Remove from the bath, shake off excess liquid, then transfer to the drying cabinet. Film and plates will normally dry in less than 5 min after this treatment.

Do not attempt to use alcohol on material which has already partially dried. Stronger alcohol mixtures may cause clouding or opalescence of the emulsion. If the negatives are very valuable, try the technique first on an unwanted negative, similarly processed.

The rapid drying technique of setting fire to an alcohol-soaked print or plate, often recommended in non-scientific photographic journals, is a useful attraction on laboratory open days, but is certainly not recommended for use in routine electron microscopy!

7.5.8 POSSIBLE DIFFICULTIES AND REMEDIES IN NEGATIVE PROCESSING

During development, negative materials will gradually blacken at the exposed areas. Negatives in a processing solution always look more dense than they really are, and the beginner, seeing a rapidly blackening negative, may be tempted to 'snatch' the negative before development is completed. This is very bad practice and will lead to poor quality images. Development should always be controlled by time and temperature; if the resultant image is too dense, the negative is overexposed, and the electron image should be recorded again with the correct exposure.

A mottled appearance on the negatives may result if the water rinse is inadequate in time or degree of agitation, or if the note concerning acid stop-baths is disregarded.

Streaks and patches of uneven density will arise if the agitation is insufficient, or is too violent. (Uneven density can also arise from incorrect setting of the EM illuminating system see § 4.2.)

Dense stress marks will appear if the emulsion has been pressed or abraded. Care must be taken during loading and unloading, and also to avoid plates sliding over one another during dish development.

Small air-bells sometimes appear on the surface of negative materials in the developer. They must be dislodged, otherwise development will be impeded at the area of contact and 'pinholes' will appear on the processed plate. A sharp tap on the developing rack or spiral will usually suffice. If air-bells persist during dish development, they can be removed with a soft, wide camel-hair brush.

Emulsions may become detached from the film or glass support if washing is very prolonged. Overnight washing, apparently attractive as a time saver, can be particularly troublesome in this respect.

Drying marks, in the form of streaks or blobs on the surfaces of the negative, sometimes appear especially if the washing water contains particulate matter and dissolved mineral salts. They can be avoided by a final rinse, after the main washing, to which a few drops of a neutral detergent liquid have been added (e.g. Teepol). Proprietary photographic wetting agents are also available for this purpose.

Squeegees and sponges are sometimes recommended for removing excess water from negatives prior to drying. They must be exceptionally clean and applied lightly, otherwise the swollen emulsion may be damaged irreparably. Sponging should not be necessary if the washing water is clean and the drying cabinet is effective.

7.5.9 INITIAL PREPARATIONS – PRINTING DARKROOM

(1) Switch on the room lighting.

(2) Switch on the safelighting. If the printing darkroom has previously been in use for non-EM work, check that the correct type of safelight screen is in position in each fitting.

(3) Set out the processing dishes (trays). Large dishes are usually left permanently in place on the wet bench.

(4) Prepare the developer. If a liquid concentrate is used, dilute the concentrate with water in the correct proportions, using the graduated measures. Adjust the temperature during preparation by mixing hot and cold water, as required. The temperature is usually set in the range 18 °C–21 °C. Add the developer to the dish. If the dish is in use for the first time, the required volume of solution must be established, and noted for future reference. A solution depth of about 2 cm is usually adequate. Set out the print forceps.

(5) Prepare the stop bath.

(6) Prepare the fixer. Adjust the temperature to be in the range 18 °C–21 °C. For processing prints, the preservation of a reasonably constant temperature throughout processing is not quite as important as for negative materials. If the temperature falls much below 18 °C, however, the solution activities will begin to fall off rapidly.

(7) Note the time and date of preparation of the fresh solutions on the darkroom notice board.

(8) Wipe up any spilt solution, and wash the hands.

(9) Check that the glazing sheet or drum is clean and polished. (Follow

the glazer manufacturer's detailed instructions for maintenance of the glazing surfaces.)

(10) Check the functioning of the enlarger, and set up the lenses and negative carriers appropriate for the work in hand.

(11) Switch on the enlarger timer and exposure meter, if fitted.

(12) Check that the guillotine is in place, for cutting test strips. A soft pencil will be required for identifying the prints.

(13) Check that the appropriate type of printing paper, in a suitable range of grades, is available in the darkroom cupboard.

(14) Switch on the drying and glazing machine.

(15) If it is intended to mount the prints immediately after processing, switch on the dry-mounting iron and the dry-mounting press, on the print finishing bench.

7.5.10 PROCESSING PRINTS – ENLARGING

When a negative is enlarged for the first time, the correct exposure for the printing paper, and the correct grade of printing paper will be unknown. The first steps in the enlarging procedure, therefore, are concerned with establishing the exposure time and the grade of paper to use. When some experience has been gained, the appropriate grade of paper can usually be selected after a visual examination of the negative. Several devices are available commercially for measuring the required exposure and considerable time can be saved by using such a device. However, beginners are recommended to follow the test-strip procedure outlined below until a little experience has been gained. The advantages, methods of use, and limitations of enlarger exposing devices will then be much more readily appreciated.

(1) Insert the negative into the negative carrier, emulsion side facing the easel. If necessary, blank off any completely clear regions surrounding the negative (see § 7.4.2.)

(2) Adjust the easel or paper holder to the required size of print.

(3) Switch the safelighting on, and the room lighting off.

(4) Switch on the enlarger light source.

(5) Set the enlarging lens to maximum aperture (i.e. diaphragm fully open).

(6) Adjust the focus of the enlarging lens, and the position of the enlarger assembly on the column, so that a focused image of the desired magnification is obtained on the easel.

(7) Adjust the lateral position of the paper holder to obtain the desired field of view and picture composition.

(8) Check the focus carefully, using a magnifier.

(9) Stop down the enlarging lens to say, two stops below maximum aperture. (This adjustment may be modified by experience.)

(10) Switch off the light source. (Some enlargers are provided with a miniature safelight screen, which can be swung over the enlarging lens, thus obviating the need to switch off when making adjustments to the paper.)

(11) Select a sheet of normal grade paper, and cut a strip about 5 cm wide and long enough to straddle a representative region of the image, in terms of densities and contrast.

(12) Place the test-strip, emulsion side up, to cover the chosen region of the image.

(13) Prepare a piece of opaque card, large enough to cover the test strip. Hold the card in the hand, at the ready.

(14) Switch on the light source, to expose the paper strip. After say, 2 sec exposure, cover about 1 cm of the width of the test-strip with the opaque card. (The card does not have to be in contact with the printing paper.) At intervals of 4, 8, 16, 32 sec, progressively cover the test strip, so that at the completion of exposure the strip will be exposed in 5 bands of 2, 4, 8, 16 and 32 sec exposure respectively.

(15) Switch off the enlarger, and remove the test strip.

(16) Slide the test strip into the developer. Using the print forceps, turn the strip over once to dislodge air-bells and ensure that the strip is uniformly wetted. Develop for 2 min (or the paper manufacturer's recommended time). Agitate steadily throughout development by gently raising and lowering alternate edges of the dish, as in negative processing by the dish method (§ 7.5.2).

(17) Lift the print out of the developer and allow a few seconds for excess developer to drain off.

(18) Transfer the print to the stop bath for 15 sec, moving the print in the bath with forceps.

(19) Lift out of the stop bath, drain, and transfer to the fixer. Agitate steadily throughout fixing. Fix for 3 min (rapid fixers) or 10 min (conventional fixers).

(20) After 30 sec of agitation in the fixer (or 3 min if conventional fixer is used) the white light may be switched on. *Before* switching on the white light check that the lid is in place on the box of printing paper, and if the lid has to be replaced, make sure that the hands are dry and clean. (If the forceps have been used correctly, the hands will be dry and clean. It is *bad practice* to paddle with the fingers in processing solutions – see § 7.4.1.)

(21) Inspect the test-strip under the white light. The print will show 5 bands of different density and contrast. One of these bands should show a good density range in which a very light part of the negative has printed as a dense black, and a dark part of the negative has printed to a shade just perceptibly darker than the white base of the paper. The exposure for this band of the test strip will be the correct exposure for the whole print. A small adjustment to the exposure time may seem desirable. If, unluckily, the test exposures have not successfully straddled the optimum exposure, the test must be repeated with an appropriate adjustment to the exposure increments.

If the correct grade of paper has not been chosen, the test strip will show a band with an adequate density range (black to white) but with few intermediate tones. In this case the paper chosen is too hard, and a softer grade is required (see § 7.2.6). Alternatively, the test strip may show a band with a good range of separable tones, but with poor contrast from one end of the tonal range to the other; in this case, the paper chosen is too soft, and a harder grade is required.

(22) If necessary, repeat the test using a different grade of paper.

(23) When the correct grade of paper and the required exposure time have been established, place a full sheet of the chosen paper in the paper holder, expose the paper, and develop the print as described for the test strip. Mark the grade of paper on the back of the print, for future reference, before processing.

(24) The white light may be switched on for inspection of the print after 30 sec of fixing but if the print is satisfactory and is required for further use, it must be fixed for the full 3 min.

Wet prints viewed under a darkroom white light have a slightly different appearance to that seen in daylight when the print is dry and glazed. This must be allowed for when judging the adequacy of exposure in the darkroom.

(25) After fixing, transfer the print to the lowest tray of the print washer. In the echelon type of washer, batches of prints are moved *up* the echelon, so that freshly introduced prints containing a high proportion of fixer do not contaminate prints which have been partially washed. Wash for 60 min.

(26) After washing, transfer the prints to a final water rinse containing a few drops of wetting agent (e.g. Shell 'Teepol' or Kodak 'Photoflo') and agitate for about 1 min. This will improve the quality of the glaze.

(27) Remove the prints from the water rinse, sponge or drain off excess water, and dry the prints with the emulsion side in contact with the glazing drum

(28) If the prints tend to curl after drying, draw the back of the print steadily and firmly over a clean smooth table edge, to flatten the print, taking care not to crack the glazed emulsion.

The tendency to curl after drying and glazing can be considerably reduced by stacking or storing prints face-to-face, in pairs.

7.5.11 PROCESSING PRINTS – CONTACT PRINTING

A contact print is made by exposing a printing paper to a light source, as in the enlarger, but with the negative held in firm contact with the printing paper, usually by means of a transparent glass plate. The glass plate, negative and printing paper can be held in a *contact printing frame* under a light source, or contact *printing boxes* may be used in which the light source is in a closed box fitted with a glass plate on the upper surface. The negative and paper are laid on the glass plate and held in contact with a pressure pad.

Test exposures are usually necessary to ensure that the exposure time and grade of paper are correct, and the processing procedure is closely similar to that for enlargements.

Contact prints are not much used in electron microscopy – the print is the same size as the negative and the opportunities for special printing techniques (see § 7.5.13) are somewhat limited – but some laboratories like to keep records of all exposures in the form of contact prints. When the negatives are small, 35 mm size for example, a large number of records can be held on a single 25 cm × 20 cm (10 in × 8 in) sheet of printing paper.

7.5.12 PROCESSING PRINTS – RAPID STABILISATION PROCESSORS

The procedure for preparing prints using automatic rapid stabilisation processors is exactly the same as for the conventional enlarging technique described in § 7.5.10, except that steps 16 to 27 concerned with developing, fixing, washing, and drying, are replaced by only two steps in which:

(a) the exposed test-strip or complete print is fed into the processor and retrieved from the processor some 15 sec later;

(b) the processed print, which is slightly damp and limp, is exposed to the room atmosphere for a few minutes, in order to dry the print.

The correct exposure for the paper must be determined, as described in § 7.5.10, and a correct contrast grade must be found.

The exposure range (contrast grade) of certain stabilisation processing papers (Kodak Ektamatic SC paper, for example) can be varied by exposing the paper to light which has passed through one of a range of special Polycontrast filters. Variation of exposure range is achieved by varying the filter.

In this way, only one type of paper has to be stocked; the appropriate filter can be brought into use on the enlarger, as required.

(This method of grade variation has also been used for conventional printing paper – see § 7.3.7.)

Stabilised prints cannot be glazed. The main advantage is rapid processing. When the immediate purpose has been served, stabilised prints can be made permanent, if desired, by fixing and washing, as described in § 7.5.10.

7.5.13 SPECIAL PRINTING TECHNIQUES

The majority of EM negatives can be printed in the straightforward manner previously described, but occasionally negatives will be encountered which, because of some unusual extremes of contrast, or unavoidable non-uniformity of exposure, or other disturbing feature, will benefit from special attention during the exposing procedure in the enlarger.

For example, a print may have a strong highlight which attracts the eye, but which is not at the intended centre of attraction from the point of view of information. The distracting highlight can be subdued by increasing the exposure of the distracting region, during enlargement.

The amount of increased exposure can be readily determined from a test-strip, or it can usually be estimated satisfactorily, when some experience has been gained. A 'burning-in' or 'printing-in' mask is required which consists of a sheet of opaque card, sufficiently large to more than cover the whole of the print. An aperture is cut in the card, to correspond in shape to the area to be printed-in. To avoid a sharp shadow edge, the card is held several centimetres above the print during the exposure, and the size of the aperture must, therefore, be smaller than the size of the highlight. Printing-in is facilitated if the enlarger is fitted with a safelight shutter which can be swung in front of the enlarging lens. The image can then be seen on the paper, without exposing the paper. When the normal print has been exposed, 'close' the shutter. Hold the mask in place, with the aperture in the correct position. 'Open' the shutter, and expose the highlight for the desired time. A slight circular oscillation of the mask during exposure will be an additional help in avoiding an unsightly shadow edge. Skilled printers will often use their hands and fingers to form a suitable aperture, but some experience is required for this method to be used with confidence. More control can be exercised if the exposure time for the printing-in is increased by stopping down the enlarging lens, after the normal print exposure has been made.

An alternative method of printing-in is to use a very small pentorch, or one of the proprietory light micro-probes used for inspection purposes in the fine mechanisms industry. The author has successfully used L-shaped pieces of Perspex (methyl methacrylate) rod, with ground ends, fastened to a hand torch, as narrow light pipes. The light beam can be 'painted' over the highlight in stages, and this method is especially useful when the highlight is rather narrow or is of unusual shape.

An opposite situation is frequently encountered where a region of the print is too dark. In this situation it is necessary to reduce the exposure to the dark area by interposing a stop made from a piece of opaque card cut approximately to the shape of the dark area, and held on a thin wire. If the normal exposure time for the print is only a few seconds, it is worth increasing the time by stopping down the lens. More control of the print 'dodging' procedure can be achieved in this way. To avoid a shadow from the wire, and a hard shadow edge from the stop, keep the stop moving with a suitable small amplitude during the time it is interposed in the light beam.

Electron diffraction patterns can often present some appreciable difficulties in printing, especially when there is a high background of inelastically scattered electrons whose intensity falls off rapidly with the angle of scattering, i.e. as the edge of the negative is approached. Often, straight printing will reveal either the central region of the pattern, or the outer regions, but not both. This is an example where compression of the density range can be usefully attempted during negative development (Farnell and Flint 1969) and where printing-in, or alternatively, dodging, can be used during printing. Circular stops will be required, and to achieve a variation in effective diameter during the dodging exposure, the stop can be moved steadily from a position close to the enlarging lens, where the masking effect on the print will be greatest, to a position close to the print, where the masking will be effective over a small central region only. Unsharp masking techniques can also be used (Hamilton 1968) as described in Beeston et al. (1972).

Compression and expansion of the density range, and local printing-in and dodging can be achieved very effectively by electronic means. The Log Etronic enlarger uses a flying-spot raster on the face of a cathode-ray tube as a light source in the enlarger. Feedback probes at the easel, and also below the negative, can modulate the light beam and control the exposure of the printing paper, element by element. The apparatus is comparatively expensive, and would be difficult to justify in a small EM department. In circumstances where there is a large and continuous throughput of 'difficult' negatives, however, the economics of the situation may show that this type of apparatus is a worthwhile investment.

The skill of the photographic technician in performing these special printing techniques should not be used as an excuse for careless workmanship in operating the electron microscope, or in processing the negative. The main purpose of the techniques is to make a reasonably good print even better, not to rescue a poor negative from the waste bin.

7.5.14 PRINT FINISHING AND PRESENTATION

Many electron micrographs are rapidly evaluated 'as received' from the printing room. Measurements may be made, for example, and then the prints are filed with an identification number or other information simply written on the back of the print. There is no point in spending time on print finishing operations if the prints are to be evaluated rapidly and filed in this way.

The majority of electron micrographs are usually evaluated in a more detailed fashion, and will often be examined frequently as an investigation progresses. A double-weight paper, to withstand the increased handling, will be an advantage for this type of print. For this more detailed and more frequent type of evaluation, it is a great convenience to have the serial number and other relevant information transferred to the front of the print after processing is completed. A writing space can be provided by setting the printing frame so that a suitable width of the paper is obscured during enlargement. The obscured strip will be white, after processing, and can be written on with a fibre-tipped pen, when dry.

There are many occasions when prints are to be evaluated more formally, perhaps outside the EM department, or are to be displayed and presented in a more permanent fashion. A critical examination of such prints may reveal some minor blemishes, not noticed in the darkroom; it is desirable to remove these blemishes especially if they are in high contrast with the image on the print.

Dust and small hairs on the negative can be particularly troublesome and will show on the print as small white spots and lines. Usually these can be eliminated almost completely by 'spotting' the defects on the print with retouching dye on a fine-pointed artist's water-colour brush. The brush should be relatively dry, with the tip drawn out into a point. Only a very small amount of dye should be picked up on the tip of the brush. Apply the dye to the defect, with a stippling action, until the defect is no longer visible at the normal viewing distance. Painting-in the defect, rather than stippling, is not recommended. Different shades of dye are available, and can be mixed if necessary, to match the tone of the area surrounding the defect.

Small black spots and lines on a print usually arise from corresponding clear defects on the negative caused by dust on the emulsion before exposure. (A surprising amount of particulate matter seems to collect in electron microscope camera chambers; this is often overlooked during routine maintenance.) Small dark defects on the print can be eliminated by scraping the defect with a scalpel. Gentle scraping of the *surface* is required; the paper base will be completely exposed to view if the action is too severe. Slight over-removal can be restored by spotting with retouching dye. Retouching the blemish on the negative itself is not advisable unless the retoucher has considerable experience.

Large area defects, perhaps arising from drying marks, are very difficult to disguise unless the retoucher is very skilled, and wherever possible, it is best to remake the negative or print if very obvious blemishes are present.

All prints, whatever their quality, will appear very much more impressive if they are mounted and titled. Slip-in overlay mounts, in which the print is sandwiched between a card frame and a backing, are satisfactory for temporary use. For more permanent display, prints are usually dry-mounted onto thick card. Proprietary mounting boards are available, but mounts can be guillotined from any suitable grade of artist's board. Various textures and colours are available, but in making a choice it should be remembered that the purpose of the mount is to emphasise the print.

A caption or title, with perhaps a magnification marker, and one or two arrows or letters emphasising the significant features on the micrograph, will add appreciably to the value and appearance of a mounted print. Captions can be typed or printed onto a separate piece of thin card, which can then be dry-mounted alongside the print. Alternatively, proprietary transferable lettering can be used (e.g. Letraset). The Letraset system contains arrows and other useful markers. Hand printing is not recommended unless the printer is an expert.

Prints intended for publication are best finished in accordance with the wishes of the publisher. In general, publishers prefer a print with good but not extreme contrast, and adequate tonal separation of the important detail. A white smooth glossy paper is preferred, as for normal electron micrographs. Annotations to be included on the print are best attached to a translucent overlay, so that the publisher's draughtsmen can print the annotations in the publisher's house-style.

7.5.15 POSSIBLE DIFFICULTIES AND REMEDIES IN PRINT PROCESSING

The use of a test-strip (see § 7.5.10) will have established the correct exposure time and contrast grade for a print. Notwithstanding this test, however, the beginner may see a rapidly blackening print in the developer, and fearing that it has been overexposed, will attempt to rescue the situation by snatching the print before development is completed. As with negative snatching, this is a very bad practice and will lead to poor quality prints. It is important that prints are fully developed for 2 min (or the manufacturer's recommended time) at the appropriate temperature. Prints always appear darker when immersed in developer and illuminated by safe-lighting; if they have been genuinely overexposed, they should be discarded, and fresh prints made. Some degree of over-development is permissible, if a print has been inadvertently under-exposed in the enlarger, but extended over-development will reduce the image quality, and may lead to fogging and overall yellow staining of the print.

Overall yellow staining and a generally weak image, can also arise from the use of exhausted or oxidised developer. Yellow stains over parts of the print can arise from insufficient agitation in the fixer, or exhausted fixer. It is important that prints are kept fully immersed during development and fixing, otherwise the projecting areas may be stained brown.

The presence of iron in the solutions or in the processing apparatus can give rise to blue stains. Use stainless steel or plastic fittings, and discard chipped enamel dishes and containers.

If it seems difficult to obtain white highlights, and the overall contrast of the prints seems unusually poor, the prints may have been fogged. Check that the paper is reasonably fresh and has been adequately stored. Ensure that flare in the enlarger is minimised by cleaning the optical components, and by masking off any large clear areas surrounding the negative. Flare can also arise by reflection of light (from the enlarger image) from nearby surfaces. If thin printing paper is used, a temporary backing sheet of black paper will ensure that light is not being reflected through the back of the print from the white easel or printing frame. Check the safelighting by resting a coin on a sheet of printing paper. Expose the paper to the safelighting for a time at least equal to the maximum normal handling time during processing, and develop the print. If a shadow image of the coin can be seen, the lighting is unsafe. If the safelighting itself appears to be in order (correct and intact screens, correct lamp, minimum distance from bench, etc.) look for a light leak into the darkroom. A door seal may be worn, or the outer door of a light-trapped entrance may have been inadvertently left open.

Unglazed patches on the print, and generally poor surface quality can usually be traced to a contaminated glazing drum. The glazing surface should never be handled with the fingers, and the glazing cloth should be washed or replaced from time to time. Check that the glazer is operating at the recommended temperature and speed.

7.6 Photographic chemicals

A very wide range of chemicals is used for photographic purposes; comprehensive catalogues are available from the photographic manufacturers, and a survey of the principal applications and methods of use appears in Crawley (1974).

Fortunately only a few chemicals are required for the production of electron micrographs. Negative developer, print developer and fixer are essential. A wetting agent is highly desirable, and if rapid processors are used, activator and stabilizer are necessary. A stop bath is desirable.

As pointed out in § 7.3.1 there is some merit in using the developer recommended by the emulsion manufacturer, but other developers can be used with excellent results. May and Baker Teknol, and Ilford PQ Universal, are examples of liquid concentrates which give excellent results with printing papers as well as negative materials (Table 7.3).

If a proprietary universal developer is not immediately available, a useful general purpose developer can be made as follows:

Universal Phenidone Hydroquinone Developer (Ilford ID62)

Sodium sulphite anhydrous	50 g
(*or* crystals)	(100)
Hydroquinone	12 g
Sodium carbonate anhydrous	60 g
(*or* crystals)	(162)
Phenidone	0.5 g
Potassium bromide	2 g
Benzotriazole	0.2 g
Water to make	1 litre

Dissolve the chemicals in the order given, and make up to 1 litre. Filter the concentrate with a cottonwool plug, before bottling.

Kodak developer D19 is recommended for use with Electron Image Plates, EM Film 4489 and also Commercial Film 4127 which is used for recording image display tubes (scanning electron microscopes, TV display monitors and image intensifiers) (Table 7.4).

TABLE 7.3

Developing conditions for three universal developers

Developer	*Dilution for*	
	Tank development EM negatives	*Dish development prints*
May and Baker Teknol	1:9	1:9
Ilford PQ Universal	1:9	1:9
Ilford ID62	1:4	1:3
develop for	4 min 18–21 °C 10 sec/min agitation	2 min 18–21 °C continuous agitation

If D19 is not immediately available as a prepared powder, it can be made according to the following formula:

D19

Metol	2 g
Hydroquinone	8 g
Sodium sulphite anhydrous	90 g
(*or* crystals)	(180)
Sodium carbonate anhydrous	45 g
(*or* crystals)	(122)
Potassium bromide	5 g
Water to make	1 litre

Start with about 750 ml of water. Add a pinch of the sodium sulphite to the water first, to minimise oxidation of the developing agent. Then add the metol and hydroquinone, followed by the remainder of the sulphite. Always add the sulphite (preservative) before the carbonate (alkali), otherwise oxidation may be accelerated. Make up to 1 litre. Filter through a cotton wool plug before bottling.

D19 is an excellent high contrast developer for general scientific purposes.

Printing papers can be processed in any universal or print developer. The main influence of the particular developer formulation will be on the general colour of the print, i.e. cool blue-black, neutral black, or warm brown-black. In general, the universal developers in Table 7.3 will give a neutral to slightly

TABLE 7.4
Developing conditions for D19 high contrast negative solutions

Negative material	*Tank development 18–21 °C 10 sec/min agitation*		*approx. speed S* (100 kV)
	dilution	*time* (min)	
Ilford EM5 plate Ilford N4E50 EM film (sheet and roll film)	1:2	4	0.3
Kodak EM film 4489 (sheet)	1:2	4	0.4
Kodak Electron Image plate	1:2	12	2.0
Kodak Electron Image plate	1:2	4	0.7
Kodak Electron Image plate	full strength + 1 g/ litre of benzotriazole	2	0.2
Kodak Fine Grain Positive film (roll film)	full strength	2	0.06

warm black at the concentration and development times shown. Colder images can be obtained by slightly reducing exposure and extending development; warmer images can be obtained by decreasing the development time or the concentration, with an increase in exposure to maintain the density. The very rich warm tones achievable with certain paper developers have limited application in electron microscopy.

For large batches of prints where the capacity of the fixer is likely to be extended to the limit, the fixer life can be prolonged by using a stop bath immediately after development. A solution of 3% glacial acetic acid in water makes a simple but very effective stop bath for papers. Concentrated glacial acetic acid should be stored as a 50% solution, to prevent solidification at low temperatures. Alternatively, proprietary liquid concentrates are available which contain a coloured indicator to show when the stop bath is exhausted. For very small batches of prints, where the fixer is likely to be discarded after use, a stop bath is not essential; a 10 sec water rinse will suffice.

Fixers are available as powders or liquid concentrates; the rapid type of fixer, usually containing ammonium thiosulphate, is now widely used (e.g. May and Baker 'Amfix', Ilford 'Hypam', Kodak 'Kodafix'). If a prepared fixer is not immediately available the following formula will give excellent results:

Acid Fixer Stock Solution

Potassium metabisulphite	50 g
Sodium thiosulphate ('hypo') crystals	500 g
Water to make	1 litre

Dilute 2:3 for prints
3:2 for negatives
Fix for 10 min at 20 °C.

A few drops of a neutral liquid detergent (e.g. Shell 'Teepol') added to a final water rinse before drying, will help to avoid drying marks. Proprietary concentrates (e.g. Kodak 'Photoflo') are also available for the purpose. Agitate for 1 min in the final rinse.

In regions where water conservation is important, the washing time for negatives and prints can be reduced considerably by using a fixer clearing bath (e.g. Kodak 'Hypo Clearing Agent'). After fixing the negatives or prints, wash thoroughly for 30 sec in running water, and transfer to a bath of hypo clearing agent. Agitate continuously for 3 min at about 20 °C. Wash thoroughly in running water for 5 to 10 min, before drying.

All processing solutions have a maximum useful life, which is determined by the amount of material processed, and by the tolerable degree of degradation or loss of quality when the solution is nearing exhaustion. It is very difficult to quantify either factor, but as a help to beginners, an approximate guide to the useful life of the type of developers and fixers used in electron microscopy is given in Table 7.5. When some experience has been gained, a change in the time of processing or in the quality of the result will be the best guide.

Processing solutions will also deteriorate in the unused condition. Developers, in particular, are susceptible to oxidation. In the small EM department

TABLE 7.5

Useful life and keeping properties of processing solutions

	Useful life – area of emulsion processed	*Keeping properties*			
		Stoppered bottle		*Tank*	*Dish*
		Full	*Half-full*		
Developer	3500 cm²/litre	6 months	2 months	3 days	1 day
Fixer	14000 cm²/litre	3 months	3 weeks	3 weeks	1 week

where the darkrooms may not be used continuously, it is likely that solutions will be discarded because of their limited keeping properties, rather than their degree of exhaustion. An approximate guide to the keeping properties of unused solutions is also given in Table 7.5. It will be seen that unused solutions keep longest in full stoppered bottles, and for the small department there is some merit in purchasing a larger number of small bottles of developer and fixer, rather than one large bottle which will eventually contain a large volume of air. Print developers in open dishes rarely keep more than 24 hr; it is best to discard print developer at the end of each working day. Negative developers will keep longer in the small tanks normally used for EM negative processing; used solutions will deteriorate rapidly, however, and in the interests of consistent high quality, negative developers are best discarded every two or three days, assuming that they have not been exhausted before this time.

Continuous replenishment of processing solutions is not worthwhile unless a large and continuous throughput of work can be maintained. Details of replenishment techniques are available from the photographic manufacturers.

7.7 External photography of the electron image

The electron image seen on the fluorescent screen of the electron microscope may be photographed through the viewing window with an external camera; several of the early ciné-films showing *in situ* dynamic experiments were recorded in this way (Von Ardenne 1943; Hirsch et al. 1957; Horne and Ottewill 1957, 1958; Horne 1959). This arrangement has the serious disadvantages that normal viewing of the image is impaired, unless the photographic apparatus is moved away from the viewing window, and the fluorescent screen has to be tilted in order to provide a flat field, with consequent distortion of the image.

These disadvantages can be overcome by photographing a transmission fluorescent screen located underneath the camera of the electron microscope. In the arrangement developed by Kenway and Anderson (1967) a transmission fluorescent screen with a blue P11-type phosphor is coupled to a 35 mm recording camera with a Dallmeyer *f*/1 copying lens of 50 mm focal length. The demagnification is 4:1, in order to match the resolution of the phosphor (about 80 μm) to the resolution of the film (about 20 μm). Kenway and Anderson showed that Ilford Pan F film, although panchromatic, has greatest sensitivity in the blue region of the spectrum, and gives optimum recording of the blue phosphor image.

The exposure time required for external recording with Pan F film is only about 1.5 times the exposure required for direct recording on an internal emulsion. The contrast of externally recorded images is comparable with that of internally recorded images, but the resolution is noticeably inferior. Kenway and Anderson concluded that the use of a fibre-optic plate would allow a resolution to be obtained comparable with that of the conventional directly exposed emulsion.

External photography can also be achieved in association with image intensifying apparatus (see § 7.8).

Although external photography has some obvious advantages and applications, it would seem unlikely that external methods will replace the highly efficient and comparatively inexpensive method of direct recording on an internal emulsion in the conventional transmission electron microscope.

7.8 Electronic display and recording systems

Following the early work of Haine et al. (1958) and the demonstration by Haine and Einstein (1960) of a practical system in which the electron image in a transmission microscope could be intensified electronically and displayed on a cathode-ray tube, several types of electronic recording and display systems have been developed. Various types of image intensifier have been investigated (Haine and Einstein 1962; Komoda et al. 1963; Premsela 1968; Brandon et al. 1970; Thomas and Danyluk 1971; English and Venables 1972; Herrmann et al. 1970; Anderson 1968) and most of the leading transmission microscope manufacturers now offer television display systems, with or without image intensifiers.

In the display system described by Van Dorsten et al. (1966), the electron image falls on a transmission fluorescent layer deposited on a fibre-optic window. The output side of the window is attached to a Plumbicon pick-up tube, and carries a conducting layer of tin oxide and a photo-conducting lead oxide layer which is scanned by the Plumbicon low-energy electron beam. The output video signal is taken from the conducting layer, and is fed into a standard video amplifier and TV display system.

In a further development of this system described by Premsela (1968), the input fibre-optic window is coupled to a self-focusing electrostatic diode image intensifier tube, and the output end of the intensifier is similarly coupled with a fibre-optic plate to a Plumbicon pick-up tube which feeds a TV display system.

The use of such an electronic system offers a number of important advantages for certain types of EM work. Clearly visible images can be obtained on the TV display (monitor) with electron image current densities as low as 10^{-14} A/cm^2. The human eye requires about 100 times more current density on the standard fluorescent screen for the same type of image to be visible. Thus, electron images which are almost invisible on the fluorescent screen, either because the scattered electron beams are inherently of very low intensity, or because the intensity is deliberately restricted to avoid radiation damage to the specimen (heating or ionisation), can be clearly revealed on the TV monitor screen, and photographed from the monitor screen. The image contrast can also be enhanced and fine structural detail from, for example, crystal lattice planes or unstained biological material, which may be extremely difficult to see on the standard EM screen, can be resolved clearly, without dim-light adaption, on the TV monitor. Premsela has suggested that the Plumbicon display system is of value in ultimate performance microscopy, when the phase structure of a supporting film, seen clearly on the monitor with good contrast, can be used in the correction of astigmatism, for precise focusing, and to detect small amounts of drift, vibration, and other image instabilities. TV display and intensifier systems are subject to noise and contrast transfer limitations; the important aspects of noise and contrast transfer are discussed by Premsela (1968) and (1972), and by Kühl (1968).

One of the important attractions of electronic display systems is that the video output signal can be fed directly to a video tape recorder (VTR). Thus, recordings of dynamic phenomena can be played back immediately without waiting for film to be processed. The video output can also be fed into standard closed-circuit television (CCTV) systems, for display to larger numbers of observers. This can be particularly valuable during training activities, lectures and demonstrations.

For certain purposes, a conventional ciné-film may be more convenient to use than a video tape. Heerschap and De Cat (1972) have described a technique for recording transient phenomena by filming the image on the output phosphor of an intensifier, and using the TV display for viewfinding and continuous observation during filming.

In several applications of the electron microscope, meaningful statistical analysis of the electron image is becoming increasingly important. The Quantimet 720 (see Appendix) is a highly developed form of image analysing computer, and this instrument can be readily interfaced to an electron microscope by means of a fibre-optic coupled Vidicon or Plumbicon pick-up tube. Crawley and Gardner (1972) have discussed the application of image analysing techniques to transmission electron microscopy.

It will be clear from the previous discussion that electronic display and recording systems have some very valuable advantages in certain specialised applications. The apparatus is comparatively expensive, however, and it seems unlikely on economic grounds that the conventional internal photographic emulsion will be displaced as a means of image recording for general purpose transmission electron microscopy.

7.9 Viewing the electron image

The complementary nature of visual inspection of the image on the fluorescent screen and internal recording was stressed in the introduction to this chapter. In the haste to operate the electron microscope and produce electron micrographs, the fact that the conditions for viewing the fluorescent screen can be optimised, just as the photographic recording can be optimised, is often overlooked.

7.9.1 THE BINOCULAR VIEWING TELESCOPE

The resolving power of the eye is about 100 μm for a moderately bright object, and the resolving power of the phosphor on a well prepared fluorescent screen is usually in the range 50–80 μm. For critical viewing, it is usual to magnify the image on the screen with a binocular telescope. In practice, a higher magnification than is strictly necessary to match the resolving powers is used, and although this gives some empty magnification, the larger image detail is easier to see. A magnification (M) of 8–10 × is recommended for use at a working distance (near point) of 25 cm.

For an extended object, i.e. an object size greater than the resolving power of the eye, the brightness of the image on the retina cannot exceed the brightness of the object, and is usually less because of absorption and scattering losses in the viewing system. With a small object, i.e. less than the resolving power of the eye, all the light will fall on a single receptive area of the retina. The apparent brightness will depend on the total luminous flux entering the eye, and is proportional to M^2. Since magnification will improve the brightness of a small source but not that of an extended source, the use of a binocular telescope will enhance the contrast of very small detail on the fluorescent screen relative to the general screen background.

It is important that the exit pupil of the binocular telescope is matched to the entrance pupil of the eye. If the eye pupil is larger than the exit pupil, the eye will not be filled with light and there will be a loss of brightness. The eye pupil is about 7 or 8 mm in diameter when viewing high resolution EM images close to maximum magnification, and about 2 mm in daylight.

The required objective lens diameter (for d eye = 7 mm and $M = 10\times$) is 70 mm. Most practical binoculars have somewhat smaller objectives, and less than optimum brightness will be obtained.

If the exit pupil of the binocular is larger than that of the eye, the effective aperture of the system will be determined by the eye pupil, rather than by the aperture of the objective lens. In this situation the resolving power of the system will be less than optimum.

Observers with myopic vision (short-sight) may prefer to wear spectacles during operation of the electron microscope, in order to see the various controls and instrument readings clearly, and it is very convenient to keep the spectacles on when viewing through the binocular telescope. Unfortunately, binoculars supplied with electron microscopes have the exit pupil too far inside the eyepiece cup, to allow satisfactory viewing with spectacles. Some binocular manufacturers will provide special eyepieces for use when spectacles are worn, with a suitable position of the exit pupil and a soft cushion to prevent scuffing of the spectacle lenses.

Traditionally, the viewing telescope is regarded as part of the basic electron microscope, but in recent years alternative binocular telescopes have been marketed as EM accessories (e.g. Olympus). The points discussed above should be borne in mind if a change of binocular is proposed.

7.9.2 THE FLUORESCENT SCREEN

At higher brightnesses, the eye has maximum sensitivity in the green region of the visible spectrum, and as object brightnesses are reduced towards the detection threshold, the peak of sensitivity moves into the blue region of the spectrum. The phosphors provided with electron microscopes are usually of the zinc cadmium sulphide type with a peak response at about 560 nm, i.e. yellow-green in colour.

Fluorescent screens are easily damaged, both mechanically and by the electron beam, especially at high electron energies where knock-on damage can be severe. The efficiency gradually decreases with prolonged electron irradiation. For optimum viewing, fluorescent screens should be renewed regularly. The electron microscope manufacturers offer a replacement service, and some manufacturers also provide instructions for making screens in the EM laboratory.

Many workers have attempted to make high resolution screens by filtering or centrifuging to remove the coarse particles of phosphor, and depositing a very thin layer. Unfortunately the light yield decreases as the phosphor thickness is reduced, and if the brightness is too low, the acuity of the eye will deteriorate and it will become impossible to see fine detail. Agar (1957) has shown that a screen brightness of at least 0.02 foot-lamberts is necessary to see an 80 μm Fresnel fringe. If much work is undertaken at very low electron intensity, and hence low screen brightness, it may be preferable to use thicker phosphors than usual in order to improve the light yield, and to accept that the resolving power will be impaired; the resolution will be impaired in any event, by the reduction in visual acuity at the low screen brightness.

The efficiency of phosphors also decreases at the higher electron energies above 100 keV, and thick phosphors are necessary for high voltage electron microscopes. Brightness has to be balanced carefully against loss of resolving power.

It is natural to think that the highest possible phosphor resolution is required, but viewing of the fluorescent screen is subject to electron noise in a similar way to the recorded image. During observation of the phosphor there is a statistical deviation in the rate of arrival of electrons, which gives rise to a time-varying contrast or scintillation on the screen. The contrast is about 0.045 for a resolution of 100 μm, an integration time for the eye of 0.1 sec, and a current density of 10^{-11} A/cm^2. Thus if the viewing conditions remain the same, an improved resolution will result in an increased noise contrast. The scintillation can be seen readily on a good screen at low current densities.

7.9.3 THE AMBIENT LIGHTING

All microscopists will be familiar with the temporary and partial loss of vision which occurs on moving from bright sunlight to a darkened room, and vice versa. The sensation produced by a given illumination depends on the current state of the eye. When the eye has been rested in the dark for a period of about 15 min or more, the eye is said to be *dark-adapted.* Objects of very low brightness can be detected, but the sensation of colour tends to disappear and the objects are discerned as different shades of grey. The resolution is very poor and the forms of objects are difficult to distinguish.

Adjustment of the eye to operating in high light intensities, or *light-adaptation*, also takes time, although this process is much more rapid than for dark-adaptation.

For almost all EM applications an intermediate level of adaptation is necessary; full dark adaptation is *not required.* The fluorescent screen will be moderately bright and the microscopist will be adapted to *dim-light.* The field of view of the eye is very large, about 160° in the horizontal plane (some 100° on the temporal side) and 120° in the vertical plane, and in a normal observer the field of view is being constantly scanned by involuntary movements of the eye muscles. The ability to discern dim objects in the field of view will be severely restricted if the eye becomes light-adapted, and since a single bright source anywhere in the field of view can cause the eye to light-adapt, it is important that such sources of *glare* are removed when the microscopist is adapted to dim-light. The scattered light in the eye from a source of glare will also tend to obscure the retinal image of other objects in the field of view.

It is necessary, therefore, that sources of light in the EM room other than the fluorescent screen, are reduced to a brightness less than but comparable with the brightness of the screen. It is unnecessary and *undesirable* for these other sources to be suppressed completely. The ability to see fine detail on the screen in dim-light conditions will not be affected by the presence of other light sources, provided that the sources are diffuse and of reasonably large area. For this reason it is usual to provide a low level of overall room lighting, from a small safelight fitting behind the electron microscope, i.e. in a position where a direct view of the safelight is obscured by the EM column. The eye is much less fatigued and has greatest sensitivity to small changes in brightness, when the whole of the field of view is illuminated at the appropriate dim-light level.

Most electron microscopes are fitted with a dimmer to control the brightness of the illuminated control panels and meters. On some microscopes, certain pilot lights are not controlled by the dimmer. This is unsatisfactory for optimum viewing, as a small intense light in the field of view is a source of glare and will catch the eye in a most irritating fashion. An attempt must be made, with the help of the manufacturer, to include all EM lights in the dimming system.

During a prolonged operating session in which the microscopist will be alternating between the EM room and the negative dark room, it is an advantage to be able to move between the two rooms without destroying the dim-light adaptation. If the rooms are not in direct communication, but are adjacent, the preservation of adaptation can be easily achieved by fitting a combined light-trap as described in Alderson (1974). A prolonged session of image viewing and recording can be very fatiguing, and it is strongly recommended that the microscopist should emerge into normal daylight after, say, an hour or so of EM operation. In this way, the eyes will be rested, and a further session of EM operation can proceed with the knowledge that, once again, critical viewing will be optimised.

In a limited number of special EM applications, where it is necessary to work with the lowest possible electron intensity, and barely perceptible screen brightness (in order to minimise radiation damage to the specimen, for example), full dark adaptation may be essential. *All* sources of light in the EM room must be extinguished, and the microscopist must allow sufficient time for full adaptation to occur. This type of work requires intense concentration and can be very fatiguing. If much of this work will be necessary, the possibility of acquiring an image intensifier should be explored.

REFERENCES

Agar, A. W. (1957), On the screen brightness required for high resolution operation of the electron microscope, Brit. J. appl. Phys. *8*, 410.

Agfa-Gevaert (1970), Personal communication, Agfa-Gevaert Limited, Great West Road, London.

Alderson, R. H. (1974), Design of the electron microscope laboratory, in: Practical methods in electron microscopy, A. M. Glauert, ed. (North-Holland, Amsterdam).

Anderson, K. (1968), An image intensifier for the electron microscope, J. scient. Instrum. (ser. 2) *1*, 601.

Beeston, B. E. P., R. W. Horne and R. Markham (1972), Electron diffraction and optical diffraction techniques, in: Practical methods in electron microscopy, Vol. 1, A. M. Glauert, ed. (North-Holland, Amsterdam).

Brandon, D. G., D. Schectman and D. N. Seidman (1970), Preliminary results with a channel plate image intensifier in the electron microscope, Proc. 7th Int. Congr. Electron Microscopy, Grenoble *1*, 343.

Burge, R. E. and D. F. Garrard (1968), The resolution of photographic emulsions for electrons in the energy range 7–60 keV, J. scient. Instrum. (J. Phys. E) *1*, 715.

Burge, R. E., D. F. Garrard and M. T. Browne (1968), The response of photographic emulsions to electrons in the energy range 7–60 keV, J. scient. Instrum. (J. Phys. E) *1*, 707.

Crawley, G. (1966), The Melico enlarging exposure meter and general considerations affecting enlarging photometry, Brit. J. Photography 28 January, 1966.

Crawley, G. (1974), The British Journal of Photography Annual, ed. G. Crawley (formerly the British Journal Photographic Almanac), Henry Greenwood & Co. Ltd., London.

Crawley, J. A. and G. M. Gardner (1972), The applications of image analysis techniques to transmission electron microscopy, Proc. 5th Eur. Reg. Conf. Electron Microscopy, Manchester.

Digby, N., K. Firth and R. J. Hercock (1953), The photographic effect of medium energy electrons, J. Phot. Sci. *1*, 194.

English, C. A. and J. A. Venables (1972), An evaluation of channel plates as image intensifiers for transmission electron microscopy, Proc. 5th Eur. Conf. Electron Microscopy, Manchester, p. 172.

Evennett, P. and K. Oates (1968), Personal communication, Department of Zoology, University of Leeds.

Farnell, G. C. (1973), Personal communication, Kodak Research Laboratories, Harrow, Middlesex.

Farnell, G. C. and R. B. Flint (1969), Low-contrast development of electron micrographs, J. Microscopy *89*, 37.
Farnell, G. C. and R. B. Flint (1973), The response of photographic materials to electrons with particular reference to electron micrography, J. Microscopy *93*, 271.
Farrand, R. (1961), To illustrate a lecture, Brit. J. Photog. April 21, 1961.
Frieser, H. and E. Klein (1958), Die Eigenschaften photographischer Schichten bei Elektronenbestrahlung, Zeit. angew. Phys. *10*, 337.
Gabor, D. (1953), The history of the development of electron microscopes, Institute of Physics, Bournemouth Meeting, May 1953 (subsequently published as: Die Entwicklungsgeschichte des Elektronenmikroskops, Elektrotechnische Zeitschrift, *78*, 1957).
Haine, M. E. and P. A. Einstein (1960), Image intensifier, Proc. 2nd Eur. Reg. Conf. Electron Microscopy, Delft, p. 97.
Haine, M. E. and P. A. Einstein (1962), Intensification of the electron microscope image using cathodo-conductivity in selenium, Proc. I.E.E., *109*, 185.
Haine, M. E., A. E. Ennos and P. A. Einstein (1958), Image intensifier for the electron microscope, J. scient. Instrum. *35*, 466.
Hamilton, J. F. (1968), The use of unsharp masking in printing electron micrographs, J. appl. Phys. *39*, 5333.
Hamilton, J. F. and J. C. Marchant (1967), Image recording in electron microscopy, J. Opt. Soc. Am. *57*, 232.
Hammond, J. H. (1957), Illustrations for lectures, Brit. J. Photog. January 18, 1957.
Hammond, J. H. (1936a), Making 2 × 2 slides, Brit. J. Photog. July 12, 1963.
Hammond, J. H. (1936b), Lantern slides – legibility and information content, Brit. J. Photog. May 24, 1963.
Hawkes, P.W. (1972), Electron optics and electron microscopy (Taylor and Francis, London).
Heerschap, M., and R. De Cat (1972), Direct filming of transient phenomena with a closed TV circuit as viewfinder, Proc. 5th Eur. Reg. Conf. Electron Microscopy, Manchester, p. 170.
Herrmann, K. M., D. Krahl and V. Rindfleisch (1970), Storage operation in an image amplification device with SEC camera tube, Proc. 7th Int. Congr. Electron Microscopy, Grenoble *1*, 339.
Hirsch, P. B., R. W. Horne and M. J. Whelan (1957), Direct observations of the arrangement and motion of dislocations in aluminium, Phil. Mag. *1*, 667.
Hochhäusler, P. (1929), Ein- und Ausführung von Platten und Filmen an Kathodenoszillographen ohne Störung des Hochvakuums, Electrotechnische Z. *50*, 860.
Horder, A. (1968), The Ilford manual of photography, A. Horder, ed. (now The Manual of Photography, Focal Press, London.)
Horne, R. W. (1959), Applications of ciné techniques to electron microscopy, Sonderdruck aus Research Film, *3*, 150.
Horne, R. W. and R. H. Ottewill (1957), Electron microscope studies on colloidal silver iodide employing cinematographic techniques, Nature, Lond. *180*, 910.
Horne, R. W. and R. H. Ottewill (1958), Examination of colloidal silver iodide by electron microscopy, J. Phot. Sci. *6*, 39.
Hurter, F. and V. C. Driffield (1890), Photo-chemical investigations and a new method of determination of the sensitiveness of photographic plates, J. Soc. Chem. Ind. *9*, 455.
Ilford, (1970), Personal communication, Technical Sales Department, Ilford Limited, Ilford, Essex.
Ingelstam, E., E. Djurle and B. Sjogren (1956), Contrast-transmission functions determined experimentally for asymmetrical images and for the combinations of lens and photographic emulsion, J. Opt. Soc. Am. *46*, 707.
Iwanaga, M., H. Ueyanagi, K. Hosoi, N. Iwasa, K. Oba and K. Shiratsuchi (1968), Energy dependence of photographic emulsion sensitivity and fluorescent screen

brightness for 100 kV through 600 kV electrons, J. Electron Microscopy *17*, 203.
Jones, R. C. (1955), New method of describing and measuring the granularity of photographic materials, J. Opt. Soc. Am. *45*, 799.
Jones, R. C. (1958), On the point and line spread functions of photographic images, J. Opt. Soc. Am. *48*, 934.
Kelly, D. H. (1960), Systems analysis of the photographic process; I. A three-stage model, J. Opt. Soc. Am. *50*, 269.
Kenway, P. B. and K. Anderson (1967), External photography of the microscope image, Proc. 25th Ann. Meeting EMSA, p. 244.
Kodak (1973), Personal communication, Industrial and Professional Sales Department, Kodak Limited, Kodak House, Hemel Hempstead, Herts.
Komoda, T., M. Oikawa, M. Hibi, and H. Kimera (1963), The development of an image intensification system for the electron microscope, Hitachi Central Research Laboratory report, 1963.
Kühl, W. (1968), Information transfer with electron microscope TV systems, Philips Electron Optics Bulletin, no. EM30, October 1968.
Marton, L. (1935), Le microscope électrique et ses applications, Rev. Opt. (Théor. Instrum.) *14*, 129.
Mees, C. E. K. and H. James (1966), The theory of the photographic process, 3rd edition (Macmillan, New York).
Oates, K. (1972), Personal communication, Department of Biological Sciences, University of Lancaster.
Premsela, H. F. (1968), A versatile TV image display system for electron microscopy using an image intensifier and a Plumbicon pick-up tube with fibre optics window, Philips Electron Optics Bulletin, no. EM 30, October 1968.
Premsela, H. F. (1972), Some aspects of the modulation transfer of the EM-TV chain, Proc. 5th Eur. Reg. Cong. Electron Microscopy, Manchester, p. 168.
Thomas, E. L. and S. Danyluk (1971), A channelplate image intensifier for the electron microscope, J. Phys. E, *4*, 483.
Turnbull, P. (1958), Legible lantern slides, Brit. J. Photog. September 19, 1958.
Valentine, R. C. (1965), Characteristics of emulsions for electron microscopy. Lab. Invest. *14*, 1334.
Valentine, R. C. (1966), Response of photographic emulsions to electrons, in: Advances in optical and electron microscopy, vol. 1, R. Barer and V. E. Cosslett, eds. (Academic Press, London and New York), p. 180.
Valentine, R. C. and N. G. Wrigley (1964), Graininess in the photographic recording of electron microscope images, Nature, Lond. *203*, 713.
Van Dorsten, A. C. (1950), Statistical effects in electron microscope image recording in relation to optimal electron optical magnification and picture quality, Proc. 2nd Eur. Reg. Conf. Electron Microscopy, Delft, *1*, 64.
Van Dorsten, A. C., P. H. Broerse and H. F. Premsela (1966), Extending the limits of visual observation and picture recording in the electron microscope, Proc. 6th Int. Congr. Electron Microscopy, Kyoto *1*, 275.
Van Dorsten, A. C. and H. F. Premsela (1960), Low voltage electron microscopy, Proc. 2nd Eur. Reg. Conf. Electron Microscopy, Delft, *1*, 101.
Van Horn, M. H. (1951), The use of film in X-ray diffraction studies, Rev. Scient. Instrum. *22*, 809.
Von Ardenne, M. (1943), Elektronenmikrokinematographie mit dem Universal-Elektronenmikroskop, Z. Phys. *120*, 397.
Von Borries, B. (1942), Über die Intensitätsverhältnisse am Über-mikroskop; I. Die Schwärzung photographischer Platten durch Elektronenstrahlen, Phys. Z. *43*, 190.
Zeitler, E. and J. R. Hayes (1965), Electrography, Lab. Invest. *14*, 586.

Chapter 8

Image Interpretation

8.1 Introduction

If one takes the term 'Image interpretation' in its broadest sense, it will embrace the whole of electron microscopy, from the preparation of the specimen, through the imaging and recording processes, and including the effects of external factors and instrumental defects. A complete discourse on these matters would certainly occupy more than one book. However, the influence of specimen preparation techniques on the appearance of the image is considered in detail in other books of this series, and many of the electron optical, photographic, and instrumental factors which influence the image have already been discussed in earlier chapters of this book. Consequently this chapter will mainly deal with the very important effects on the interpretation of the image of instrumental settings, and of instrumental defects and ambient conditions.

8.2 The effect of instrumental settings on the interpretation of the image

8.2.1 THE EFFECT OF COHERENCE

The coherence of the illumination has very little practical effect on the image unless there is a substantial contribution from phase contrast. Thus, for all structures greater than about 1 nm in size, the coherence is relatively insignificant. Where phase contrast is dominant, the coherence is very important, as shown in § 3.8.

8.2.2 THE EFFECT OF FOCUS SETTING

The setting of the objective lens current for focusing the image has very little significance for interpreting images formed by amplitude contrast; the main effect of an out-of-focus image is some loss in sharpness of detail. The significant effects arise from phase contrast and are mainly therefore confined to image detail of 1 nm and below. The most obvious effect is the appearance of the bright (underfocus) or dark (overfocus) fringes at the edge of a feature of an object. These intensity changes must not be interpreted as due to

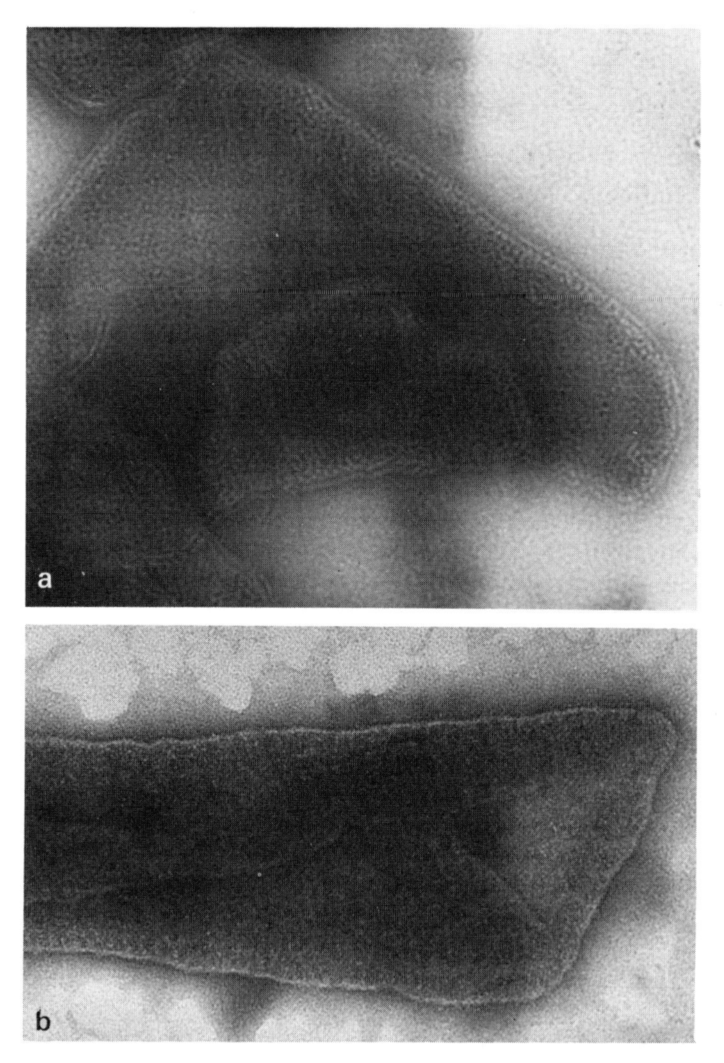

Fig. 8.1. Negatively-stained preparation of bacterial cell walls. (a) Overfocused micrograph. (b) In-focus micrograph. Magnification 120,000 ×.

thickness or density changes in the specimen; in particular the overfocused fringe must not be interpreted as a biological membrane. For example, Fig. 8.1a shows an overfocused micrograph of a negatively-stained preparation of bacterial cell walls, and Fig. 8.1b shows an in-focus micrograph of the same preparation. An additional 'membrane' appears to be present at the surface of the cell walls in (a).

When interpreting finer substructures, however, the effects may be more subtle and less easy to differentiate from genuine specimen structure (Chescoe and Agar 1966; Haydon 1969). The grain of the specimen support film changes appreciably in appearance according to the focal setting of the objective lens. These changes are well illustrated by the appearance of the film structure in Fig. 4.11. Near focus, very little structure can be seen, but the size of the grain increases as the objective lens is further underfocused, and for overfocused conditions the contrast reverses. In the absence of an edge, where a Fresnel fringe may be seen, it can be quite difficult to decide whether the setting is under or over focus. An example of these difficulties in interpretation is illustrated in Fig. 8.2.

In a micrograph of a dispersion of ferritin particles on a carbon film at high magnification, with the objective lens near focus, very little structure can be seen (Fig. 8.2a). When the objective lens is underfocused (Fig. 8.2b) a strong granular structure appears in the supporting film, and the ferritin particles now appear to have internal structure. This micrograph, taken in isolation, might indeed represent true internal structure; the comparison with the in-focus micrograph shows, however, that any contrast arises solely from phase effects, and that, in this preparation, no significant amplitude contrast exists for any internal structure. (There may indeed be true amplitude contrast in some other ferritin preparations). If the negative of Fig. 8.2b is now printed with one small region double-exposed (by light), one obtains Fig. 8.2c in which there appears to be a new ferritin particle (arrow). This illustrates that structure in regions of higher photographic density tends to be differentially identified by the eye as significant. This enjoins particular caution in interpreting any sort of 'dot' structure in particles, membranes or indeed any regions of a section or particle preparation which have a high density on the photographic plate. If a through-focal series of micrographs is available, of course, one can see whether the structure changes with the phase structure of the support film (or embedding material) or whether it has some residual amplitude contrast.

For very thin specimens wholly dependent on phase contrast the situation is very difficult. Referring to the contrast graphs of § 3.8, reproduced here as

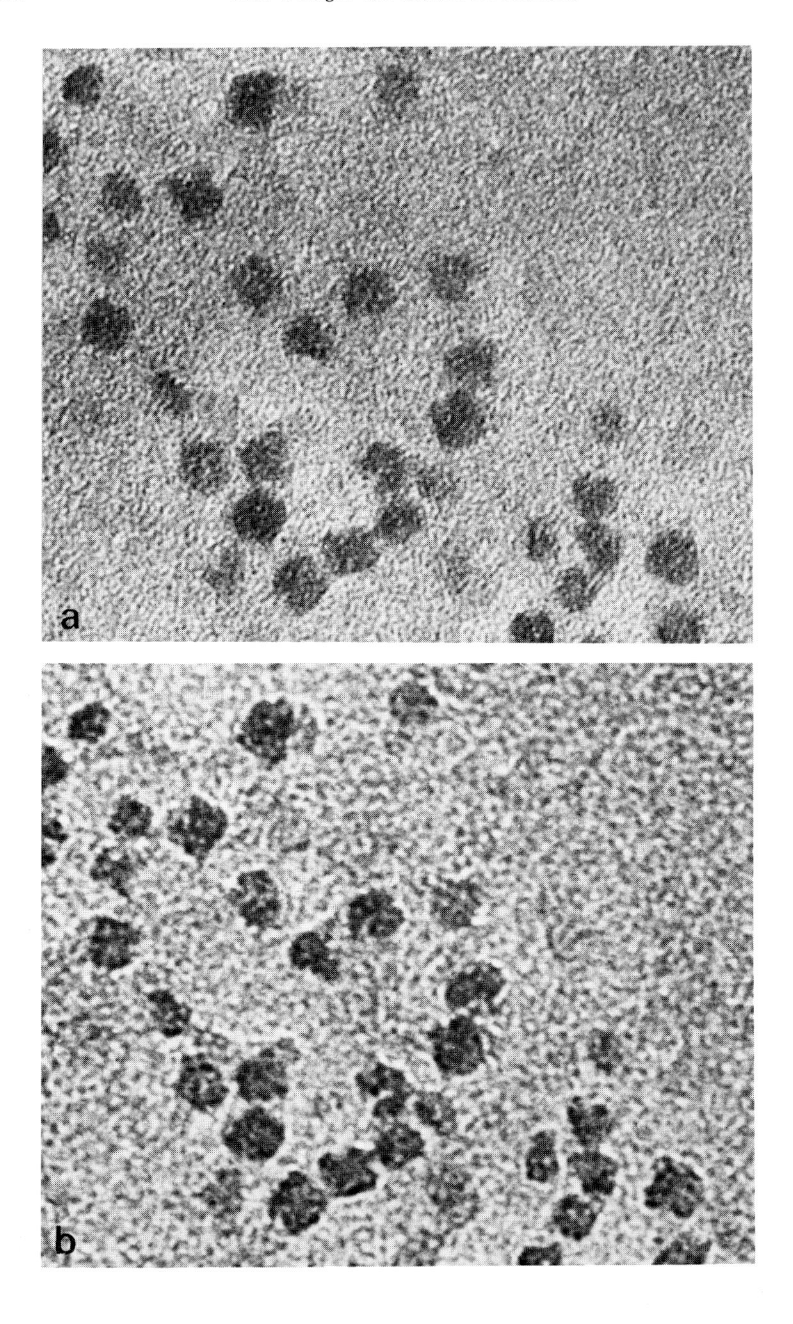

Fig. 8.2 a-b

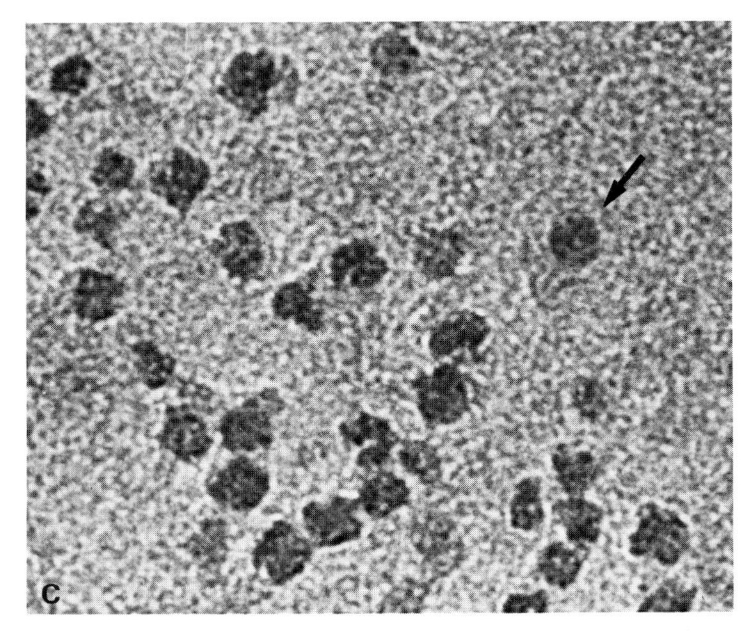

Fig. 8.2 c

Fig. 8.2. (a) Ferritin particles on a carbon film. Near focus. (b) Same field as 8.2a. Underfocused. (c) Same field as 8.2b. Printed with a double exposed region (arrow) on the printed area formed by illuminating this area twice. Magnification 670,000 ×

Fig. 8.3, which relate phase contrast to the degree of defocus of the objective lens and to the size of the structure, it can be seen that, for a given degree of defocus, there will be some structure sizes in high contrast, while others are in low contrast. In the case of particular settings of objective lens focus (e.g. 78 nm underfocus for the particular lens illustrated), there may be a broad band of structure sizes for which the phase contrast is near maximum (0.45 nm–1 nm in this case). The interpretation of structure in this size range is then reasonably straightforward. But at a different degree of defocus (e.g. 234 nm) there may be three regions of zero or very low contrast, and a highly selective picture of the structure would emerge. If, for instance, some high frequency component of the wave spectrum defining an object were suppressed (Fig. 8.4), it could give rise to a quite erroneous particle shape as deduced on 'logical' grounds and one might conclude that the particle had a hollow centre.

The conclusion must be that any attempt to interpret structure smaller than

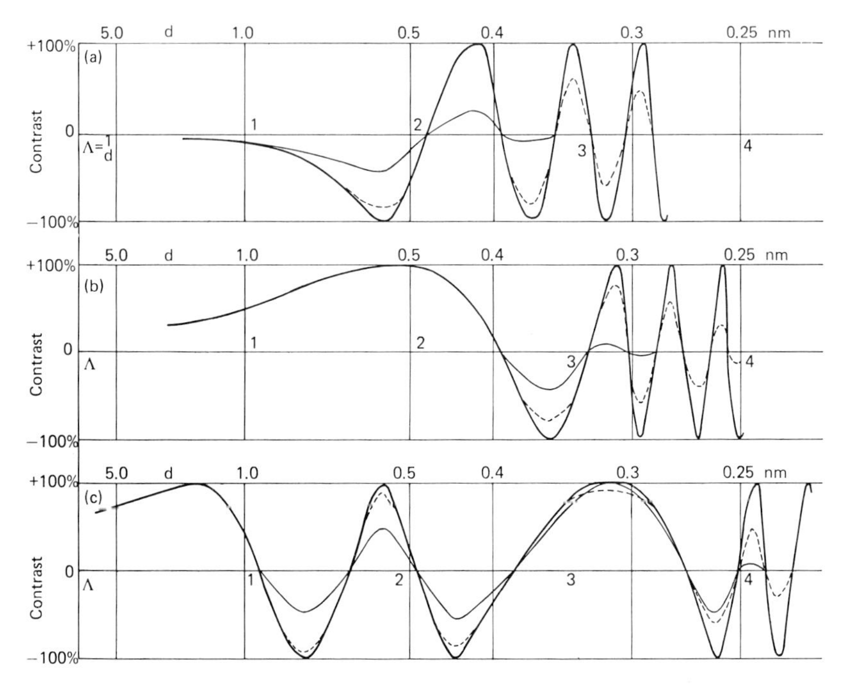

Fig. 8.3. Plot of phase contrast as a function of structure size. (a) Objective lens in focus. (b) Objective lens 78 nm underfocus. (c) Objective lens 234 nm underfocus.

1 nm in size based on a micrograph depending on phase contrast, is exceedingly hazardous. If the conditions can be very closely defined (as for the 78 nm defocussing described above), then there may be more confidence in the results, but they ought to be backed up by optical diffraction analysis of the micrographs. This reveals very clearly whether there is selective filtering of structure size, and what dominant structure sizes have been selected. The evidence should also be checked by strioscopy, if possible, where the absence of the unscattered electron beam prevents the phase contrast effects and simplifies the interpretation.

It should perhaps be repeated that these extreme difficulties in interpretation will only be encountered with very thin specimens.

8.2.3 EFFECTS OF ILLUMINATION APERTURE AND SUPPLY STABILITY

A further examination of Fig. 8.3 shows that the effect of a finite illumination aperture (thin line, for an aperture of 10^{-3} radians) is to reduce the contrast to a negligible level very quickly. Thus, when the condenser lens is operated

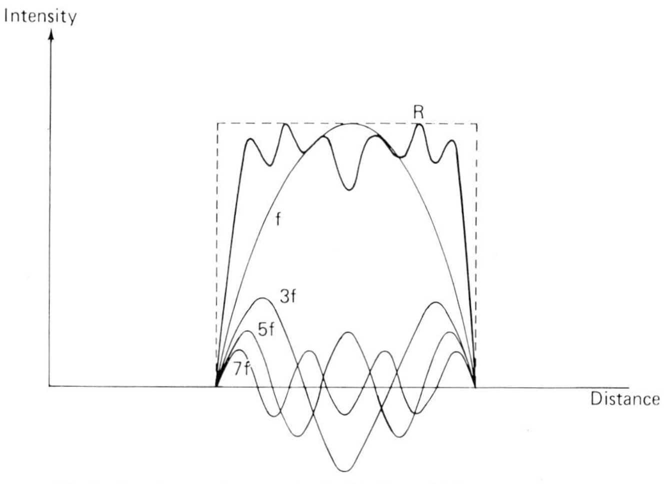

(a) Profile due to frequencies f, 3f, 5f and 7f

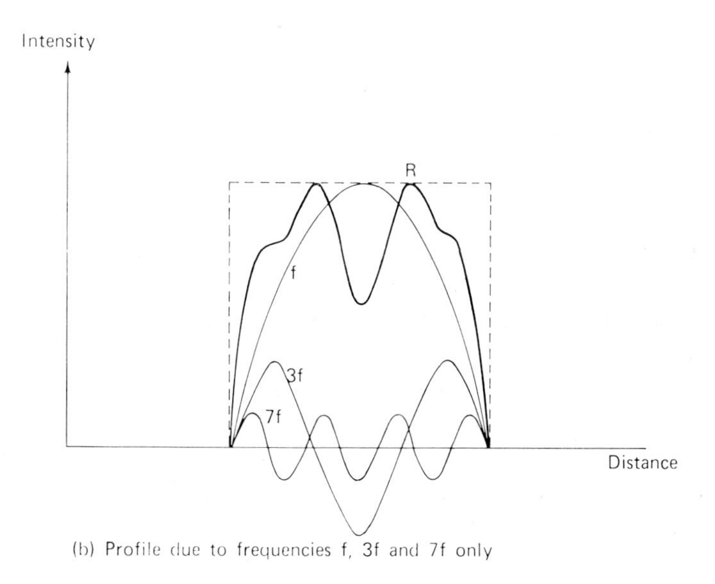

(b) Profile due to frequencies f, 3f and 7f only

Fig. 8.4. A cylindrical particle: (a) with several component wave frequencies giving an approximate match to its shape; (b) with one higher frequency suppressed. Note the strong dip in the resultant intensity profile of this image which might be interpreted as a hollow centre to the particle.

near focus, the phase effects will be much less dominant, and many of the ambiguities may be reduced (but the total contrast may become too low to be perceptible as well!). It can be seen, conversely, that very fine structure will have imperceptible contrast with focused illumination (assuming a normal size of condenser aperture), even if the microscope is nominally capable of resolving the structure. This loss of contrast of fine structure may also lead to very 'soft' images of easily resolved structure. When calculating the effective illumination aperture, it should be remembered that most powerful objective lenses have at least some 'prefield' (or condenser effect) which can increase the effective illumination aperture (a factor of 1.25 is quite common).

The supply stability has a much less serious effect on the loss of contrast of the phase patterns, (dotted line, Fig. 8.3). However, the graph is drawn for an instrument with good stability and with a very low chromatic aberration coefficient. One would rapidly lose contrast if the supplies were not performing well. An instrument with a higher chromatic aberration coefficient would also be significantly more sensitive to supply stability in determining image contrast.

8.2.4 SPECIAL EFFECTS WITH CRYSTALLINE MATERIAL

The study of thin foils of metals and alloys has revolutionised our understanding of the science of materials. The images formed are due to a series of diffraction phenomena which require a complete treatise to describe them (Hirsch et al 1965; Howie 1974).

If one considers only the extension of lattice images already mentioned, to two or three dimensional arrays, the problem of interpretation becomes very complex because the scattered electrons have selective phase relationships which combine in ways which depend critically on the plane of focus of the objective lens. This means that regular structures may be seen in the image which bear no simple relationship at all to the original structure. It is of course possible to analyse them if the focus conditions are accurately known and controlled, but a single micrograph at an unknown focal setting may be totally misleading.

Where the image is of a single set of crystal planes, the basic plane spacing will usually be obtained, although even here, a different focal setting may yield half-spacings (Komoda 1964b). It is mainly when two or three lattice spacings interact that the most complex patterns arise (Komoda 1964a; Cowley and Iijima 1972). Fig. 8.5 shows a through-focal series of micrographs of a complex oxide of titanium and niobium and illustrates vividly the interpretation difficulties. Biological specimens can be subject to similar

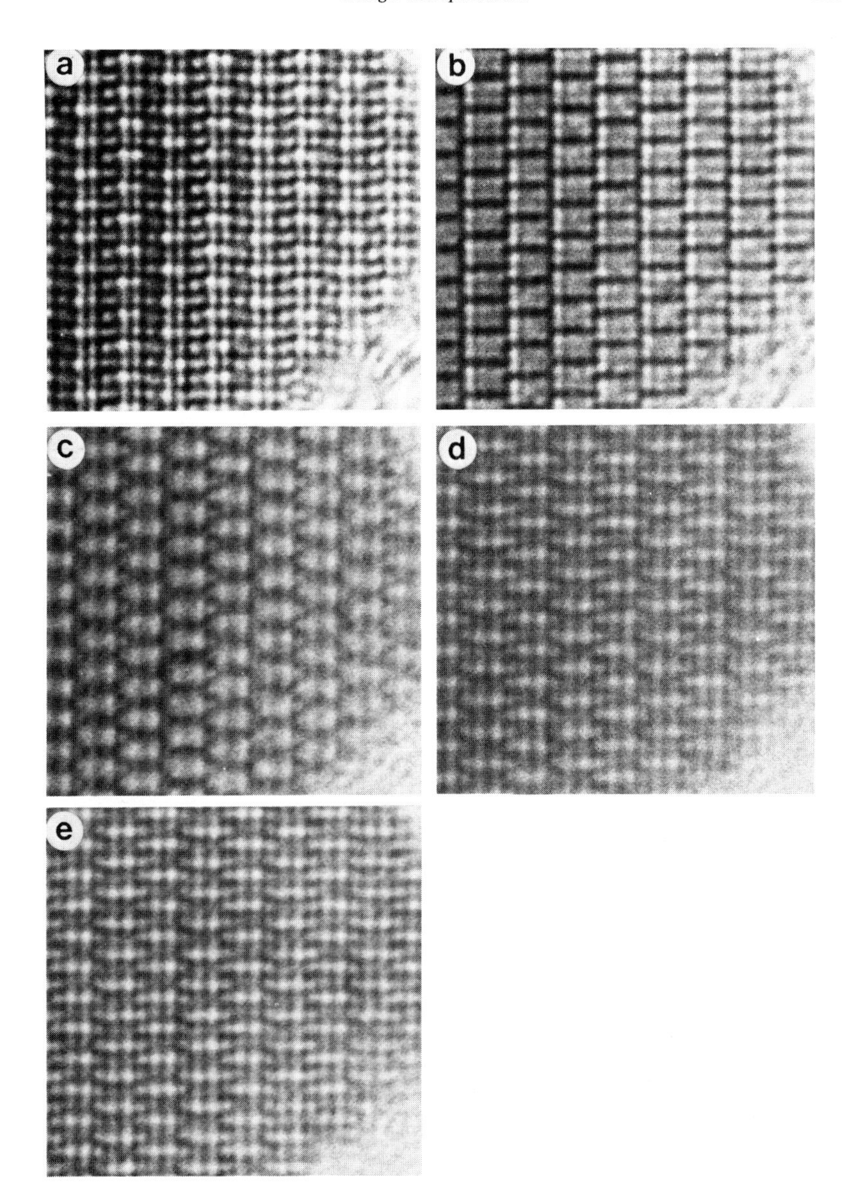

Fig. 8.5. Through-focus series of micrographs of complex titanium niobium oxide ($Ti_2\,Nb_{10}\,O_{29}$) (after Cowley and Iijima 1972). Defocus distance Δf_0 equal to (a) -160 nm (b) -96 nm (c) -24 nm (d) zero (e) $+24$ nm (positive defocus is overfocus). Magnification 6,700,000 ×.

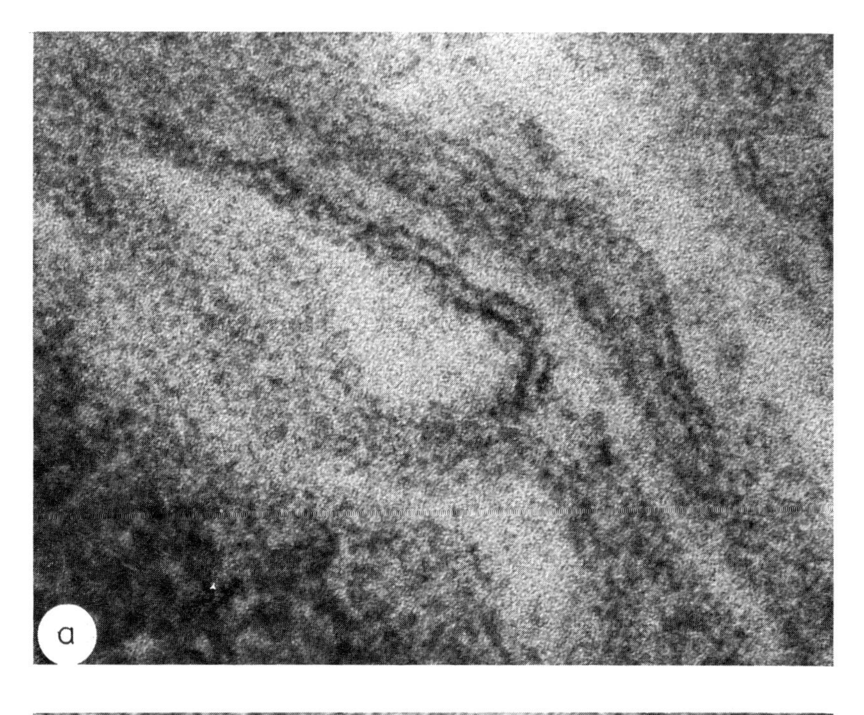

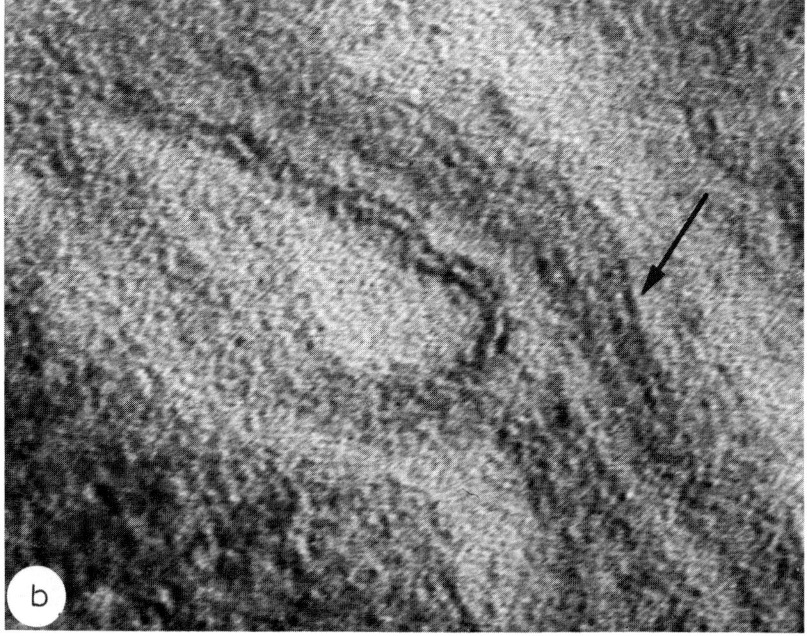

Fig. 8.6 a-b

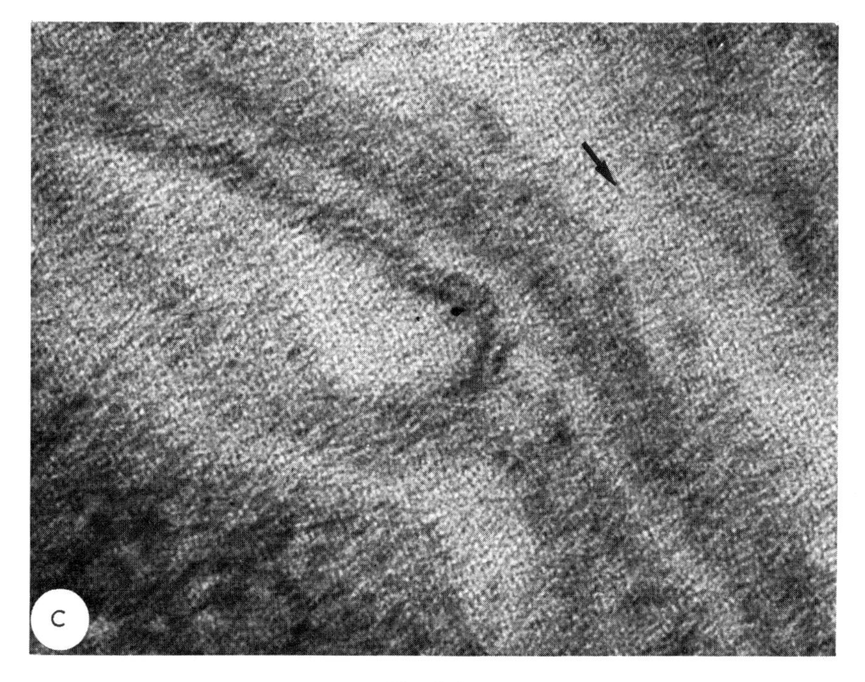

Fig. 8.6c

Fig. 8.6. Series of micrographs of a section of bacterial cell wall. (a) Astigmatism corrected, in focus. (b) With astigmatic objective lens imaged near one of the focal lines. (c) As (b) but focused near the circle of least confusion. Magnification 400,000 ×.

difficulties of interpretation (Johnson and Sikorski 1962; Johnson and Crawford 1973).

8.2.5 THE EFFECT OF ASTIGMATISM

Where there is residual astigmatism in the objective lens, a round object point is imaged into two focal lines some distance apart along the axis of the instrument. At any one focal line, a round object point is rendered as a line, of length determined by the degree of the astigmatism. The overall effect on the image is to produce a directional structure running parallel to the direction of the focal line – and this structure can be rendered at right angles by adjusting the strength of the objective lens until the other focal line is imaged. A very clear illustration of this is seen in Fig. 4.12a.

By way of further illustration, Fig. 8.6a shows part of a bacterial cell wall photographed with a corrected objective lens, while Fig. 8.6b shows the same area, but with a considerable amount of residual astigmatism and with one of the focal lines imaged. The direction of the linear structure is indicated by an arrow; it runs right across all structure in the picture. Where it might be dangerous from the interpretation point of view, is where this structure crosses a membrane (arrow), since it greatly modifies the appearance. Fig. 8.6c is a micrograph taken with the same amount of astigmatism but

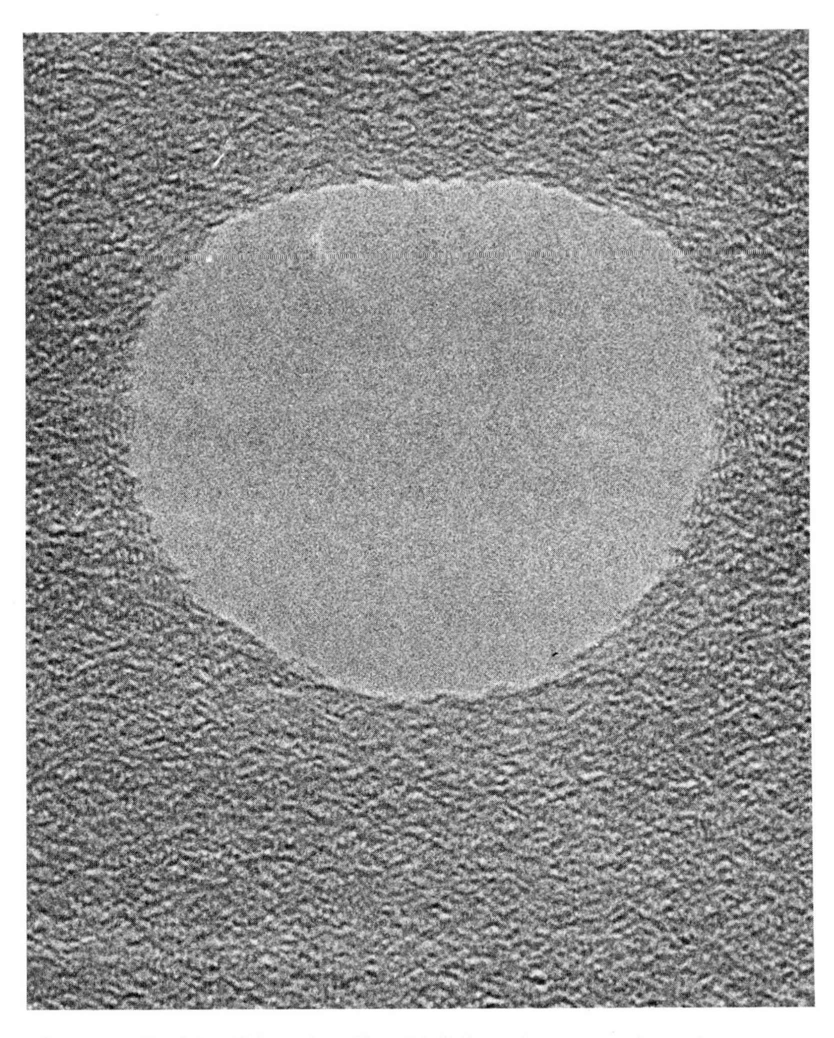

Fig. 8.7. Thin carbon film with hole, to demonstrate 'tartan' structure with slight astigmatism. Magnification 2,100,000 ×.

with the image near the circle of least confusion. The directional structure has mainly disappeared, but each image point is enlarged, so that all the fine detail is lost. In some regions (arrow) of this micrograph a kind of tartan structure is visible in the background; this is characteristic of an image between the focal lines in an astigmatic lens. If this kind of structure is detected in a picture, the micrograph should be rejected. In a higher magnification picture of a hole in a carbon film (Fig. 8.7) this tartan structure is again seen. The amount of astigmatism is very small (note the small asymmetry of the Fresnel fringe), but the tendency to directional structure can still be detected.

As a protection against interpretation errors because of astigmatism, the setting of the objective lens corrector should be carefully checked before any high resolution micrograph is recorded. A through-focal series of micrographs will very quickly reveal astigmatic effects, and will aid in distinguishing directional structure in the micrograph from that caused by specimen stage drift (§ 8.3.1), vibration of the column (§ 8.3.2), electrical instabilities (§ 8.3.3) or external fields (§ 8.3.4).

8.3 Directional structure in the image

8.3.1 THE EFFECT OF SPECIMEN DRIFT

The main causes of specimen drift have been described in § 5.4, and it is unlikely that gross drift will pass unnoticed on the fluorescent screen, especially if the viewing telescope is used for critical examination of the image. A very small drift, however, may not be detected on the screen, and will appear on the photographic plate as a uni-directional blurring of image detail. It is possible in some circumstances for a very small blurring to be wrongly interpreted as a linear or orientation effect in the object structure, and vice versa, it could be said that certain types of extraction replica or thin foil images showing oriented lamellar precipitates have the appearance of a blurred image!

8.3.2 EFFECT OF MECHANICAL VIBRATION

The various sources of mechanical vibration (§ 5.8.1) result in relative motion between different parts of the column of the electron microscope and this shows up as an oscillatory motion of the image. If the image on the final screen is viewed by telescope the unidirectional blurrings of the image can be detected, and it can quickly be related to the orientation of the stage motions by making small movements of the stage drives. Attempts can then be made to damp the vibrations, or exaggerate them, so as to confirm that the blurring is indeed due to this cause. This will usually be adequate to locate the source and direction of the vibration, and aid in correcting the problem.

8.3.3 ELECTRICAL INSTABILITIES BEYOND DESIGN TOLERANCES

The appearance of the image in the presence of supply instabilities and illumination tilt was described in § 5.3. When there is a fault condition and the supply stability is worse than the normal figure, the effect will intensify. In acute cases it will be fairly easy to detect whether the blurring of the image is radial (HT instability) or rotary (objective lens instability) and whether it changes in magnitude across the screen (as the distance from the current or voltage centre changes).

8.3.4 AMBIENT FIELD

A varying magnetic field above the tolerance level for the microscope will cause an image vibration in a direction perpendicular to the field. The effect will be uniform across the field of view, and in this respect will be similar in appearance to the effect of mechanical vibration.

In the absence of any test equipment, the amount of blur should be checked at a given magnification at different kilovoltages. If the source of trouble is a field, the blur will be a minimum at the top kilovoltage. If the source is a mechanical vibration, the magnitude will remain unchanged with kilovoltage.

8.3.5 INTERPRETATION OF DIRECTIONAL STRUCTURE

It can be seen from the foregoing paragraphs that a number of causes may contribute to directional structure in the image. When one is confronted with the effect on a photographic plate, it may be difficult to decide what the source of the blur is, because, as described above, the effects can really only be sorted out by dynamic observations on the fluorescent screen.

However, for the purposes of interpreting a micrograph already recorded, it is generally possible to decide if directional structure is real or not by examining the background structure of the support film where no specimen is present. If this structure shows directional properties, the micrograph should be rejected. If not, the specimen structure is probably genuine, and attempts should be made to confirm the observations by experimenting with an appropriate amount of defocus to see if the contrast of the structure can be maximised.

8.4 The effect of contamination

At moderate magnifications, the principal deleterious effect of electron-induced contamination on the specimen is one of inconvenience; the specimen becomes coated with a patch of carbonaceous material, about the size of the focused electron beam, and the resulting electron micrographs will show. dark diffuse patches which spoil the appearance of the image.

At higher magnifications where fine detail is being examined, it will often be found that the contrast in the fine detail, perhaps already uncomfortably low, will be almost eliminated by the addition of the layer of contamination. Thus, interpretation becomes more difficult and information may be completely lost. Furthermore, the deposition of contamination will usually result in a thickening of the specimen and consequent loss of resolution due to chromatic effects. For example lattice images less than 1 nm in spacing may be impossible to record unless the anti-contaminator devices (see § 6.14) are performing effectively.

The layer of contamination is in tension, and will cause the specimen on which it is deposited to bend or buckle. With an amorphous specimen (a replica or thin section of biological material) the bending is normally of no consequence and is often unnoticed. With a crystalline specimen, however, the bending is immediately revealed by the movement of the diffraction contours. The fact that the specimen is buckled can be an advantage in the examination of crystalline foils, but continuous movement of the contours arising from a steadily increasing thickness of contamination can make interpretation difficult if not impossible. Freedom from contamination is especially important where a large number of micrographs or diffraction patterns are required from one particular region of the specimen. This situation arises in tilting experiments on layered materials, and in the examination of crystalline material, where corresponding micrographs and diffraction patterns are required for a number of different specimen orientations.

From the interpretation point of view, freedom from specimen contamination is now an essential requirement of the electron microscope.

8.5 The effect of etching

Attempts have been made (see § 6.14) to eliminate or control contamination by the introduction of oxidising gases but one of the important difficulties of this method is that some of the specimen material may be removed as well as the contamination. The removal of specimen material by an oxidising mechanism is known as 'etching'. Unfortunately, etching can sometimes occur in a normal electron microscope where gases are not deliberately introduced. The appearance of the image is marred by the rather coarse appearance of the etched portion (Fig. 8.8), but more important, if the specimen material is removed, so is the information! Attempts have been made to remove organic specimen material selectively, in the manner of metallographic etching, but control of the selective removal is extremely difficult and no convincing interpretation of the appearance of organic etched material has yet been presented.

The principal source of oxygen at the specimen, in an electron microscope which is otherwise performing normally, is a vacuum leak in the specimen chamber. Unfortunately the high-vacuum gauge is usually located close to the diffusion pump, for convenience in mounting (see § 2.13.2) and a small leak into the specimen chamber may pass undetected. At pressures in the region of 10^{-3} Torr or less, the gas flow in molecular (i.e. in straight lines, not turbulent or viscous flow), and if the leak is such that the gas flow is beamed directly at the specimen region, etching effects can be particularly serious. Fortunately, leaks of this type are a rare occurrence, but when they do occur, location of the leak usually requires the use of sophisticated leak detection apparatus.

In an investigation of residual gas reactions in the electron microscope Hartman et al. (1968) showed that the action of water and the electron beam on organic specimens results in the removal of oxidisable material (hydrogen and carbon) by reactions similar to the water gas–equation of the form:

$$C_nH_{2n+2} + nH_2O \rightleftarrows nCO + (2n+1)\,H_2$$

the electrons supplying sufficient energy to drive the reaction to the right. The etching effects obtained by Hartman et al. are identical with those found when water has not been deliberately introduced, and a vacuum leak cannot be found.

Now, it is well known in vacuum engineering practice that water vapour is often present in large quantities in an unbaked metal vacuum system, and is one of the last vapours or gases to be pumped away. It is not surprising, therefore, that a high partial pressure of water vapour should be present in the electron microscope, especially since an obvious source of water vapour – the photographic material – is deliberately introduced into the instrument. Water vapour may also be introduced via the specimen and camera air-locks especially if the electron microscope is not fitted with an effective air-drying tube, or dry-nitrogen inlet system (see § 2.13.2).

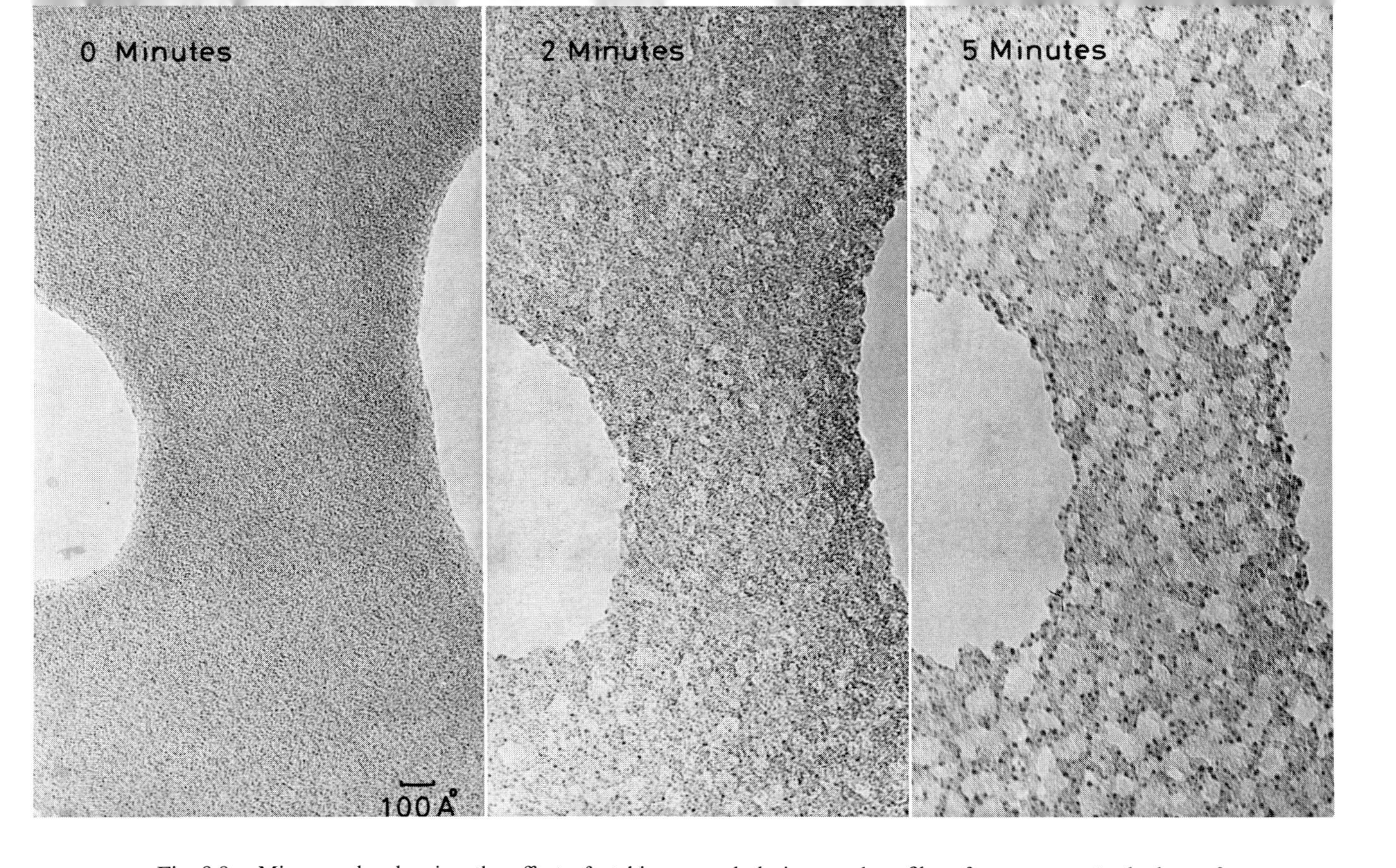

Fig. 8.8. Micrographs showing the effect of etching on a hole in a carbon film after exposure to the beam for: (a) 0 minutes; (b) 2 minutes; (c) 5 minutes.

The anti-contaminator device will trap some of the water vapour quite effectively (at liquid nitrogen temperature the water vapour pressure is less than 10^{-9} torr) but if the temperature of the anti-contaminator device is allowed to rise appreciably (water vapour pressure is 1 torr at $-20\,°C$) or the electron beam melts the ice, a very high water vapour pressure will exist in the specimen region and gross etching may occur. In these circumstances, meaningful interpretation of the image will not be possible. Every effort should be made to exclude water vapour and oxidising gases from the electron microscope by proper attention to the dessication of the photographic material (see § 7.3.5) and the maintenance of the air-inlet and vacuum seals (see § 2.13.2).

8.6 The effect of radiation damage

The results of Williams (1972) discussed in Chapter 6 (Fig. 6.6) show how quickly radiation damage can change the appearance of a specimen, and how the operator may not be aware of these changes. Other experiments, such as shadowing a specimen before and after irradiation by the electron beam, have shown that biological material may flatten (losing much of its mass) after irradiation.

Thus electron radiation may cause loss of fine structure, change in shape, and change in dimension. An associated change, sometimes observed, is the increase of granular structure in embedding plastics after intense irradiation. Some materials are very susceptible to this defect. Such granularity may mask fine structure which would otherwise be observed.

While these very serious effects must be borne in mind when attempting to interpret micrographs of fine structure in viruses and proteins, it is also true to say that very little of the coarser structure, particularly in stained biological sections, is likely to have been modified in any way which significantly affects its interpretation, provided that any dimensions are recognised as approximate due to possible shrinkage.

8.7 Analysing the source of image defects

Taking together all the background information provided, one can now summarise the various image defects due to instrumental causes, in terms of the appearance of the image. This summary is presented in Table 8.1, which gives references to the sections where further information may be found.

TABLE 8.1

Analysis of image defects in electron micrographs

Image appearance	*Likely cause*	*Remedy*
Coarse structure	Out-of-focus image	Focus critically with viewing telescope
	Astigmatism	Correct astigmatism; check centring of objective aperture
	Radiation damage	Use minimum exposure technique
	Etching	Check vacuum leaks and desiccation of plates and drying tube on air inlet
	Contamination	Check that anti-contaminator is cooled and not touching other parts of the specimen chamber
Linear structure	Astigmatism	Re-correct astigmatism
	Specimen drift	Check points discussed in § 5.4 and § 8.3.1
	Mechanical vibration	Check points discussed in § 5.8.1
	Ambient field	Check points discussed in § 5.8.2 Try changing kilovoltage
	Misalignment of illumination tilt	Check points discussed in § 5.2
	Charging effects in column	Clean sensitive parts (see § 5.7)
	Genuine line structure in the specimen	Rejoice!
Lack of sharpness towards the edges of the plate	Thick specimen (chromatic change of magnification)	Reduce specimen thickness Increase kV Use higher magnification
	H.T. instability	Check supplies; gun cleanliness
	Objective lens instability	Check supplies
Variation of sharpness across the plate	Tilted specimen (particularly if thick)	Check specimen tilt
	Misalignment of illumination tilt	See § 5.2
Uneven intensity on the plate	Uneven thickness of specimen	Try a different field of view
	Filament setting below saturation	Saturate filament (§1.6)
	Structural defect in filament if it is saturated	Change filament (§1.6)
	Image distortion	See § 5.9.2. Check calibration of magnification
	Misalignment of illumination	See § 4.2
Defects on photo print	Photographic enlarger	Check original plate for same defects

8.8 General precautions in interpretation of electron micrographs

The very great clarity of detail rendered by an electron microscope tends to invest electron micrographs with a credibility which is not always justified. It has to be remembered that the visualisation of micro-structures so small takes one far beyond the normal range of the senses, and into a region where new forces become dominant (nuclear forces, Van der Waals forces, surface tension) compared with those commonly encountered in the macro-world. It is vital that observations with an electron microscope be correlated wherever possible with those obtained by other techniques (prior observation with a light microscope, X-ray diffraction information, crystallographic information, chemical information, biological information; checks by optical diffractometers).

Where independent checks are difficult, the parameters of the specimen preparation and of the operating condition of the electron microscope should be varied to provide corroborative evidence. It must always be remembered what an exceedingly small sample is being examined (about 10^{-15} gm of material in the field of view at 100,000× magnification). It is very easy to photograph something atypical!

REFERENCES

Chescoe, D. and A. W. Agar (1966), Interprétation des micrographies en très haute résolution, J. Microscopie *5*, 91.

Cowley, J. M. and S. Iijima (1972), Electron microscope image contrast for thin crystals, Zeit. Naturf. *27a*, 445.

Hartman, R. E., R. S. Hartman, and P. L. Ramos (1968), Residual gas reactions in the electron microscope: Qualitative observations on the water gas reaction, Proc. 26th Ann. Meeting EMSA, p. 292.

Haydon, G. B. (1969), Visualisation of substructure in ferritin molecules: an artefact, J. Microscopy *69*, 251.

Hirsch, P. B., A. Howie, R. B. Nicholson, D. W. Pashley and M. J. Whelan (1965), Electron microscopy of thin crystals (Butterworths, London).

Howie, A. (1974), Application of wave theory in electron microscopy, in: Practical methods in electron microscopy, A. M. Glauert, ed. (North-Holland, Amsterdam), in preparation.

Johnson, D. J. and D. Crawford (1973), Defocusing phase contrast effects in electron microscopy, J. Microscopy *98*, 313.

Johnson, D. J. and J. Sikorski (1962), Molecular and fine structure of alpha-keratin, Nature, Lond. *194*, 31.

Komoda, T. (1964a), Resolution of phase contrast images in electron microscopy, Japan J. appl. Phys. *3*, 122.

Komoda, T. (1964b), On the resolution of lattice imaging in the electron microscope, Optik *21*, 93.

Williams, R. C. (1972), Personal communication.

Chapter 9

Future developments of the electron microscope

For a period of almost twenty years from 1949, the multipurpose 100 kV electron microscope has been the standard instrument. During this period, the resolution improved and specimen facilities were greatly extended but the mode of operation and type of application were practically unaltered.

Towards the end of this period, however, new trends were becoming apparent, which will lead in the 1970s to a much more diverse range of instruments than were current during the 1960s. Apart from a continued improvement in instrumental resolution five main trends can be discerned:

(1) The appearance of simplified instruments.
(2) The use of higher accelerating voltages.
(3) The development of analytical facilities.
(4) The wide-spread use of scanning microscopes for surface studies.
(5) The development of scanning transmission electron microscopes.

9.1 Simplified electron microscopes

It seems self-evident that the versatile research instruments are not really needed to carry out the great bulk of routine investigations where ultimate resolution is not called for and where complex treatment stages are irrelevant. Over the years, a whole series of less expensive or simple electron microscopes has been produced but many have not been very successful and still only command a fraction of the total electron microscope market. The reason for this seems to lie in the choice of compromises which must be made in order to effect savings in the cost.

On the one hand, very little saving is possible if the accelerating voltage is not reduced below 100 kV. This means that a cheaper instrument is almost certainly unsuitable for most metallurgical studies and is therefore designed for biologists and for industrial control applications. On the other hand, the loss of penetrating power with reduced voltage sets a lower practical limit to the operating voltage of 50–60 kV for most purposes.

Most biological sections are of such a thickness that chromatic aberration limits the achieved resolution to about 1.5 nm; and indeed it is questionable whether there is much meaningful structure below this level because of changes occurring during specimen preparation and during examination in the electron microscope. However, it is obviously important that the instrument should not limit the resolution too drastically and 1 nm seems to be the poorest acceptable resolution, while there is considerable pressure to have a guarantee of 0.7 or 0.8 nm resolution from the 'simple' instruments.

This requirement immediately defines other instrumental specifications. A resolution of 0.7–1.0 nm requires an objective astigmatism corrector to be fitted, and the maximum screen magnification must be at least 100,000 × (and preferable 150,000 ×) in order to allow adequate astigmatism correction. Since the minimum magnification should be near 1000 × (full screen picture) the imaging system must have three fully controllable lenses. With a top magnification as high as 100,000 ×, an efficient illuminating system is needed to yield enough brightness for comfortable viewing. Hence condenser lenses are required. If only one condenser lens is used, the heating of the specimen, and consequent thermal and radiation damage, will normally be excessive for the resolution aimed at. So a double condenser lens is called for.

Finally, as far as control is concerned, the instrument has to be designed so that it is relatively foolproof, since it may be used by less skilled operators than the research instruments. It will be particularly necessary to interlock the vacuum system and to provide very simple camera operation. Such interlocking can usually only be obtained by relatively complex mechanisms or circuits. Thus, the logic of the simple instrument leads to a quite sophisticated design, and any attempt to omit a number of the features mentioned above will result in an instrument which fails to carry out its purpose.

The first instrument to gain rather wide acceptance for routine use in biological investigations was the Zeiss EM 9 (Fig. 9.1). This instrument has only one condenser lens, but a telefocus gun goes some way to give the necessary extra protection to the specimen. The top magnification of 60,000 × is rather marginal for accurate correction of astigmatism for the rated performance. Otherwise it is a very compact instrument, easy to use,

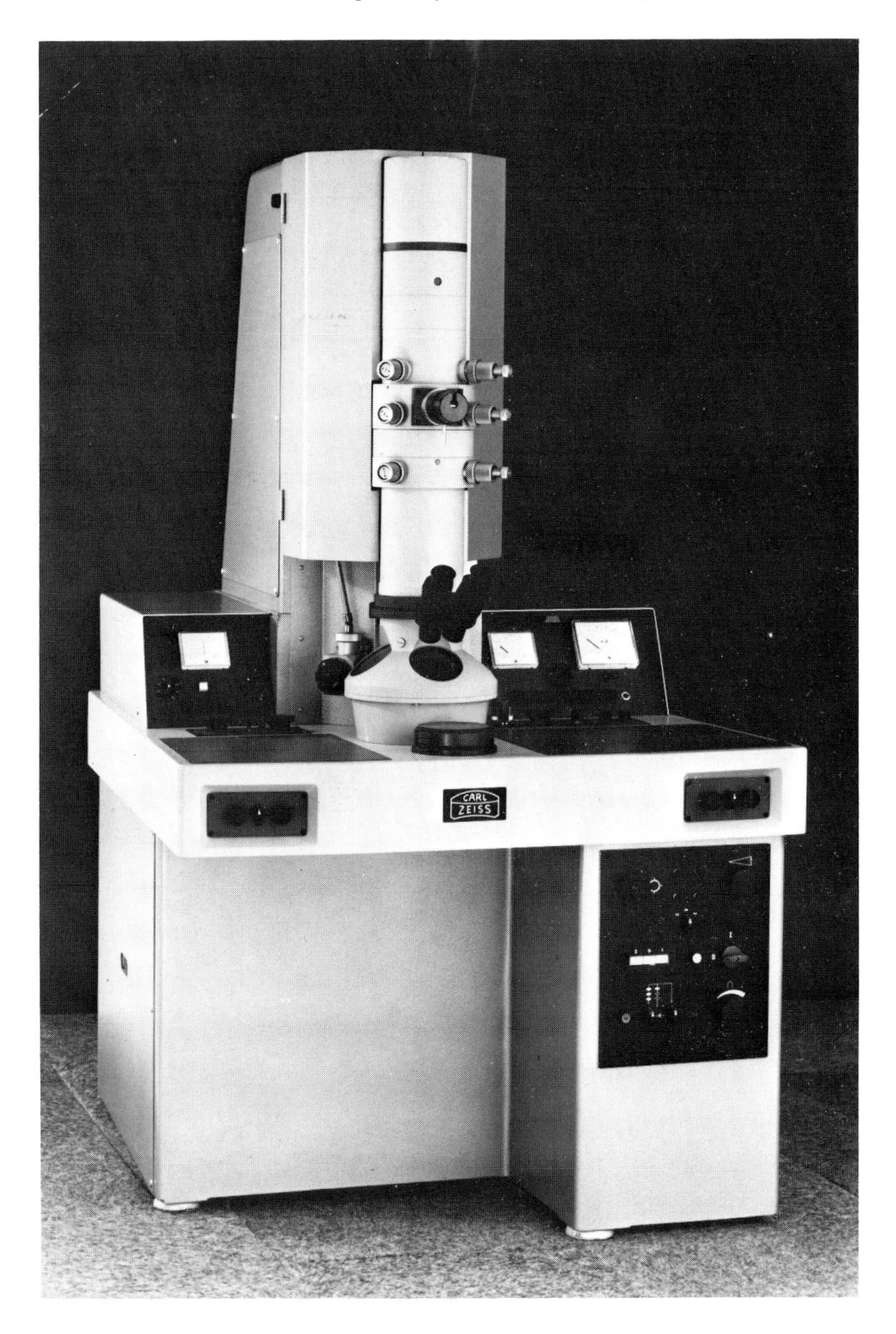

Fig. 9.1. Zeiss EM 9S Electron Microscope for biological studies.

with a minimum of controls and a very simple specimen exchange mechanism on a single rod. Its sales bear witness to its acceptability for the work it was designed to do.

It was followed some time later by the AEI Corinth 275 which has a rather more comprehensive specification (and therefore higher price) and meets all the requirements listed above. Fig. 9.2 shows this instrument, which is unconventional in having the column inverted and supported from the ground. This gives it good mechanical stability and excellent viewing arrangements, but this is only practicable for an instrument with a relatively short column, as otherwise the operator could not sit in comfort. A feature of these simplified instruments is that they are much smaller than the research instruments, the whole of the supplies being contained within the small desk. This factor is of increasing importance in these days of pressure on laboratory space.

Fig. 9.2. AEI Corinth 275 Electron Microscope for biological studies.

The only other niche for a simplified instrument would seem to be for those who will accept an instrumental resolution of 1.5–2.0 nm and possibly a lower voltage of 40–50 kV. The relaxation of ultimate resolution would permit savings in both the imaging and illumination systems and hence a significant saving in cost. It might paradoxically be less easy to operate, since the required cost savings might preclude the sophistication of vacuum interlocks and some of the lens linkages.

In spite of the lack of enthusiasm for simple instruments in the past, it now appears certain that they will command an increasing share of the market in the future, because economic factors will demand that research grants are only used to purchase instruments of adequate sophistication for the task. There is no doubt that a high proportion of all biological and industrial control studies could be carried out quite adequately with the simpler instruments now available (see Appendix).

9.2 High accelerating voltages

9.2.1 ADVANTAGES OF HIGHER ACCELERATING VOLTAGES

There are a number of advantages stemming from the use of higher voltages for accelerating the electron beam (Dupouy and Perrier 1962; Cosslett 1969).

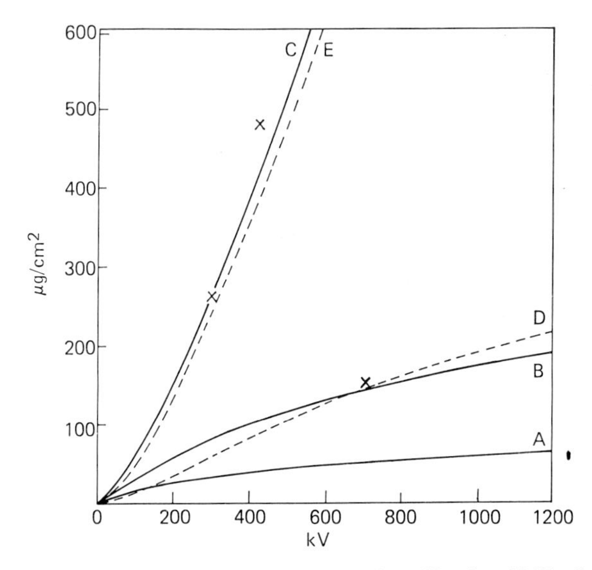

Fig. 9.3. Thickness–kilovoltage relationship for carbon (Cosslett 1969). Curves A, B, C for $I/I_0 = e^{-4}$ when $\alpha_0 \rightarrow 0{,}1 \times 10^{-3}$ and 5×10^{-3} respectively.

9.2.1a *Greater penetration of the object*

Since the electron scattering cross-section of a specimen decreases as the energy of the incident electrons is raised, it follows that the specimen becomes more transparent to the electron beam. Thus, thicker specimens can be examined at higher voltages than at the conventional 100 kV. Cosslett (1969)

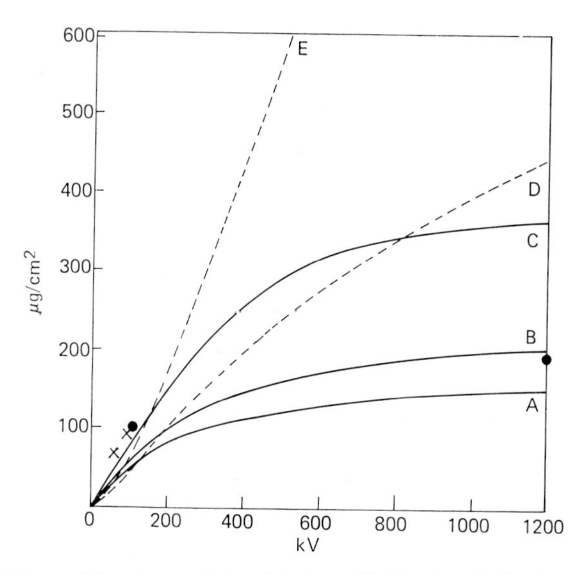

Fig. 9.4. Thickness–kilovoltage relationship for gold (Cosslett 1969). Curves A, B, C for $I/I_0 = e^{-4}$ when $\alpha_0 \rightarrow 0,1 \times 10^{-3}$ and 5×10^{-3} respectively.

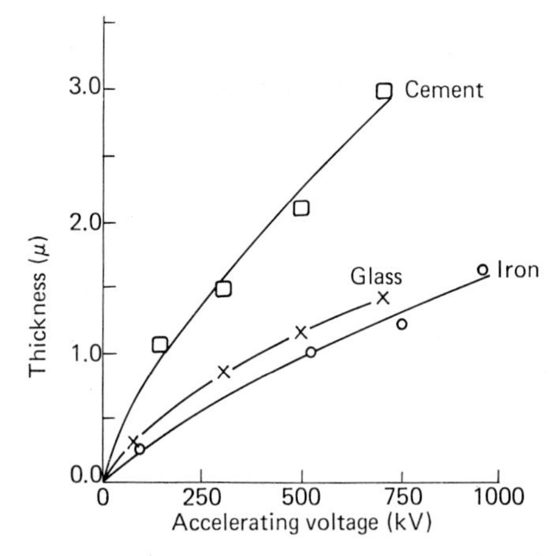

Fig. 9.5. Thickness of specimens giving good average penetration for different kilovoltages (after Hale and Henderson-Brown 1970).

calculated that the amount of extra penetration depended strongly upon the aperture angle used in the objective lens. Fig. 9.3 reproduced from his paper shows the relationship between film thickness and kilovoltage at a constant transmission for carbon. Curve A is for a vanishingly small objective aperture, B for $\alpha_0 = 1 \times 10^{-3}$ radian and C for $\alpha_0 = 5 \times 10^{-3}$ (an aperture angle commonly used in 100 kV microscopy). Fig. 9.4 shows the same curves for gold. It will be seen that there is a flattening of the curve C for gold, although in the case of carbon, the increase of penetration with kV is practically linear, under these imaging conditions. The practical results of Hale and Henderson-Brown (1970) (Fig. 9.5) for iron, cement and glass confirm these findings.

9.2.1b *Better electron diffraction information*

Not only do more electrons penetrate the specimen but since the scatter angles decrease with increased kilovoltage, the electron diffraction pattern contains many more orders of diffraction than at lower voltages. Furthermore, since the specimen is often thicker, it is possible to image the very informative Kikuchi line patterns when the crystalline material is thick enough.

A further advantage springs from the small angles of scatter. Since the electrons now enter the objective lens at much smaller angles, the displacement of the beam from the true image points due to spherical aberration of the objective lens, is greatly reduced. Hence, the limitation to the accuracy of selected area diffraction which is to an area of about 1 μm diameter at 100 kV, is much smaller at 1000 kV, when a selected area down 50 nm diameter may be accurately imaged. For a fuller discussion of the potentialities of electron diffraction at high voltages see Beeston et al. (1972).

9.2.1c *Better dark field imaging*

This is closely connected with the diffraction effects described above. Since the diffracted beams are now near the objective lens axis, they can be selected for dark field imaging by displacement of the objective aperture. The rays do not suffer large off-axial aberrations, as happens at 100 kV, and therefore good quality dark field images can be obtained from different reflexions very quickly and easily at high voltage.

9.2.1d *Radiation effects*

These have already been discussed at some length in § 6.13. The reduced cross-section for scattering at higher voltages results in less interaction between the electron beam and the specimen, and hence ionisation damage is reduced at higher accelerating voltages. Unfortunately, the improvement is not linear and present evidence suggests an improvement by about a factor of 3 by increasing the voltage from 100 kV to 650 kV (Kobayashi and Sakaoku 1965; Kobayashi and Ohara 1966). Later experiments tend to confirm these figures.

Conversely, the use of higher accelerating voltages brings many more materials into the range where displacement damage (due to atomic nuclei being moved by beam interaction) may occur, and this damage has been observed in many metal specimens.

9.2.1e *Resolution*

The high resolution obtainable with a high voltage microscope is due to two factors. Most specimens examined in a conventional instrument are thick enough to cause considerable chromatic losses to the electron beam and this is the main limitation to resolution (see § 3.2). The much lower energy losses in a high voltage instrument represent a very small relative loss compared with the incident energy and may result in a factor of 20 improvement in resolution in a thick specimen in a 1000 kV microscope compared with one operating at 100 kV (Dupouy and Perrier 1966). This is a major advantage of high voltage microscopy. In addition, the effective wavelength of electrons decreases with increased accelerating voltage; it is 0.00087 nm at 1000 kV compared with 0.004 nm at 100 kV. This results in a theoretical resolution limit of about 0.1 nm instead of 0.2 nm and, if other limiting factors can be overcome, this may be a crucially important factor. An analysis by Riecke (1970) suggests that to resolve 0.1 nm it will be essential to use an accelerating voltage of 1000 kV.

9.2.2 OPERATIONAL ADVANTAGES OF HIGH VOLTAGES

As a result of the technical features listed above there are certain clear advantages to both materials scientists and biologists from the use of high voltage microscopy. For the materials scientist, the thicker specimens which can be examined will generally bring the specimen thickness into the region where bulk properties can be demonstrated. The properties of the thin metal foils which can be examined at 100 kV are strongly modified by the two surfaces of the foil. Both precipitation and dislocation arrays are affected. Also the improved dark field and diffraction facilities both contribute to additional understanding of materials problems. The thicker foils are much easier to prepare than the very thin ones required for normal microscopy and much greater areas can be penetrated by the beam. This improves both the representativeness of the areas viewed and interpretation of the results observed.

Furthermore, since bulk properties reside in the specimens under examination, it becomes useful to attempt dynamic experiments within the microscope. These are further facilitated by the larger space available in the objective lens in the larger instrument, making feasible much more complex treatment stages than can be made for 100 kV instruments. Caution must be exercised in the choice of accelerating voltage, as it will often be desirable to restrict the voltage to just below the threshold value at which displacement damage starts to occur (about 500 kV for copper; over 1000 kV for uranium). On the other hand, valuable new information about the effects of radiation damage on reactor materials can be obtained with a high voltage microscope; it can show results in a few hours which would take a year to develop in a reactor (Makin 1971).

Biologists are finding great interest in the thicker sections which can be examined with good resolution. They were formerly obliged to try to reconstruct three-dimensional structure by the very tedious procedure of cutting serial thin sections, identifying corresponding areas in successive sections and then superposing and reconstructing the structures. By using thick sections ($\frac{1}{2}$–3 μm) they can record stereo photographs which immediately yield three-dimensional information in a readily comprehended form (Ris 1969; Porter 1969; Hama and Nagata 1970). In specimens consisting mainly of membraneous structure, much thicker specimens may be examined.

A good survey of the situation was provided by the papers presented at the EMCON meeting in Manchester, 1972. For relatively complex structures, section thicknesses of 0.5 to 1.0 μm were found to give good results for examination at 1 MV (Cox and Juniper 1973; Glauert and Mayo 1973). If material is selectively stained so that only a few components in the cell have high contrast, much thicker sections can usefully be examined. Rambourg et al. (1973) used section thicknesses in the range 3 μm to 7 μm for examining the Golgi apparatus at 1 MV. In a comprehensive survey of the techniques which might be used in high voltage microscopy, Favard and Carasso (1973) found that there was a limiting thickness of about 10 μm for Araldite sections beyond which the plastic is cratered by the action of the electron beam. They showed, however, that sections up to 10 μm in thickness can be readily examined at 3 MV.

9.2.3 HIGH VOLTAGE INSTRUMENTS

It is clear that there are good scientific reasons for building electron microscopes operating at voltages higher than 100 kV. In the actual choice of voltage however, factors of costs and convenience may weigh more heavily than the purely scientific ones. The lowest 'high voltage' offered in a commercial instrument is 200 kV. Operation at this voltage gives roughly a factor of two in the thickness of specimen penetrated compared with 100 kV. It therefore simplifies specimen preparation (particularly for thin metal specimens) and is valuable for studying particles which do not yield diffraction patterns at 100 kV. Such instruments are therefore useful in extending the range of observation but they do not really change the type of observation that is possible; for example they do not get into the range of specimen thickness where bulk properties are exhibited.

There are other commercial instruments rated at 650 kV, 1000 kV and 1200 kV and these will be considered together since they have many features in common. Fig. 9.6 shows a high voltage microscope (the AEI EM7) (Agar et al. 1970). This instrument operates up to 1200 kV and has the following main features, which are typical of high voltage electron microscopes.

(a) High Voltage Set. The high voltage generator and accelerator are housed in tanks containing insulating gas at high pressure (4 atm) (Reinhold 1969). These tanks are relatively massive and contribute considerably to the weight and height of the equipment. However, they enable the high voltage stability to be independent of the environment in the microscope room. The set can be operated at any voltage between 100 kV and 1200 kV to enable comparative experiments to be carried out, or radiation thresholds to be determined.

(b) Microscope Column. This also is massive, being 18.5 in (47 cm) in diameter and accommodating four imaging lenses as well as two condenser lenses. It is considerably higher than a conventional microscope column (9 feet to the top of the condenser lenses). Consequently the controls at the upper part of the column cannot be reached by the seated operator and remote control is required.

(c) The electron optical controls are very similar to those for a conventional 100 kV instrument and no difficulty is experienced in changing from one instrument to another. In fact, it has been possible to build in convenient controls which make this instrument simpler to operate in many ways than a 100 kV instrument.

(d) The very penetrating X-rays generated by the electron beam when they strike parts of the microscope column, require the construction of complicated shielding so as to permit safe operation.

These instrumental features result in very massive installations, requiring

Fig. 9.6. AEI High Voltage Electron Microscope Type EM7.

special accommodation and relative high running costs. There is very little saving in building requirements or running cost between the designs at present available for 650 kV and 1200 kV, although of course there is a saving in capital cost of the instrument. For a number of metallurgical problems, 650 kV may be adequate since in some dynamic experiments it will be desirable in any case to work below the threshold for radiation damage. On the other hand, it looks as if 1200 kV is required for some experiments in physics and for any studies of radiation damage effects; and for many biological experiments the penetration of suitable thick specimens seems to require at least 800 kV and preferably 1000 kV.

In an effort to overcome the problem of very expensive buildings for a high voltage instrument, Le Poole et al. (1970) have designed a 1000 kV instrument with a double-folded column so that the whole installation could be accommodated in a room of normal height (Fig. 9.7). This is achieved by curving the electron beam from the accelerator through 180° into the bottom of the microscope column; after the beam has passed through the intermediate lens it is again sharply folded by magnetic prisms to pass through the projector lens on to the viewing screen and photographic plate. Good progress has been made in solving the formidable technical problems involved and it will be interesting to see if the instrument can be developed to provide the operating simplicity and range of operation required by a typical group of users. If these problems are solved, the instrument would overcome one of the major objections to high voltage instruments.

One of the earliest hopes pinned on high voltage electron microscopy was that it may make possible the examination of living material by enclosing it in a small cell containing water or other environment in which life processes could continue. It now looks as if this hope may not be realised as the radiation dose imparted to the specimen in only a brief examination is usually lethal, and is at least so high that any phenomena observed would be heavily modified by the effects of radiation. Studies by Glaeser (1972) show that even under optimal recording conditions, the minimum electron dose is likely to exceed the lethal level for all except the largest organisms, and these can be examined under a light microscope.

However, there are important uses to be made of a closed cell, since it permits the examination of material in the wet state and this may avoid many drying and chemical artefacts. Matricardi et al. (1972) have already obtained diffraction patterns from wet samples of catalase and have shown that the shrinkage observed with specimens dehydrated by normal preparation methods has been avoided.

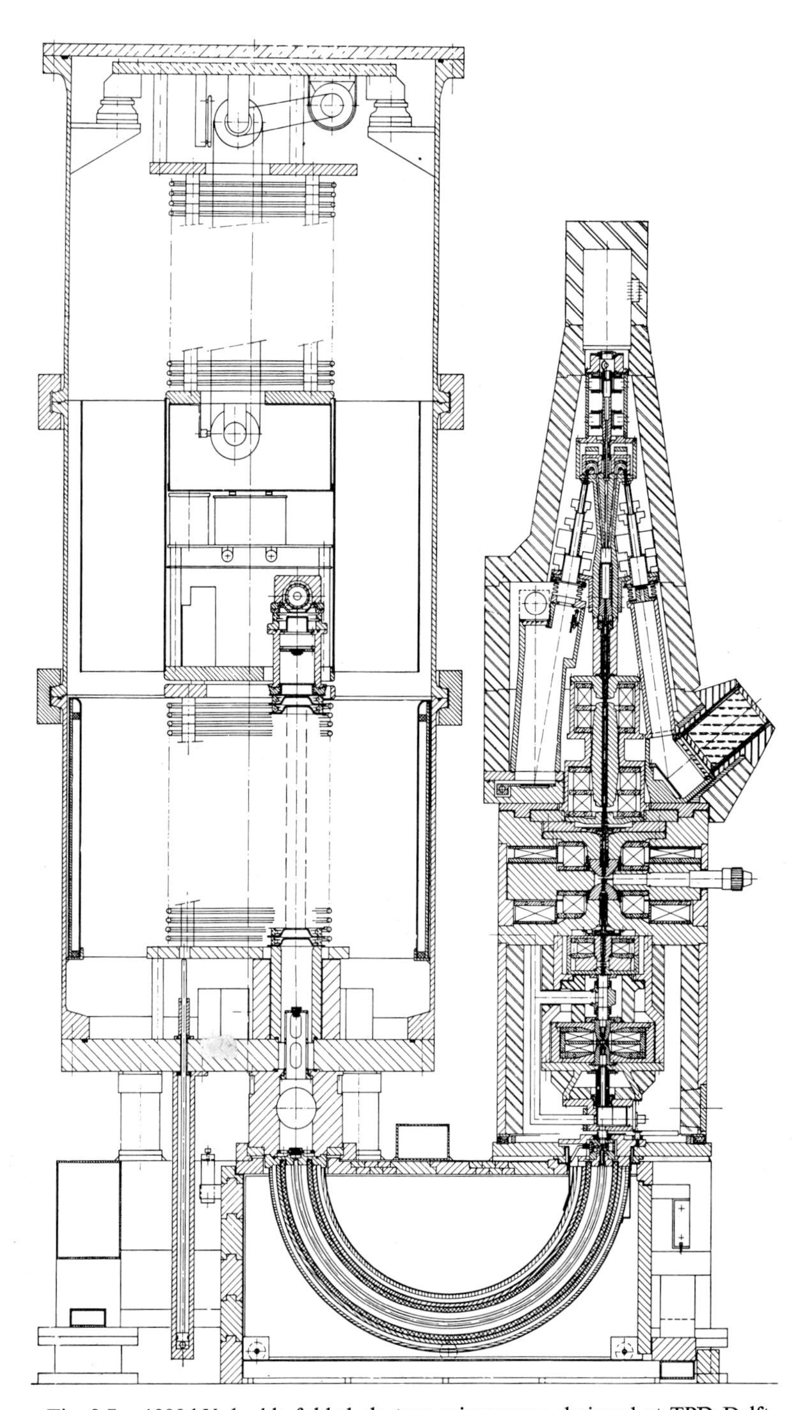

Fig. 9.7. 1000 kV double folded electron microscope, designed at TPD Delft.

9.2.4 SPECIAL TREATMENT STAGES

The advent of high voltage microscopes has made for new experimental possibilities within the instrument. Because all the dimensions of the instrument are scaled up, there is quite a large space in the objective lens gap, and this permits the introduction of much more complex mechanisms than are possible in a 100 kV microscope. It becomes possible to treat the specimen (heating, cooling, straining) with a considerable degree of precision, and combine this treatment with good specimen tilt facilities. For the first time, controlled straining experiments are possible within an electron microscope; indeed, it is only because specimens with bulk properties can be examined that such experiments are worth doing.

Amongst the most interesting developments have been the environmental stages. Two different designs have been evolved. One, by Allinson (1970) is contained within a rod specimen holder; the end of which is shown diagramatically in Fig. 9.8. The nose cone A spring-loads the specimen rod R

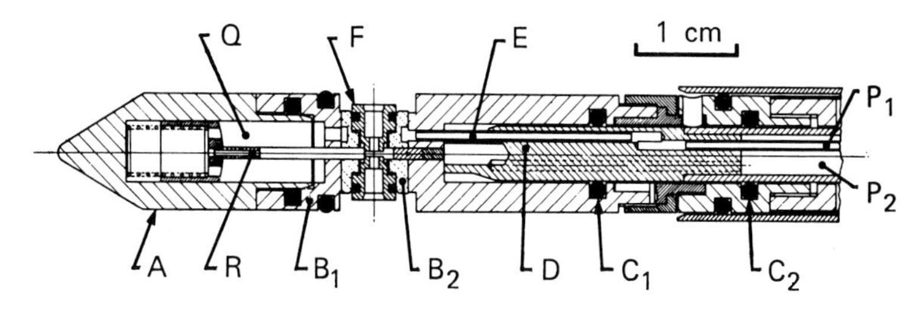

Fig. 9.8. End portion of NPL environmental cell for the EM7 high voltage microscope (Allinson et al. 1970).

against the sealing tube D. The gas cell is contained in B_2 by the window sealing mounts F. Gas or vapour is introduced through capillary pipe P_1 through the sliding capillary E to the chamber Q, from whence it passes over the specimen and out by pipe P_2. C_1 and C_2 are sliding seals. The windows are corundum crystals ion-thinned to about 200 nm. The gap between the windows can be varied in the range 0.1 to 0.5 mm. A special feature of such a design is that the specimen may be removed from the microscope and kept within the controlled atmosphere for a considerable time, and re-examined at a later time when the reaction has had time to develop. It is being used in this way in a study of the hydration of cement.

Swann and Tighe (1971) have adopted a different principle (Fig. 9.9). A section of the objective lens gap is isolated by small apertures above and

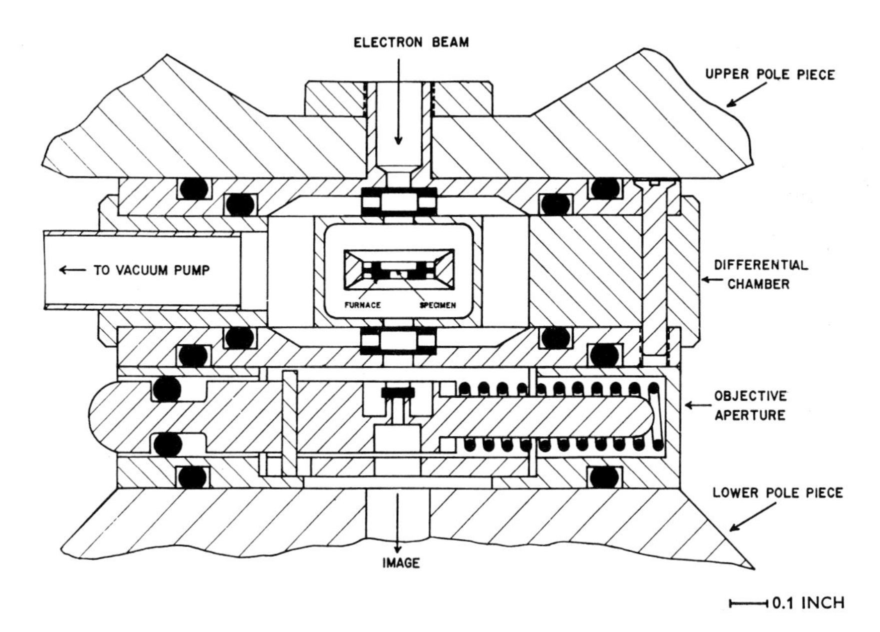

Fig. 9.9. Gas reaction chamber for the EM7 high voltage electron microscope. This gas chamber spans the gap between the upper and lower pole pieces. The controlled atmosphere fills the central reservoir and surrounds the specimen. Small apertures above and below this volume restrict the leakage of gas to the main microscope volume. A differential pumping tube provides additional pumping at each end of these apertures. The objective aperture is hermetically sealed into the assembly (Swann and Tighe 1971).

below it, and an ordinary specimen rod can be introduced into this space. There is a relatively large reservoir of gas connected to the region between the pole pieces so that the pressure can be precisely controlled. The small apertures restrict the rate of loss of gas to the microscope vacuum, and supplementary pumping is provided for the specimen chamber. Once this special section has been inserted into the objective lens (an operation taking less than an hour), any other of the standard treatment stages (heating, cooling etc.) may be introduced through the air-lock in the normal way. The gas path length is longer (1.5 mm or 5 mm) than in the Allinson stage, but it has been found possible to operate at a pressure of 300 torr in air, or at atmospheric pressure in helium, or in helium saturated with water vapour. Typical results are shown in Fig. 9.10 and illustrate the good results obtained in a helium atmosphere at atmospheric pressure. A detailed study of the electron transmission through gases at different pressures and operating voltages has been made by Swann and Tighe (1972).

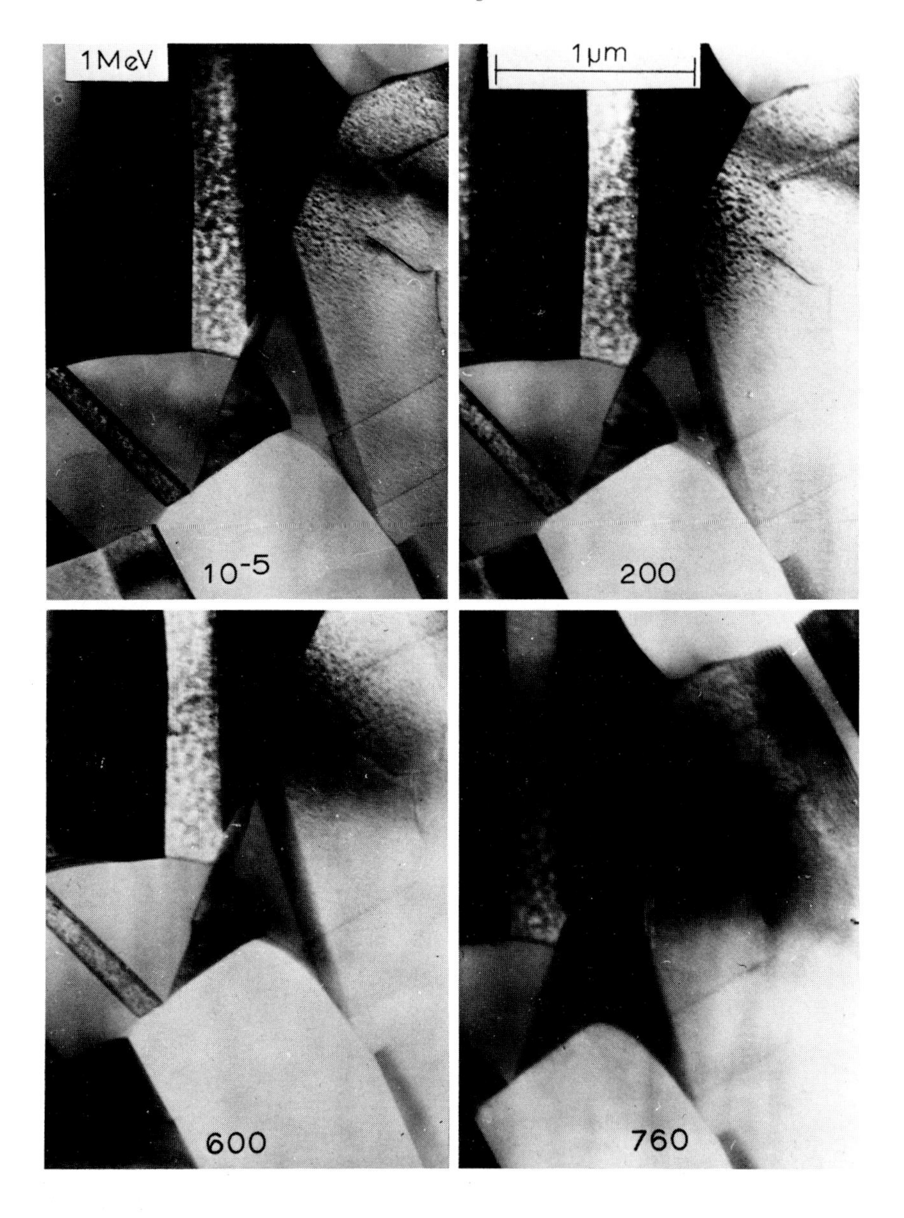

Fig. 9.10. Stainless steel specimen in environmental cell with different pressures of helium within the cell. Cell thickness 1 mm. Voltage 1000 kV. Pressures are expressed in torr (Swann and Tighe 1971).

Obviously, stages of this kind will be of great importance in biological studies, and particularly for examinations in the wet state.

9.2.5 VOLTAGES ABOVE 1 MV

Already, two instruments have been built to operate at 3000 kV, one by Dupouy et al. (1970) and the other by Sugata et al. (1970) and Ozasa et al. (1970). The Toulouse instrument has a massive column nearly 1 m in diameter and 3.9 m high (Fig. 9.11). These instruments have shown the continued improvement in resolution of thick specimens due to the reduction of chromatic defects. They also permit experiments on threshold voltages and on electron scattering of interest to solid state physicists. It is not yet clear how much more value they will prove to have in the study of metals and biological structures.

There is some interest in even higher voltages – at 10 MV, there would be a further useful gain in lifetime of sensitive materials to irradiation, and such an instrument might be important from such a standpoint. An instrument to operate at this voltage is under construction in Japan.

9.3 Analytical facilities for electron microscopes

9.3.1 SELECTED AREA ELECTRON DIFFRACTION

The first analytical studies using the electron microscopes were possible more than 20 years ago when selected area electron diffraction facilities (§ 3.10.1) were included in the conventional transmission electron microscope (Ruska 1954; Haine and Page 1956). See Beeston et al. (1972) for a full account of the technique.

9.3.2 X-RAY MICRO-ANALYSIS

There followed various attempts to provide facilities for elemental analysis by extracting the characteristic X-rays generated as the electron beam interacted with the specimen. Katagiri and Osaza (1967) attached a crystal spectrometer to the specimen chamber of an Hitachi instrument, and the specimen was raised from its normal position in order to extract the X-rays. This system suffered from a rather poor X-ray collection efficiency and also from the fact that a large area of the specimen was irradiated and hence there was insufficient precision in defining the area of analysis. An improved technique suggested by Anger et al. (1968) consisted of sinking the specimen deeper into the objective lens field, so that the upper part of the lens field acted as a third condenser lens and reduced the cross-section of the incident

Fig. 9.11. 3 MV electron microscope at Toulouse (Dupouy et al. 1970). Note the remote control rods for adjustments to the upper part of the column.

beam. This system improved the precision of the analysis, but involved tilting the specimen and also some rather unfavourable geometry for the X-ray spectrometer, which was attached to the outside of the microscope column. This placed the detection limit for chromium at about 10^{-16} gm in a probe of diameter 0.1 μm.

The next development arose from the pioneer work of Duncumb (1963) who built a combined electron microscope and microanalyser and showed how valuable an electron microscope could be if more sophisticated X-ray detection facilities were available. A further improvement by Cooke and Openshaw (1969) resulted in the combination of a conventional high resolution electron microscope with high quality X-ray detection facilities; an air-cored mini-lens was added, which permitted a much smaller region of the specimen to be analysed. This combined instrument (the AEI EMMA-4) (Fig. 9.12), permits analysis of elements from sodium to uranium from a region of 100–200 nm diameter, with detection limits of the order of 10^{-18} g, with local elemental concentrations of less than 1 in 1000. In addition an electron microscope image can be obtained with a resolution of 1 nm and selected area diffraction from a region 50 nm in diameter. In principle the method could be used for quantitative analysis; and this is already possible for some specimens.

Another approach to analytical electron microscopy has been the fitting of solid state energy dispersive analysers onto the column of conventional instruments. The fine electron probe is obtained by the system of Anger et al. (1968) described above. It is still difficult to place the detector close to the specimen because of space restrictions due to the objective lens itself, but qualitative analyses are possible if there is a reasonable concentration of the element of interest in the analysed region. The non-dispersive system is attractive because of the simultaneous detection of all elements and the graphical display available. Developments in solid state detectors are now permitting a much smaller size for the detector head. This permits a closer approach to the specimen, and so the sensitivity of the method is improving rapidly. It has become an attractive addition to a standard electron microscope, although it must at present be regarded as a qualitative technique only.

Analytical electron microscopy by X-rays will be of greatly increasing importance in the future. It already merited a symposium at Nottingham in 1971 (Agar et al. 1972) followed by a symposium session at the European Congress for Electron Microscopy in Manchester, 1972, and a rising proportion of electron microscopes can be expected to have some analytical facilities in the years ahead.

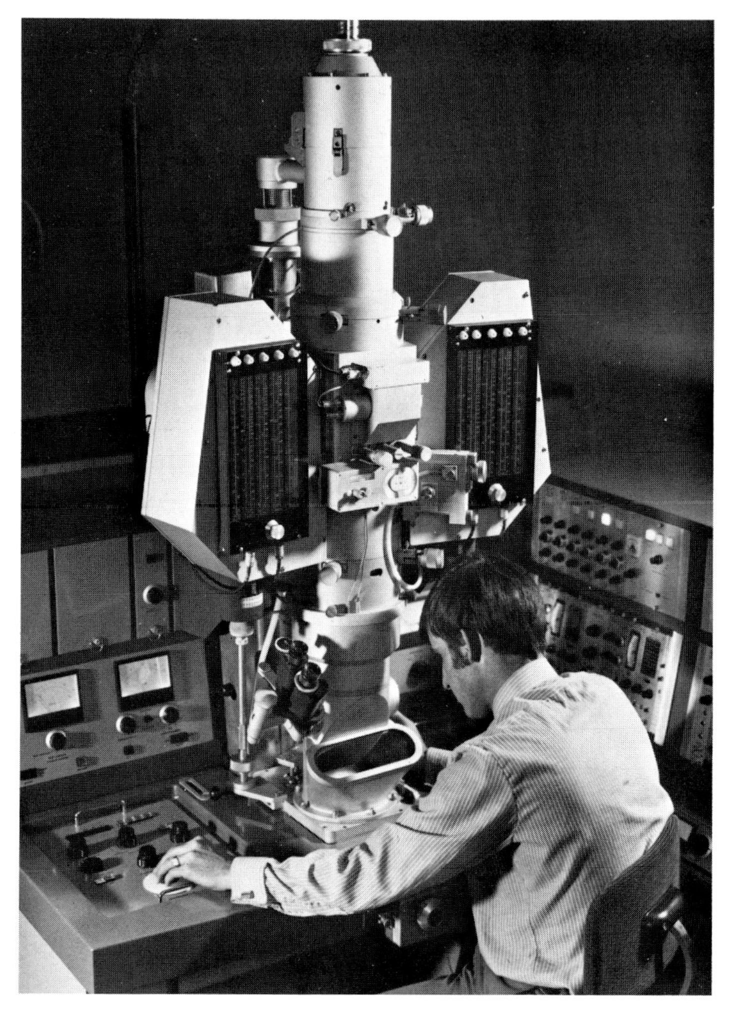

Fig. 9.12. The AEI EMMA – 4 Analytical Electron Microscope. Fully focusing vacuum spectrometers are attached to either side of the column.

9.3.3 ENERGY ANALYSIS

It was very early suggested that a measurement of characteristic energy losses in a specimen might yield a microanalysis technique (Hillier and Baker 1944). Since then a number of different experimental arrangements have been devised to achieve this end (e.g. Crick and Misell 1971).

One such arrangement (an energy selecting microscope) allows the whole of the electron image to pass through an energy filtering system and results in an image free of energy-loss electrons or made up only from electrons of some specific narrow band of energy loss. Instruments of this type have been described by Boersch (1949), Watanabe (1951) and Castaing and Henry (1962). There are considerable design difficulties because a good energy discrimination demands small apertures in the system and this restricts the field of view. This is a basic problem in the conventional electron microscope where the whole image is formed simultaneously over a considerable area.

An alternative system is to form the electron image in the normal way and to select a small area (or line) of the image which is then subjected to energy analysis either by a retarding field filter, or by some spectrometer arrangement (an energy analysing microscope).

Such a system gives a fine spectrum of energy losses for a small selected

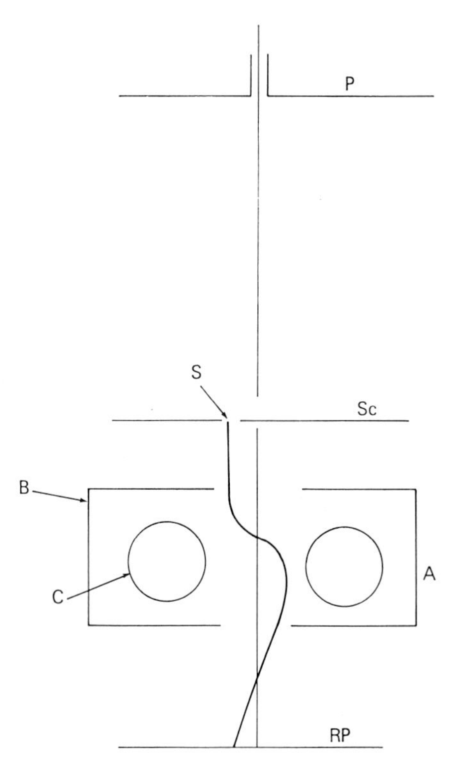

Fig. 9.13. The energy analysing system for an electron microscope after Cundy et al. (1966). P is the projector lens S the selecting slit in the viewing screen, Sc. The Mollenstedt analyser A consists of an earthed box B containing cylinders C at the high voltage of the electron gun. The energy spectrum appears on the recording plate RP.

region of the specimen. A considerable amount of work has been done with an electrostatic cylinder lens first described by Möllenstedt and Rang (1951) and further developed by Cundy et al. (1966); the experimental arrangement of Cundy et al. (Fig. 9.13) permits an energy resolution of 1.5 eV to be attained, and the region from which analysis is carried out is only about 10 nm in width – much smaller than for any other technique so far described. However, the analysed region is not known with precision, so the method is principally of use to study transitional regions (e.g. grain boundaries, large precipitates), where the exact location of the selected region in a direction perpendicular to the line of analysis is not critical. The technique has been used to study grain boundary segregation in an aluminium – 7% magnesium alloy (Cundy et al. 1968a), silicon inclusions in nickel (Cundy and Grundy

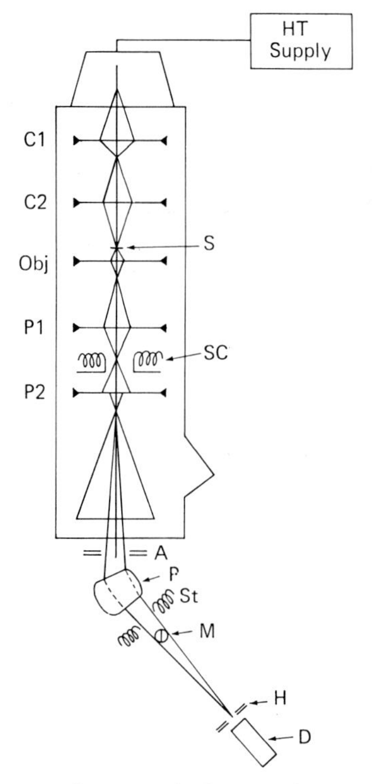

Fig. 9.14. Magnetic energy analyser attached to an electron microscope (after Wittry 1969). The standard microscope with lenses C1, C2 Obj, P1 and P2 has the specimen at S and special scan coils SC between the projector lenses. The beam for analysis is selected by the aperture A, dispersed in the prism P and shaped by the stigmator St to fall on the slit H. Electrons are counted by the scintillation counter D. The modulation coils M are to counteract stray magnetic fields.

1966), aluminium – 4% copper (Cundy et al. 1968b), silica particles in silicon (Ditchfield and Cullis 1970) and glass (Cook 1971). Quantitative analysis with an accuracy of about 1% by wt appears to be feasible.

Further work by Whelan (1974) has resulted in a new design of energy selecting microscope which shows an energy filtered image on the screen, while allowing an energy spectrum to be recorded with an aperture-counting system beneath the screen. This is achieved by sweeping the HT so that the energy window sweeps along the spectrum. The dispersion is linear up to high energy losses and has now superseded the old energy analysing system described above.

A further technique has been described by Wittry (1969) who pointed out the limitations of the Cundy method for binary systems. Wittry placed a magnetic analyser below a conventional electron microscope thus simplifying the construction (Fig. 9.14). The scan across the energy spectrum is achieved by sweeping the high voltage supply through an appropriate voltage range, so that all the loss electrons in turn are imaged along the instrument axis and without chromatic errors. The optics of the analysing system are then fairly simple. Wittry has used this electron spectrometer mainly to study the larger energy losses due to the K ionisations which produce X-rays. It turns out that the spectrometer is particularly sensitive to the ionisation losses of light elements which are the most difficult to analyse by X-rays, so it has considerable promise as a complementary technique to that of X-ray analysis. It works with high sensitivity only with very thin specimens however. As pointed out by Wittry et al. (1969) the smaller plasmon losses can be used to measure the thickness of the specimen under examination.

So far, none of these energy analysis methods have been used in a commercial microscope, but they may well be applied in the scanning transmission electron microscope (§ 9.4.2).

9.3.4 SCANNING ELECTRON DIFFRACTION

The technique of scanning electron diffraction was first refined and applied by Grigson (1962) and further developed by Tompsett et al. (1969) in a special electron optical system. If the system is attached to an electron microscope, it becomes a very powerful analytical tool (Wittry et al. 1969; Barden and Craig-Gray 1970; Beeston et al. 1972).

If an electron diffraction pattern, formed in the usual way, is scanned across a small aperture at the bottom of the instrument, by means of special coils built into the electron microscope column, the electron diffraction information can be collected sequentially in a series of line scans. This

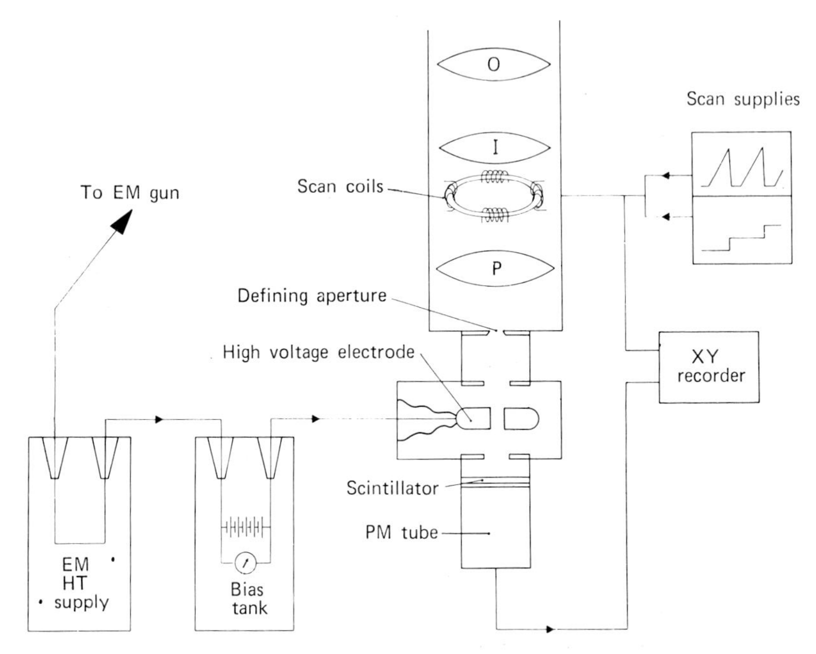

Fig. 9.15. Scanning electron diffraction attachement for an electron microscope (after Barden and Craig-Gray 1970).

electrical detection allows information to be recorded in the period of the scan, that would take maybe one minute to record on a photographic plate. It may therefore be used to record and follow transitory phenomena such as re-crystallisation of a specimen. Fig. 9.15 shows schematically the system developed by Barden and Craig-Gray (1970).

The electron beam passing the selector aperture may also be energy filtered so that any electrons which have lost energy are removed. It therefore becomes possible to record a diffraction pattern free of background scatter and hence to measure true intensities from the pattern. This renders possible the accurate analysis of thin films of material which would not yield an X-ray diffraction pattern because of lack of material. In addition the electron diffraction pattern is collected in a fraction of the time which would be required for X-ray measurements, even if they were possible. Alternatively, the voltage on the filter can be set to accept electrons of a defined energy loss.

9.4 Scanning transmission electron microscopy

Scanning reflection electron microscopes are now so important that they will be discussed in a complete book in this series (P. Echlin and G. R. Booker, in preparation).

It is not proposed to discuss these instruments in any detail, but for an understanding of the importance of the Crewe-type scanning transmission electron microscope the question of information gathering must be briefly discussed.

A scanning electron microscope uses the lenses in a demagnifying mode to form an exceedingly small electron probe. This is scanned in a raster over a specimen surface, and either back scattered or collected electrons, or secondary electrons from the specimen surface, are detected and displayed on a television tube scanned in synchronism. In this way, a picture of the surface structure can be built up. Alternatively, if the specimen is thin, the electrons transmitted may be collected and displayed, and a picture similar to that from a conventional transmission microscope is obtained.

Neglecting all the other detail about the instrument, one can see that, to a first approximation, the resolution obtained should be determined by the size of the electron probe formed at the specimen (neglecting electron spread in the specimen). There is nothing in principle to prevent the formation of exceedingly small electron probes. The real limit to performance, however, stems from the numbers of electrons in the beam, and the need to be able to recognise real structure against a background of noise. If there are N electrons reaching one image point, the statistical fluctuations will amount to $\sqrt{N}$, so the signal-to-noise-ratio is $N/\sqrt{N}$. It has been shown that, in order to be sure an absolute minimum signal-to-noise level for *detection* of detail against a statistically varying background is 5 to 1; 10 to 1 would be more acceptable. In order to obtain a good quality image (such as with a transmission electron microscope) a figure of 100 to 1 would be more appropriate.

Considering first, a good quality image, a signal-noise ratio of 100:1 requires 10^4 electrons per image point. In order to achieve a good display, 1000 image points in 1000 lines would be required: 10^6 image points.

Thus 10^{10} electrons are required, with a charge of

$$10^{10} \times 1.9 \times 10^{-19}\ \text{C} = 1.6 \times 10^{-9}\ \text{C}.$$

If the beam probe size is 1 nm $= 10^{-7}$ cm dia,

$$\text{charge density onto specimen} = \frac{1.6 \times 10^{-9}}{10^{-14}}\ \text{C/cm}^2 = 1.6 \times 10^5\ \text{C/cm}^2.$$

The maximum brightness for a tungsten filament gun at 100 kV is about 10^5 A/cm^2/sr and the maximum solid angle achievable (due to lens aberrations, an aperture of 10^{-2} radian) is $\pi \times 10^{-4}$ sr.

Achievable current density in the probe is therefore

$$10^5 \times \pi \times 10^{-4}\ \text{A/cm}^2 = 31.4\ \text{A/cm}^2.$$

The time for the charge density of 1.6×10^5 C/cm^2 to be built up with this probe is about 5000 sec. This is a hopelessly long exposure time. A barely acceptable picture, at the limit of signal-to-noise ratio, would be obtained in just under 1 min – again, hardly tolerable with the supply and mechanical stabilities available. This indicates why a scanning microscope based on a heated tungsten filament is limited in resolution to about 7 nm.

By using a lanthanum boride emitter, the brightness of which can be 10 times greater than that of a heated tungsten cathode, the resolution could be improved to about 2 nm.

The field emission cathode has a very much higher brightness – a factor of 10^3–10^4 better than a tungsten filament. This makes very high resolutions of 0.2 to 0.3 nm achievable with short exposure times.

In order to understand various claims of better resolutions than are deduced here, one should remember that savings in exposure time can be made by reducing the area of the specimen imaged, and by reducing the number of image points. Also, a noisier picture may be accepted than that given by a transmission electron microscope of comparable resolution. This enables some remarkable figures to be claimed for scanning instruments with a tungsten filament. But the only way of fully overcoming the difficulties is by using the field emission gun.

9.4.1 THE SCANNING TRANSMISSION ELECTRON MICROSCOPE (STEM)

The scanning transmission electron microscope (STEM) is the name given to the type of electron microscope devised by Crewe (1966) (see also Crewe et al. 1968; Crewe et al. 1970) and it must be clearly distinguished from two other modes of microscopy sometimes also described in this way.

(a) A normal scanning reflection microscope which yields a surface picture may be modified so that the electron detector is placed behind the specimen. If the specimen is thin enough, the detector will then collect electrons modified by passage through the specimen and will yield a scanning micrograph recorded in the transmission mode of operation. The probe size of 10–20 nm is the same as used in the scanning reflection mode.

(b) If a conventional transmission electron microscope is equipped with scanning coils in the illuminating system, and the specimen is moved to the centre of the objective lens, then a fine scanning probe of 2 nm diameter may be obtained by employing part of the objective lens field as a third condenser.

An electron detector placed at the bottom of the instrument will record information which may be displayed as a scanning transmitted image (Koike et al. 1970).

Both systems (a) and (b) have a conventional oil diffusion-pumped vacuum system and a simple electron detector. They both use a tungsten emitter as the electron source, and this severely limits the current which can be obtained in a fine electron probe. Both therefore lack the high beam current density and the ultra-high vacuum system implicit in the Crewe-type instrument and in order to avoid confusion, it has been suggested that they are not known as STEM instruments (Agar 1972) even though they do provide a transmission form of scanning electron microscope.

The importance of these transmission scanning modes of operation lies in the ability to penetrate specimens two to three times thicker than with a conventional transmission microscope (TEM) for a comparable resolution (at 100 kV). (This is because the chromatic losses in the specimen are unimportant in the scanning mode because there are no lenses after the object.) On the other hand the low current in the probe means that, to obtain a high resolution picture, the exposure times must be very long or the field of view (number of image points) restricted.

9.4.2 THE STEM INSTRUMENT

The instrument developed by Crewe has several important characteristics. A schematic layout of the instrument is shown in Fig. 9.16. The electron source is a field emission gun. This is a very fine pointed single crystal of tungsten in a field strong enough to draw electrons from the tip by field emission. The electron source size from such a gun is only of the order of 3 nm and is of very high brightness (Cosslett and Haine 1954). The very small source size means that one stage of demagnification will yield a scanning spot of the order of 0.3 nm at the specimen, and thus makes possible a very high resolution scanning instrument (because the high brightness does not limit the information collection as does a tungsten emitter). Naturally, to yield such a high resolution, the instrument has to be used in the transmission mode, and the specimen must be very thin.

The fine electron probe is scanned in a raster over the specimen surface by appropriate scan coils, and electrons transmitted by the specimen are collected on a signal detector whose output is fed through signal amplifiers to a cathode ray tube display scanned in synchronism with the electron beam. Various special signal collection arrangements can be made. The elastically scattered electrons are mostly scattered through relatively large angles, and can be collected on an annular detector. The inelastically scattered electrons, contained within the central cone of the beam, may be energy analysed to separate them from the unscattered beam, before being detected. The different signals may be viewed sequentially or electrically combined; thus, the ratio of the elastic to the inelastic signals is proportional to the atomic number of the region being examined. Dark and bright field pictures may be selected at will.

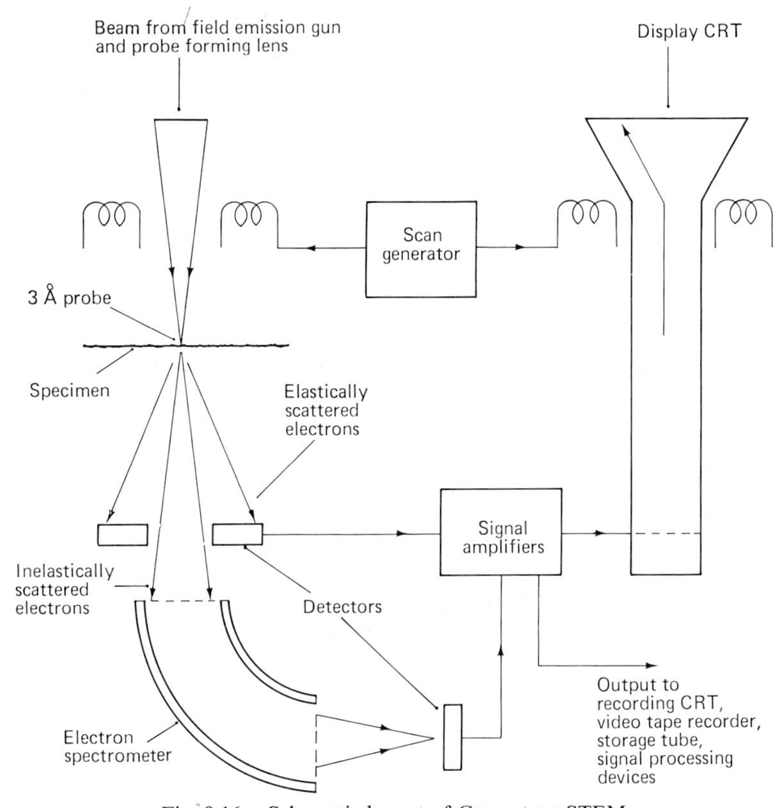

Fig. 9.16. Schematic layout of Crewe-type STEM.

Since the scanning beam gives rise to sequential signals from the specimen, and the beam is axial, the filter system is much simpler than that required in a normal transmission electron microscope, where a whole area of the specimen is imaged simultaneously and the energy analysis facilities described in § 9.3.3 become easy to provide.

The scanning transmission electron microscope requires an ultra-high vacuum in the electron gun in order to achieve stable field emission conditions (a vacuum better than 10^{-10} Torr). Since the specimen area must be at a good vacuum, the problems of contamination of specimen and apertures are no longer so serious (except in so far as the specimen itself contributes) and this is a major improvement.

There is, in theory, an exact correspondence between the performance of this type of microscope and a conventional transmission instrument; the ultimate resolution for instance, will be the same. However, there is an improved electron collection efficiency in the STEM, since the collection angle is no longer limited by lens aberrations (there are no electron lenses after the specimen). Also the contrast conditions can be modified much more easily so that fine structure near the resolution limit can be expected to be rendered in better contrast.

9.4.3 ANALYTICAL FACILITIES IN STEM

Most advanced STEM designs include an energy filter after the specimen so that the image can be processed in various ways. Thus they are already equipped for measuring electron energy loss spectra. Some instruments have Auger analysis facilities fitted also. In spite of these built-in facilities for analysis, it seems quite possible that analysis by electron energy loss may advance relatively slowly, compared with X-ray microanalysis, which is very much better documented. The main limitation in existing analytical electron microscopes such as EMMA is the beam current available in a small probe – this limits the X-ray count and hence the volume of material from which an analysis can be obtained. X-ray analysis in a STEM should make possible analysis from regions 10 nm in diameter. This would be particularly interesting in many biological problems, provided that the intense irradiation did not sublime away the material to be analysed.

One important point concerns the range of operation of STEM. The objective lens may be operated at a longer focal length in order to provide room for reflexion specimens, and provided the apertures are correctly chosen, there is no important loss in performance. However, the field emission gun will only yield a very small spot, and any attempt to obtain a large probe from such a gun will lead to a large loss in intensity. For probe sizes greater than 100 nm, a thermionic emitter remains more efficient. The STEM therefore will remain a high performance instrument and will not supplant the existing transmission and scanning instruments for routine work, (or an EMMA for analysis of areas greater than 100 nm). It seems possible that a scanning microscope with a lanthanium boride emitter will take over the important role of an instrument for the analysis of thicker specimens where the ultimate resolution is not called for.

The first commercial models of STEM instruments are now appearing, although with rather limited resolution at present. When the high resolution models appear in a year or two, they will open a new era in electron microscopy for sophisticated research work. It is not yet clear how much of the work at present handled by high performance conventional transmission electron microscopes will come to be transferred to STEM instruments. For a number of years the STEM will be more expensive and considerably more complex to operate. It seems likely that economics will generally be the deciding factor in the division of roles between these new instruments and the established transmission microscope.

REFERENCES

Agar, A. W. (1972), Nomenclature of scanning electron microscopes. Proc. R. microsc. Soc. *7*, 287.

Agar A. W. et al. (1972), Proceedings of first symposium on biological applications of combined electron microscopy and X-Ray probe microanalysis, Nottingham, Micron *3*, 81.

Agar, A. W., G. Browning, J. L. Williams, J. Davey and K. Heathcote (1970), A new 1000 kV electron microscope. Proc. 7th Int. Congr. Electron Microscopy, Grenoble *1*, 115.

Allinson, D. L. (1970), Environmental cell for use in a high voltage electron microscope. Proc. 7th Int. Congr. Electron Microscopy, Grenoble *1*, 169.

Anger, K., K. H. Hermann, G. Kempf and H. Neff (1968), X-ray microanalysis of thin foils using a very fine electron beam. Proc. 4th Eur. Reg. Conf. Electron Microscopy, Rome *1*, 255.

Barden, L. F. and J. Craig-Gray (1970), A scanning electron diffraction attachment with energy filter. Proc. 7th Int. Congr. Electron Microscopy, Grenoble *1*, 181.

Beeston, B. E. P., R. W. Horne and R. Markham (1972), Electron diffraction and optical diffraction techniques, in: Practical methods in electron microscopy, Vol. 1, A. M. Glauert ed. (North-Holland, Amsterdam).

Boersch, H. (1949), Ein Elektronenfilter für Elektronenmikroskopie und Elektronenbeugung, Optik *5*, 436.

Castaing, R. and L. Henry (1962), Filtrage magnétique des vitesses en microscopie électronique, C. r. Acad. Sci. Paris *255*, 76.

Cook, R. F. (1971), Combined electron microscopy and energy loss analysis of glass, Phil. Mag. *24*, 835.

Cooke, C. J. and I. K. Openshaw (1969), Combined high resolution microscopy and X-ray microanalysis. Proc. 28th Ann. Meeting EMSA Houston, p. 522.

Cosslett, V. E. (1969), High voltage electron microscopy, Q. Rev. Biophysics *2*, 95.

Cosslett, V. E. and M. E. Haine (1954), The tungsten point cathode as an electron source. Proc. 3rd Int. Congr. Electron Microscopy, London, p. 639.

Cox, G. and B. Juniper (1973), The application of stereo-micrography in the high voltage microscope to studies of cell-wall structure and deposition. J. Microscopy *97*, 29.

Crewe, A. V. (1966), Scanning electron microscopes: is high resolution possible? Science *154*, 729.

Crewe, A. V., M. Isaacson and D. Johnson (1970), A simple multipurpose scanning microscope. Proc. 7th Int. Congr. Electron Microscopy, Grenoble *1*, 209.

Crewe, A. V., J. Wall and L. M. Welter (1968), A high resolution scanning electron microscope, Rev. Scient. Instrum. *40*, 241.

Crick, R. A. and D. L. Misell (1971), The application of energy loss analysis to microanalysis, Phil. Mag. *23*, 763.

Cundy, S. L. and P. J. Grundy (1966), Combined electron microscopy and energy analysis of an internally oxidised Ni and Si alloy, Phil. Mag. *14*, 1233.

Cundy, S. L., A. J. F. Metherall and M. J. Whelan (1966), An energy analysing electron microscope, J. scient. Instrum. *43*, 712.

Cundy, S. L., A. J. F. Metherall, M. J. Whelan, P. N. T. Unwin and R. B. Nicholson (1968a), Studies on segregation and the initial stages of precipitation at grain boundaries in an aluminium 7 wt% magnesium alloy with an energy selecting microscope. Proc. R. Soc. A *307*, 267.

Cundy, S. L., A. J. F. Metherall and M. J. Whelan (1968b), Microanalysis of Al and 4 wt% Cu by combined electron microscopy and energy analysis, Phil. Mag. *17*, 141.

Ditchfield, R. W. and A. G. Cullis (1970), Identification of impurity particles in epitaxially grown silicon films using combined electron microscopy and energy analysis, Proc. 7th Int. Congr. Electron Microscopy, Grenoble *2*, 125.

Duncumb, P. (1963), A combined microscope microanalyser, Proc. Symp. X-Ray Optics and Microanalysis, Stanford U.S.A. H. H. Pattee, V. E. Cosslett, and A. Engstrom, eds. (Academic Press, New York), p. 431.

Dupouy, G. and F. Perrier (1962), Microscope électronique fonctionnant sous une tension d'un million de volts, J. Microscopie *1*, 167.

Dupouy, G. and F. Perrier (1966), Microscopie électronique variation de l'aberration chromatique en fonction de l'énergie des électrons, J. Microscopie *5*, 369.

Dupouy, G., F. Perrier and L. Durrieu, (1970), Microscopie électronique 3 millions de volts, J. Microscopie *9*, 575.

Favard, P. and N. Carasso (1973), The preparation and observation of thick biological sections in the electron microscope, J. Microscopy *97*, 59.

Glauert, A. M. and C. Mayo (1973), The study of three dimensional structural relationships in connective tissues by high voltage electron microscopy, J. Microscopy *97*, 83.

Glaeser, R. M. (1972), Personal communication.

Grigson, C. W. B. (1962), On scanning electron diffraction, J. Electron Control *12*, 209.

Haine, M. E. and R. S. Page (1956), A new universal electron microscope of high resolving power: Metro Vick Type EM6, Proc. 1st Eur. Reg. Conf. Electron Microscopy, Stockholm, p. 32.

Hale, K. H. and M. Henderson-Brown (1970), Some applications of high voltage electron microscopy to the study of materials, Micron *1*, 434.

Hama, K. and F. Nagata (1970), Stereoscope observation of the biological specimens by means of high voltage microscope. Proc. 7th Int. Cong. Electron Microscopy Grenoble, *1*, 461.

Hillier, J. and R. F. Baker (1944), Microanalysis by means of electrons, J. appl. Phys. *15*, 663.

Katagiri, S. and S. Osaza (1967), X-ray microanalysis attachment for electron microscope, J. Electron Microscopy *16*, 120.

Kobayashi, F. and M. Ohara (1966), Voltage dependence of radiation damage to polymer specimens, Proc. 6th Int. Congr. Electron Microscopy, Kyoto *1*, 579.

Kobayashi, F. and K. Sakaoku (1965), Irradiation changes in organic polymers at various accelerating voltages, Lab. Invest. *14*, 1097.

Koike, H., K. Ueno, and M. Watanabe (1970), Scanning device combined with conventional electron microscope. Proc. 7th Int. Congr. Electron Microscopy, Grenoble *1*, 241.

Le Poole, J. B., A. B. Bok, and P. J. Rus (1970), A compact 1 MV-electronmicroscope. Proc. 7th Int. Congr. Electron. Microscopy, Grenoble *1*, 113.

Makin, M. J. (1971), Studies of irradiation damage in the HVEM, Jernkont. Ann. *155*, 509.

Matricardi, V. R., R. C. Moretz and D. F. Parsons (1972), Electron diffraction of wet proteins: catalase, Science *177*, 268.

Möllenstedt, G. and O. Rang (1951), The electrostatic lens as a high resolution volocity filter, Z. angew. Phys. *3*, 187.

Ozasa, S., Y. Kato, H. Todokoro, S. Kasai, S. Katagiri, H. Kimura, E. Sugata, K. Fukai, H. Fujita and K. Ura (1972), 3 million volt electron microscope, J. Electron Microscopy, *21*, 109.

Porter, K. R. (1969), Biological applications of high voltage electron microscopy, Micron *1*, 229.

Rambourg, A., A. Marrand and M. Chrétien (1973), Tridimensional structure of the forming face of the Golgi apparatus as seen in the high voltage electron microscope after osmium impregnation of the small nerve cells in the semilunar ganglion of the trigeminal nerve, J. Microscopy *97*, 49.

Reinhold, G. (1969), Design problems of electron accelerators for high voltage electron microscopes. In First National Conference on Current Developments in High Voltage Electron Microscopy, Monroeville, Pa. U.S.A.

Riecke, W. D. (1970), On the problem of attaining atomic resolution by using a transmission electron microscope. Proc. 7th Int. Congr. Electron Microscopy, Grenoble *1*, 11.

Ris, H. (1969), The use of the high voltage microscope for the study of thick biological sections, J. Microscopie *8*, 761.

Ruska, E. (1954), Ein Hochauflösendes 100 kV Elektronenmikroskop mit Kleinfelddurchstrahlung. Proc. 3rd Int. Congr. Electron Microscopy, London, p. 673.

Sugata, E., K. Fukai, H. Fujita, K. Ura, B. Tadano, H. Kimura, S. Katagiri and S. Ozasa (1970), Project for construction and application of 3 MeV electron microscope, Proc. 7th Int. Congr. Electron Microscopy, Grenoble *1*, 121.

Swann, P. R. and N. J. Tighe (1971), High voltage microscopy of gas oxide reactions, Jernkont. Ann. *155*, 497.

Swann, P. R. and N. J. Tighe (1972), Voltage and pressure dependence of the electron transmission through various gases. Proc. 5th Eur. Reg. Conf. Electron Microscopy, Manchester, p. 436.

Tompsett, M. F., D. E. Sedgewick and J. St. Noble (1969), A versatile high energy scanning electron diffraction system for observing thin film growth in ultra high vacuum and in a low gas pressure, J. scient. Instrum (Series 2) *2*, 587.

Watanabe, H. (1951), Velocity analyser and electron microscope, J. Electron Microscopy *4*, 24.

Whelan, M. J. (1974), Private communication.

Wittry, D. B. (1969), An electron spectrometer for use with the transmission electron microscope, Brit. J. appl. Phys. (J. Phys. D) Ser. 2, *2*, 1757.

Wittry, D. B., R. P. Ferrier, and V. E. Cosslett (1969), Selected area electron spectrometry in the transmission electron microscope. Brit. J. appl. Phys. (J. Phys. D) Ser. 2 *2*, 1767.

Appendix

List of suppliers

N.B. The following list includes the names of suppliers known to the authors at the present time. They will be interested to hear of names and addresses of other suppliers for inclusion in later editions.

Apertures

Agar Aids
66a Cambridge Road
Stansted
Essex, CM24 8DA
England

Ernest F. Fullam Inc.
P.O. Box 444
Schenectady
New York 12301
U.S.A.

Ladd Research Industries Inc.
P.O. Box 901
Burlington
Vermont 05402
U.S.A.

Polaron Equipment Ltd.
21, Greenhill Crescent
Holywell Industrial Estate
Watford
Hertforshire WD1 8XG
England

(See also: Transmission Electron Microscope Manufacturers)

Calibration and test specimens

Agar Aids
66a Cambridge Road
Stansted
Essex, CM24 8DA
England

Balzers Union AG
Zubehöre für die Elektronenmikroskopie
Postfach 75
FL 9496 – Balzers
Fürstentum Liechtenstein

Ernest F. Fullam Inc.
P.O. Box 444
Schenectady
New York 12301
U.S.A.

EM Aids
Chestnut House
72, Dragon Road
Winterbourne
Bristol
England

Ladd Research Industries Inc.
P.O. Box 901
Burlington
Vermont 05402
U.S.A.

Polaron Equipment Ltd.
21, Greenhill Crescent
Holywell Industrial Estate
Watford
Hertfordshire WD1 8XG
England

Calibration service

Alan W. Agar
Agar Aids
66a Cambridge Road
Stansted
Essex, CM24 8DA
England

Densitometers

Joyce Loebl and Co. Ltd.
Princesway
Team Valley
Gateshead
Co. Durham NEll 9BR
England

Technical Operations Inc.
40 South Avenue
Northwest Industrial Park
Burlington
Massachusetts 01803
U.S.A.

Enlargers

Bessler Photo Marketing Co.
219 South 18th Street
East Orange
New Jersey 07018
U.S.A.

De Vere (Kensington) Ltd.
Thayers Farm Road
Beckenham
Kent BR3 4NB
England

Durst Iac.
P.O. Box 445
39100 Bolzano
Italy
or
Johnsons of Hendon Ltd.
(UK agents for Durst)
Hendon Way
London NW4
England

or
Ehrenreich Photo-Optical Industries Inc.

Photo-Technical Products Division
(USA agents for Durst)
Garden City
New York 11533
U.S.A.

Ernst Leitz, GmbH.
P.O. Box 2020
633 Wetzlar
Germany
or
E. Leitz Inc.
Link Drive
Rockleigh
New Jersey 07647
U.S.A.
or
E. Leitz (Instruments) Ltd.
48 Park Street
Luton
Bedfordshire LU1 3HP
England

LogEtronics Inc.
7001 Loisdale Road
Springfield
Virginia 22150
U.S.A.

Simmon Omega Inc.
P.O. Box 1068
Woodside
New York 11377
U.S.A.

Enlarger focusing aids

Thomas Instrument Co. Inc.
331 Park Avenue South
New York 10
U.S.A.

Enlarger photometers and timers

Medical and Electrical Instrumentation Co. Ltd (MELICO)
301a Finchley Road
London NW3
England

General microscope accessories

Agar Aids
66a Cambridge Road
Stansted
Essex, CM24 8DA
England (UK agents for Ladd)

Balzers Union AG
Zubehöre für die Elektronenmikroscopie
Postfach 75
FL 9496 – Balzers
Fürstentum
Liechtenstein

BEEM
P.O. Box 132
Jerome Avenue Station
Bronx
New York 10468
U.S.A.

EM scope Laboratories Ltd.
Kingsnorth Industrial Estate
Wotton Road
Ashford
Kent TN23 2LN
England

Ernest F. Fullam Inc. (EFFA)
P.O. Box 444
Schenectady
New York 12301
U.S.A.

or
Graticules Ltd.
(UK agents for EFFA)
Sovereign Way
Tonbridge
Kent
England
or
Touzart and Matignon
(French agents for EFTA)
8, rue Eugène Hénaff
94400 Vitry sur Seine
France

Ladd Research Industries Inc.
P.O. Box 901
Burlington
Vermont 05402
U.S.A.
or
Agar Aids (UK agents for Ladd)

Ted Pella Company
P.O. Box 510
Tustin
California 92680
U.S.A.

Polaron Equipment Ltd.
21, Greenhill Crescent
Holywell Industrial Estate
Watford
Hertfordshire WD1 8XG
England
or
Polysciences (USA agents for Polaron)

Polysciences Inc.
Paul Valley Industrial Park
Warrington
Pennsylvania 18976
U.S.A.

or
Polaron (UK agents for Polysciences)

TAAB Laboratories
52 Kidmore End Road
Emmer Green
Reading
Berkshire
England

or
Ebtec Corp.
(US agents for TAAB)
120 Shoemaker Lane
Agawam
Massachusetts 01001
U.S.A.

Image analysing equipment

Cambridge Instrument Co. Ltd.
(IMANCO)
Melbourn
Royston
Hertfordshire SG8 6EJ
England
or
Cambridge Instrument Co. Inc.
40 Robert Pitt Drive
Monsey
New York 10952
U.S.A.

Joyce Loebl and Co. Ltd.
Princesway
Team Valley
Gateshead
Co. Durham NE77 9BR
England

Technical Operations Inc.
40 South Avenue
Northwest Industrial Park
Burlington
Massachusetts 01803
U.S.A.

Photographic chemicals

Agfa-Gevaert NV
Verkooporganisatie Benelux
Septestraat 27
B 2510 Mortsel
Antwerp
Belgium
or
Agfa-Gevaert Ltd.
Great West Road
Brentford
Middlesex
England
or
Agfa-Gevaert Inc.
275 North Street
Teterboro
New Jersey 07608
U.S.A.

Alpha Photo Products Inc.
1101 Grove Street
Oakland
California 94607
U.S.A.

Eastman Kodak Company
343 State Street
Rochester
New York 14650
U.S.A.

Ilford Limited
Ilford
Essex IG1 2AB
England

Johnsons of Hendon Ltd.
Hendon Way
London NW4
England

Kodak Limited
P.O. Box 66
Hemel Hempstead
Hertfordshire HP1 1JU
England

May and Baker Ltd.
Rainham Road
Dagenham
Essex RM10 7XF
England

Photographic processing and finishing equipment

Agfa-Gevaert NV
Verkooporganisatie Benelux
Septestraat 27
B 2510 Mortsel
Antwerp
Belgium
or
Agfa-Gevaert Inc.
275 North Street
Teterboro
New Jersey 07608
U.S.A.
or
Agfa-Gevaert Ltd.
Great West Road
Brentford
Middlesex
England

Alpha Photo Products Inc.
1101 Grove Street
Oakland
California 94607
U.S.A.

BEEM
P.O. Box 132
Jerome Avenue Station
Bronx
New York 10468
U.S.A.

Bessler Photo Marketing Co.
219 South 18th Street
East Orange
New Jersey 07018
U.S.A.

David Allan (Dallas Products) Ltd.
Mark Road
Hemel Hempstead
Hertfordshire
England

Eastman Kodak Company
343 State Street
Rochester
New York 14650
U.S.A.

Ilford Limited
Ilford
Essex IG1 2AB
England

Johannes Bockemühl (JOBO)
5285 Derschlag
Köln
Germany

Johnsons of Hendon Ltd.
Hendon Way
London NW4
England

Kodak Limited
P.O. Box 66
Hemel Hempstead
Hertfordshire
England

Photographic sensitive materials

Agfa-Gevaert NV
Verkooporganisatie Benelux
Septestraat 27
B 2510 Mortsel
Antwerp
Belgium
or
Agfa-Gevaert Inc.
275 North Street
Teterboro
New Jersey 07608
U.S.A.
or
Agfa-Gevaert Ltd.
Great West Road
Brentford
Middlesex
England

Eastman Kodak Company
343 State Street
Rochester
New York 14650
U.S.A.

Fujii Photo Film Co. Ltd.
3,2-chome
Ginza-Nishi
Chuo-ku
Tokyo
Japan

Ilford Limited
Ilford
Essex IG1 2AB
England

Kodak Limited
P.O. Box 66
Hemel Hempstead
Hertfordshire
England

Konishiroku Photo Industrial Co. Ltd.
Tokyo
Japan

Specimen grids

Ernest F. Fullam Inc. (EFFA)
P.O. Box 444
Schenectady
New York 12301
U.S.A.

Graticules Ltd (MAXTAFORM)
Sovereign Way
Tonbridge
Kent
England

Ladd Research Industries Inc.
P.O. Box 901
Burlington
Vermont 05402
U.S.A.

LKB-Produkter AB
S-161 25
Bromma 1
Sweden
or
LKB Instruments Ltd.
LKB House
232 Addington Road
South Croydon
Surrey CR2 8YD
England
or
LKB Instrument Inc.
12221 Parklawn Drive
Rockville
Maryland 20852
U.S.A.

Mason and Morton Ltd.
M & M House
Frogmore Road
Hemel Hempstead
Herts., HP3 9RW
England

Ted Pella Company
P.O. Box 510
Tustin
California 92680
U.S.A.

Polaron Equipment Ltd.
60–62 Greenhill Crescent
Holywell Estate
Watford
Hertfordshire
England

Polysciences Inc.
Paul Valley Industrial Park
Warrington
Pennsylvania 18976
U.S.A.

Smethurst High-Ligh Ltd (ATHENE).
420 Chorley New Road
Bolton
Lancashire BL1 5BA
England
(ATHENE also marketed by Agar Aids)

TAAB Laboratories
52 Kidmore End Road
Emmer Green
Reading
Berkshire
England

VECO Zeefplatenfabriek NV
Postbus 10
Karel van Gelreweg 22
Eerbeek (GLD)
Netherlands

Stereo viewers

Agar Aids
66a Cambridge Road
Stansted
Essex, CM24 8DA
England

C. F. Casella Co.
Regent House
Brittania Walk
London N1 7ND
England

Ernest F. Fullam Inc.
P.O. Box 444
Schenectady
New York 12301
U.S.A.

Ladd Research Industries Inc.
P.O. Box 901
Burlington
Vermont 05402
U.S.A.

E. Marshall Smith Ltd.
64-74 Norwich Avenue
Bournemouth
Hampshire
England

Polaron Equipment Ltd.
21, Greenhill Crescent
Holywell Industrial Estate
Watford
Hertfordshire WD1 8XG
England

Thin film apertures

Ebtec Corporation
C. W. French Division
120 Shoemaker Lane
Agawam
Massachusetts 01001
U.S.A.

EM Aids
Chestnut House
72, Dragon Road
Winterbourne
Bristol
England

Polaron Equipment Ltd.
21, Greenhill Crescent
Holywell Industrial Estate
Watford
Hertfordshire WD1 8XG
England

Transmission electron microscopes

AEI Scientific Apparatus Ltd.
Barton Dock Road
Urmston
Manchester M31 2LD
England
or
Kratos Ltd.
(formerly AEI Scientific Apparatus Ltd.)
500 Executive Boulevard
Elmsford
New York 10523
U.S.A.

Hitachi Ltd.
4,1-chome
Marunouchi
Chiyoda-ku
Tokyo
Japan
or
Perkin-Elmer Ltd. (Hitachi)
Post Office Lane
Beaconsfield
Buckinghamshire
England
or
Perkin-Elmer Corporation (Hitachi)
P.O. Box 10920
Palo Alto
California 94303
U.S.A.
or
Perkin-Elmer Corporation (Hitachi)
Norwalk
Connecticut 06856
U.S.A.

JEOL Ltd.
1418 Nakagami Akishima
Tokyo 196
Japan
or
JEOL (UK) Ltd.
JEOL House
Grove Park
Edgeware Road
Colindale
London NW9
England
or
JEOL USA Inc.
477 Riverside Avenue
Medford
Massachusetts 02155
U.S.A.

F. G. Miles Engineering Ltd.
Old Shoreham Road
Shoreham-by-Sea
Sussex BN4 5FL
England

NV Philips Gloeilampenfabrieken
Analytical Equipment Department
Eindhoven
Netherlands
or
Philips Electronic Instruments
750 South Fulton Avenue
Mount Vernon
New York 10550
U.S.A.
or
Pye-Unicam Ltd. (Philips)
Analytical Department
York Street
Cambridge CB1 2PX
England

Siemens Aktiengesellschaft
Messgerätewerke
1000 Berlin 13
Postfach 140
Germany
or

Siemens (UK) Ltd.
Great West House
Great West Road
Brentford
Middlesex
England
or
Siemens Corporation
186 Wood Avenue
Iselin
New Jersey 08830
U.S.A.

Carl Zeiss
7082 Oberkochen
Germany
or
Degenhardt and Co. Ltd (Zeiss)
31–36 Foley Street
London W1P 8AP
England
or
Carl Zeiss Inc.
444 Fifth Avenue
New York
New York 10018
U.S.A.

International Scientific Instruments Ltd.
Exeter Road
Newmarket
Suffolk
England

Subject index